Vits

On Writing a Column

A puzzled man making notes . . . drawing sketches in the sand, which the sea will wash away. WALTER LIPPMANN

I got to seeing that column as a grave, twenty-three inches long, into which I buried part of myself every day. DON MARQUIS

The true appeal of this job is that I can do it when I'm sitting. MIKE ROYKO

It's as easy as opening a vein, and letting the words bleed out, drip by drip. RED SMITH

People think I went crazy on account of Eleanor Roosevelt. That wasn't it at all. It began when I went cosmic. It finished when I began writing on Monday to be printed on Friday. WESTBROOK PEGLER

I am frankly a hired Hessian on the typewriter, and I have never pretended to be anything else. DAMON RUNYON

The amazing thing is that something this much fun isn't illegal. GEORGE WILL

In writing a column, you create a personality—and you become very quickly a prisoner of that person who is the column. RUSSELL BAKER

I don't think I've ever written a column I really liked. HERB CAEN

Pundits, Poets, and Wits:

An Omnibus of American Newspaper Columns

Gathered, Annotated, and Introduced by Karl E. Meyer

New York Oxford

Oxford University Press

Oxford University Press

Oxford New York Toronto
Delhi Bombay Calcutta Madras Karachi
Petaling Jaya Singapore Hong Kong Tokyo
Nairobi Dar es Salaam Cape Town
Melbourne Auckland

and associated companies in
Berlin Ibadan

First published in 1990 by Oxford University Press, Inc.,
200 Madison Avenue, New York, New York 10016

First issued as an Oxford University Press paperback, 1991

Library of Congress Cataloging-in-Publication Data
Meyer, Karl Ernest.
Pundits, Poets, and Wits / Karl E. Meyer.
p. cm.
1. American newspapers—Sections, columns, etc.—Letters to the editor.
2. American newspapers—Sections, columns, etc.—Advice.
3. American newspapers—Sections, columns, etc. I. Title.
PN4888.L47M4 1990 070.4'4—dc20 89-36472 CIP

ISBN 0-19-506063-6
ISBN 0-19-507137-9 (pbk.)

10 9 8 7 6 5 4 3 2 1

Printed in the United States of America

These pages are regarded as an extension of the copyright page.

The editor and publisher gratefully acknowledge permission to reproduce the following copyright material. While every effort has been made to secure permission, we may have failed in a few cases to trace copyright holders. We regret any inadvertent omissions.

The Estate of Franklin P. Adams and Esther Root Adams for permission to reprint poems by Franklin P. Adams.

The Estates of Joseph and Stewart Alsop and the Lois Wallace Literary Agency for three columns originally appearing in the *New York Herald Tribune.* Reprinted by permission of the Wallace Literary Agency, Inc.

The American Play Company and Sheldon Abend for three columns by Damon Runyon originally syndicated by King Features.

Dave Barry for two columns originally appearing in the *Miami Herald.*

The Bergen (N.J.) Record for three columns by William Caldwell originally appearing in *The Record.*

John Morton Blum for an October 25, 1940, letter by Walter Lippmann appearing in *Public Philosopher: Selected Letters of Walter Lippmann* (Ticknor and Fields, 1985).

Erma Bombeck and the Aaron M. Priest Literary Agency, Inc. for two syndicated columns by Ms. Bombeck.

Jimmy Breslin for two columns originally appearing in the *New York Daily News.*

The Estate of Heywood Broun and William Cooper Associates, Inc. for permission to reprint four columns by Heywood Broun.

Art Buchwald for three columns originally appearing in the *Washington Post.*

William F. Buckley, Jr. for three columns originally appearing in the *Washington Star.*

Herb Caen for three columns originally appearing in the *San Francisco Examiner.*

Alistair Cooke and Alfred A. Knopf for a BBC "Letter from America" broadcast in 1969 and reprinted in *The Americans* (Alfred A. Knopf, 1979).

Dallas Times Herald for two columns by Molly Ivins.

Enoch Pratt Free Library for two columns by H. L. Mencken originally appearing in the *Baltimore Sun.* Reprinted by permission of the Enoch Pratt Free Library in accordance with the terms of the will of H. L. Mencken.

Farrar, Straus & Giroux for "A Toast to Harlem" and "There Ought To Be a Law" from *The Best of Simple* by Langston Hughes. Copyright © 1961 by Langston Hughes. Renewal copyright © 1989 by George Houston Bass. Reprinted by permission of Hill and Wang, a division of Farrar, Straus & Giroux, Inc.

Richard Goldhurst for material by Harry Golden originally appearing in the *Carolina Israelite* and *Only in America* (World, 1958, reprinted by Greenwood Press, Westport, Conn., 1988).

Lewis Grizzard and the *Atlanta Journal-Constitution* for two columns originally appearing in the *Atlanta Constitution.*

Pete Hamill for three columns originally appearing in the *New York Post.*

Harper Collins for permission to reprint "Hot Weather" from *One Man's Meat* by E. B. White. Copyright 1944 by E. B. White. Reprinted by permission of Harper Collins.

Harvard College for four columns by Walter Lippmann, used with the permission of the President and Fellows of Harvard College.

I.H.T. Corporation for three columns by Dorothy Thompson, originally appearing in the *New York Herald Tribune* in 1938; I.H.T. Corporation. Reprinted by permission.

Kansas City Star for three columns by Bill Vaughan, reprinted by permission, *Kansas City Star*.

Murray Kempton for permission to reprint three columns originally appearing in the *New York Post*.

King Features for two columns by Walter Winchell originally appearing in the *New York Daily Mirror*. Reprinted by permission.

Little, Brown & Company for "We All Had a Finger on the Trigger," appearing in *In a Time of Torment, 1961–1967* by I. F. Stone, copyright © 1967, 1966, 1965, 1964, by I. F. Stone.

Judith Martin for permission to reprint "Miss Manners" columns © 1979, 1980, 1981, 1982 United Feature Syndicate, Inc. originally appearing in the *Washington Post*, syndicated by United Features and reprinted in *Miss Manners' Guide to Excruciatingly Correct Behavior* (Macmillan, 1982).

Mary McGrory for two columns originally appearing in the *Washington Star*, and one column appearing in the *Washington Post*.

National Association for the Advancement of Colored People for permission to reprint poems by Dorothy Parker.

Newsweek for one column by George Will from *Newsweek*, © 1981, Newsweek, Inc., all rights reserved. Reprinted by permission.

New York Post for three columns by Jimmy Cannon. Reprinted by permission.

New York Times for three columns by Russell Baker, copyright © 1981, 1982, 1984, by The New York Times Company, reprinted by permission.

for two columns by John Leonard, copyright © 1976–1977 by The New York Times Company, reprinted by permission.

for three columns by James Reston, copyright © 1971 by The New York Times Company, reprinted by permission.

for two columns by Red Smith, copyright © 1980, 1982, by The New York Times Company, reprinted by permission.

for one column by Will Rogers, copyright 1922 by The New York Times Company, reprinted by permission.

for two columns by Anna Quindlen, copyright © 1986 by the New York Times Company, reprinted by permission.

for three columns by William Safire, copyright © 1971 by The New York Times Company, reprinted by permission.

Ernie Pyle Fund of the Scripps-Howard Foundation for five columns by Ernie Pyle originally syndicated by the Scripps-Howard Newspapers.

Mike Royko for three columns originally appearing in the *Chicago Tribune*.

Scripps-Howard Newspapers for two columns by Westbrook Pegler originally syndicated by the Scripps-Howard Newspapers.

Calvin Trillin for permission to reprint three articles originally appearing in *The Nation*: "Navasky as Male Model," © 1981 by Calvin Trillin; "Pinko Problems," © 1982 by Calvin Trillin; "The Mailbag," © 1982 by Calvin Trillin.

Nicholas von Hoffman for one column originally appearing in the *Washington Post*.

Washington Post Writers Group for two columns by Ellen Goodman, copyright © 1979, 1980, The Boston Globe Newspaper Company, Washington Post Writers Group. Reprinted with permission.
for two columns by George Will originally appearing in the *Washington Post*, 1983, 1984.

Joel White for permission to reprint "Maine Speech" from *One Man's Meat* by E. B. White. Copyright 1944 by E. B. White.

For Shareen

Acknowledgments

This venture had its formal inception at Yale University in spring 1983, when Davenport College kindly invited me to conduct a seminar on the American newspaper column. The students who assisted in the initial forays were Judy Doctoroff, Sarah Duff, Kevin Fahey, Jennifer Fetter, William Guest, Melissa Marvin, Douglas Sun, Jill Greenwald, and Diana West. Thanks are also owed to colleagues who addressed my students: Anthony Lewis, Russell Baker, Sidney Schanberg, and Robert Semple. I have been under continuous obligation to a multitude of libraries, notably the New York Public Library, the Library of the General Society of Mechanics and Tradesmen, the Firestone Library at Princeton University, and the New York Society Library. Particular thanks are due to the Pequot Library of Southport, Conn., and the Weston (Conn.) Public Library. In 1987–88, I was a New York Times-Duke University Fellow, and was given the luxury of maximum access to two fine libraries; for this I am indebted to my publisher, Arthur Ochs Sulzberger, and my editor, Jack Rosenthal. Without the support and understanding of my editors and colleagues this book would have been unthinkable. A very special debt is owed to Madlyn Dowling and Rosemary Shields for their aid and forbearance. Vital research assistance was provided by Rachel Weintraub. For help in tracking down E. B. White's early work I thank D. K. Duncan of the *Seattle Times* and Jeff Lampe, editor of *The Cornell Sun.* Colin Campbell of *The Atlanta Constitution* dug out valuable material on Joel Chandler Harris. Many columnists obliged with their time for interviews, and I want especially to thank Mike Royko of *The Chicago Tribune* and Herb Caen of *The San Francisco Chronicle.* My sister, Susan E. Meyer, and her partner at Roundtable Press, Marsha Melnick, gave valuable advice. As the widow of a columnist, my mother, Dorothy Meyer, was in every sense a begetter of this project. Sheldon Meyer, my editor at Oxford University Press (and no relation), helped the author keep his head above a sea of papers, as did Leona Capeless. Finally, the fact that this volume was finished without also finishing its compiler, is owed to the affectionate support and intelligent assistance of my wife, Shareen Blair Brysac.

Preface

This compendious book is to my knowledge the first of its kind, an anthology of the American newspaper column from Franklin and *The Federalist* to Mencken, Don Marquis and Miss Manners. Given the surfeit from which to choose, I have had to shear away ruthlessly, at the risk of slighting scores of deserving candidates, some of them friends and colleagues. I have tried to stick pretty faithfully to a few rules.

- A column is a signed article of moderate length appearing at more or less regular intervals in newspapers or magazines. But I am not a strict constructionist, and have dipped into the rich loam of personal journalism, by the prophets and cranks whose uninhibited utterances fixed the tradition on which the column rests.

- Banished without mercy are columns that rely mostly on misspellings for comic vitality, thus eliminating a prolific contingent of "phunnymen" like Artemus Ward and Josh Billings. They belong to history rather than letters, but for the sake of completeness I list many of them in a special appendix, or as they would have it, a speshul appendicks.

- Columnists who write in dialect are passed over unless redeemed by superior art, which seems to me true of Mr. Dooley, Uncle Remus, Major Downing, Will Rogers, Milt Gross, the unjustly forgotten Kurt M. Stein and James Russell Lowell in his *Biglow Papers.*

- Political columns for the most part perish, in E. B. White's phrase, like snakes with the setting sun, their bite vanishing with the controversies that provoked them. There are, nevertheless, exceptions: columnists who confound the rule by virtue of superlative vision (Lippmann), grace under deadline pressure (Kempton, Reston and

McGrory), pugnacity and erudition (Buckley and Will), and vituperative energy (Brann the Iconoclast and Pegler). All of the above and few more are represented.

• For thematic reasons, I have given preference to columns about columnizing, and to those illustrating concern among craftsmen for their essential tool, *viz.*, the English language.

• Columns are reproduced as they originally appeared, with cuts shown by ellipses. Obvious misprints have been silently corrected. Where columns have been tidied up for subsequent inclusion in a book, I have given their authors benefit of second thoughts.

This is not a random scrapbook. In the course of five years' scavenging, I found that the column has a richer history and more revealing tradition than I originally expected. For good reasons: journalism has flourished in these United States as in few other places. Before the Revolution, the printshop was a nursery of liberty and colonial newspapers formed a unifying bond. The power, and ubiquity, of the press was constitutionally confirmed in the First Amendment's express prohibition of any Federal interference. As the new Republic grew from infancy to puberty, newspapers took on wonderfully diverse missions: circus and schoolroom, arbiter of language and stepladder to letters, voice of faction and conscience, distiller of visions and how-to tutors of manners and morals. All this, I hope, will become evident in the pages that follow.

Contents

The Forthright Estate

A newspaper is a court
Where every one is kindly and unfairly tried
By a squalor of honest men . . .

STEPHEN CRANE (1899)

Pick up a newspaper anywhere in the United States and you will be addressed by insistent strangers known generically as columnists because each occupies, more or less, a column of newsprint. The term also has an apt association with the I-shaped pillar from which the columnist can harangue the populace—as was literally the case with Simeon Stylites of Syria, who spent thirty years preaching from a column until his death in 459 A.D. He has come to be the patron of an unruly calling.

Some columnists are local institutions, a few so popular that they are said to "own" their city. Many more are syndicated, so that if you travel around the country their eyes seem to follow you from passport-style photographs used to relieve the grayness of Op-Ed pages. Judgments differ on how much influence senior figures wield. Writing in 1975, when he himself was still a Washington columnist, Marquis Childs guessed that his colleague James B. Reston of the *New York Times* had roughly the power of three United States Senators. Three seems a lot. Still, a columnist of rank like Reston enjoys an obvious advantage over a Senator. Members of Congress are elected, columnists are more like life peers in Britain's House of Lords. This helps explain the resentment on the part of successive Presidents over the need to placate these potentates; there is no two-term limit for a Scotty Reston.

Of modern Presidents, none was more transfixed by columnists than Lyndon Baines Johnson. He courted them incorrigibly with lavish flattery and shoulder-squeezing bonhomie. Once during a White House news session, he saw Walter Lippmann in the throng; rushing over and grasping him, the President all but shouted: "This man here is the greatest journalist in the world, and he's a friend of mine!" That was in early 1964, when a good many people would have agreed with Johnson. Lippmann had been sought out by nearly every President since Theodore Roosevelt;

he was but twenty-six when he drafted Woodrow Wilson's famous Fourteen Points. In 1958, Nikita Khrushchev invited the columnist and his wife, Helen, to a hunting lodge in the Soviet Urals to discuss ways of thawing the cold war (the phrase was introduced by Lippmann). "Today and Tomorrow" appeared three times a week in more than 200 newspapers, and by common consent Lippmann's lucid treatises were read and pondered by an influential elite that no President could ignore. In his years at the pinnacle, Lippmann was interviewed annually by CBS television, an eminence accorded to no other journalist, before or since. In his reckoning of columnists, Marquis Childs described Lippmann as Supreme Pontiff.

Alas, the honeymoon ended badly. That any journalist might differ with his policies purely on the merits was a notion that Johnson never seriously entertained. He had courted Lippmann's support ardently, attending the columnist's seventy-fifth birthday party and instructing his aides to show him drafts of forthcoming Presidential speeches. But Lippmann was a sphere-of-influence conservative who at first doubted then dissented from the Administration belief that vital United States interests were at risk in Vietnam. As "Today and Tomorrow" became increasingly critical, the President became testier, to the point of having his staff comb through old columns in quest of blunders by the Supreme Pontiff—these were then maliciously read aloud at White House dinner parties. By this time, the White House was scoffing that Lippmann had reverted to isolationism, prompting an impassioned reply, *ex cathedra,* that conveys something of Lippmann's grand manner:

> A mature great power will make measured and limited use of its power. It will eschew the theory of a global and universal duty which not only commits it to unending wars of intervention but intoxicates its thinking with the illusion that it is a crusader for righteousness, that each war is a war to end all war. Since in this generation we have become a great power, I am in favor of learning to behave like a great power, of getting rid of the globalism which would not only entangle us everywhere but is based on the totally vain notion that if we do not set the world in order, no matter what the price, we cannot live in the world safely. If we examine this idea thoroughly, we shall see that it is nothing but the old isolationism of our innocence in a new form. Then we thought we had to preserve our purity by withdrawal from the ugliness of great power politics. Now we sometimes talk as if we could preserve our purity only by policing the globe.

As such admonitions became more insistent, communications ceased between President and Pontiff, and soon senior officials boycotted social events at which Lippmann might be present. On May 25, 1967, Lippmann wrote his farewell column, explaining that after thirty-six years of "Today and Tomorrow" he no longer felt the need to find out "what the blood pressure is at the White House, and who said what and who saw whom and who is listened to and who is not listened to." Two days later, the Lippmanns sailed for Europe, departing permanently from a capital that had grown too small for a columnist and President.

The point of the story is that it could have occurred in no other country. Where else does any chief of state care that much about the judgment of a journalist? And Walter Lippmann was scarcely an isolated case; there was also his longtime rival, Joseph Alsop.

The Indochina war that Lippmann opposed, Alsop supported, indeed even helped start by hectoring successive Administrations to take up France's burden after the fall of Dien Bien Phu in 1954. Lippmann and Alsop appeared on alternate days in the *Washington Post,* and their clash provided something like the ongoing duel between Government and Opposition benches in the House of Commons. But Alsop had an insider's advantage: he was the good friend and Georgetown neighbor of John F. Kennedy, on whom the columnist had a profound influence. A firm believer in global policing, Alsop became certain in the late 1950s that a deadly "missile gap" was widening, and that the United States was lagging perilously behind the Soviet Union in the production of Intercontinental Ballistic Missiles. Senator Kennedy seized on the issue, and brushed aside President Eisenhower's protestations that the "gap" was grievously overstated. When John Kennedy assumed office in 1961, he found that Ike was right. It made no difference. The new Administration proposed and won approval for the biggest peacetime increase in defense spending, thereby propelling an arms race that came to feed on itself without journalistic assistance. It is true that Alsop was not alone in pressing the missile issue on Kennedy, but the columnist knew better than others how to get under a politician's skin by turning a strategic argument into a test of manhood. Lyndon Johnson was once heard to murmur as he ordered 50,000 more troops to Vietnam, "There, that should keep Joe Alsop quiet for a while."

Alsop's younger brother Stewart exerted a different kind of influence, less that of a public philosopher or needling debater than of a shrewd family lawyer dispensing good sense. Stewart initially teamed up with Joseph when the Alsops launched "Matter of Fact" in 1946, then parted a decade later and eventually began his own column in *Newsweek,* occupying the pulpit that had been created for Lippmann in 1962. In a turbulent era, Stewart Alsop had the knack of catching a mood and guessing what it meant with but-of-course reasoning. Thus he took correct measure of student antiwar demonstrations and forecast their demise when Richard Nixon reinstated student draft deferments. During Watergate, he came up with a sound explanation for the strange behavior of those embroiled in the scandal:

> My theory derives from the peculiar relationship between two minority categories in the human race—the crazy-brave and the phony-tough.

> Most people who have been in a war, and a lot of people who haven't, have come across specimens of both breeds. The crazy-brave, who are a lot rarer than the phony-tough, are always doing crazy things that ought to get them killed, or at least maimed, but nothing seems to happen to them. They also exercise a kind of hex or double whammy on the phony-tough, and they keep getting the phony-tough into terrible trouble.

The Watergate crazy-brave was G. Gordon Liddy, who offered to kill himself when the break-in misfired, and who threatened to kill a wide-eyed Jeb Stuart Magruder if his hand was not instantly taken from Liddy's shoulder. Thus, reasoned Stewart Alsop, the phony-toughs who dominated the Nixon White House were too smart to give an unambiguous green-light to the harebrained idea of breaking into Democratic headquarters, but were too craven to stand up to the crazy-brave. Similarly, Lt. Col. Oliver North was the crazy-brave Marine years later who hexed his phony-tough White House superiors during the Iran-contra affair—or so theorized two professors who had clipped and saved the Alsop column and quoted its argument in a *Washington Post* Op-Ed article that began: "One of the enduring frustrations of journalism is the short half-life of the column, with the result that insights provided by the occasional exceptional column are eventually lost to future generations."

I understood their lament. My father, Ernest L. Meyer, wrote a column six days a week for fifteen years, first for the Madison, Wisconsin, *Capital Times,* then for the *New York Post.* In Wisconsin, my father was telegraph editor as well, and wrote his daily piece after locking up the final afternoon edition. He used a copy pencil for a longhand draft, then typed hunt-and-peck to exact space. He did something different every day, mixing commentary with light verse, parodies or recollections of his boyhood in Milwaukee, his first newspaper job (in Warden, Washington, for $4 a week plus board), his student years at the State University or whatever. Wherever he went, the column went along: on European trips, summers in Cape Cod, visits to Washington or Hollywood or with Frank Lloyd Wright at Taliesin. Like many other columnists, he delighted in exploiting his wife and progeny—my brother Leonard, sister Susan, and me—as walk-on extras whose continual prattling provided the ever-essential "peg."

After my father died in 1952, I came into possession of his scrapbooks, the potter's field of journalism. Some years later, having made my own way into the same business, I went carefully through these scrapbooks, better able to judge the skill and stamina that went into writing a fixed quota of literate prose every day, with no time allowed for indisposition. And it saddened me, not just on my father's account, to reflect how much first-rate work vanished into the common dustbin.

So I decided to attempt an anthology. My first surprise was the sheer number of columnists. By my reckoning, the feature syndicates and major newspapers distribute the offerings of two hundred columnists, most of them writing two or three times a week. Then there are the local columnists—light essayists, gossip-mongers, specialized writers on chess or cooking, sports columnists, cracker-barrel wits, political pundits, or whatever—on one thousand seven hundred daily newspapers, two thousand four hundred weeklies, fourteen thousand newsletters, four thousand seven hundred magazines, three thousand specialized international publications, nine hundred ethnic and foreign-language periodicals, and nobody-knows-how-many student newspapers, daily and weekly. In America, it is fair to say, at least fifteen thousand columnists compose short signed articles at regular intervals—a truly vast garrulous army feeding on the depleting forests of Canada.

It is not only a ubiquitous calling, but is deemed reputable by exalted political figures. The two Presidential Roosevelts, Calvin Coolidge, Barry Goldwater and Ronald Reagan have this much in common with Eleanor Roosevelt, Jeane Kirkpatrick, Fiorello LaGuardia, Henry Wallace, Harold Ickes, and Fighting Bob LaFollette: each sought for a while to improve the American mind through the medium of a newspaper or magazine column (in Mrs. Roosevelt's case, she did both simultaneously, writing for the Scripps-Howard newspapers and *Woman's Home Companion*).

Interestingly, the column continues to flourish, notwithstanding the supposed decline of the print medium and the rise of electronic journalism. Television, if anything, has given syndicated columnists more influence plus a celebrity status useful in commanding vast speaking fees. During the reign of radio, Walter Winchell and Drew Pearson hawked their revelations on weekly broadcasts, and Franklin Pierce Adams enjoyed an autumnal renown as a panelist on "Information, Please," an early-day quiz program stressing wit more than prizes. But this only feebly adumbrated the star status of columnists who, beginning in the 1970s, became fixtures on current-affairs talk shows: George Will, Carl Rowan, Tom Wicker, William F. Buckley, Ellen Goodman, to name a few. Thus the shrinkage of a literate audience has been offset by the gain in the columnist's reach and influence.

Columnists, in short, have benefited the land of the free by contributing as Lippmann did to what President Bush has called "the vision thing." More commonly, they have been energetic advocates exerting influence through the quality of their arguments, and their independence from an electorate that has learned to put up with them. They have from time to time added sparkle and sense to the national discourse, disseminating phrases that quickly became commonplace (egghead, middle America, cold war, the Reagan Doctrine). And having been given a license to be rude, reckless, silly, and prejudiced, the columnist can in a single effusion entertain at breakfast and provide at dinner the stuff for post-prandial argument.

Yet nowhere could I find a connected history of the newspaper column, or any account of the principles that might justify so self-indulgent an institution.

II

> The liberty of the Press is the Foundation of all our other Liberties, . . . , and whenever the Liberty of the Press is taken away, either by open Force or any little, dirty infamous Arts, we shall immediately become as wretched, as ignorant, and as despicable SLAVES, as any one Nation in all Europe.
>
> *The New-York Weekly Journal,* January 14, 1734

In colonial America, the column's begetter was a sixteen-year-old printer's apprentice, a "Bookish Lad" bred in Boston when it comprised 12,000 souls. Benjamin Franklin sketches himself vividly in his autobiography—his fondness for argument, his vegetarianism, his early reading of Defoe and John Bunyan's works; and then

> About this time I met with an odd Volume of the *Spectator*. I had never before seen any of them. I bought it, read it over and over, and was much delighted with it. I thought the writing excellent & wish'd if possible to imitate it. With that View, I took some of the Papers, & making short Hints of the Sentiment in each Sentence, laid them by a few Days, and then without looking at the Book, try'd to compleat the Papers again, by expressing each hinted Sentiment at length & as fully as it had been express'd before, in any suitable Words that should come to hand. Then I compar'd my Spectator with the Original, discover'd some of my Faults & corrected them.

It was a propitious encounter. *The Spectator* had flourished in the age of Queen Anne, after Parliament repealed press licensing laws and a hundred gazettes flowered. Affluence, liberty, and literacy combined to create an audience in sitting-room and coffee house for elegant satire in a mannered style. Setting the pace was *The Spectator,* edited by Addison and Steele in a fruitful run from 1711 to 1714, yielding no less than six hundred and twenty-four essays, none inconsiderable. So good was the original outpouring that one may say of the column that it has been in retreat since its invention.

Thus smitten, young Franklin tried to emulate his masters in an ingenious way. He was apprenticed to his brother James, who had recently established Boston's second newspaper, the *New-England Courant.* Rightly fearing his brother would object to publishing anything he knew to be by Benjamin, young Franklin resorted to ruse. He slipped an article under the printing-house door as if it were an unsolicited offering from a young widow. Franklin's pen-name, Silence Dogood, was a mischievous poke at the divine, Cotton Mather, whose works included *Silentarius* and *Bonifacius or Essays to Do Good.*

The trick succeeded, resulting in fourteen Dogood letters from April to October 1722. When the secret of their authorship finally came out, it so vexed James Franklin that his apprentice had to leave town to seek his fortune elsewhere, and he soon wound up in Philadelphia. Yet in this first rough draft the precocious Benjamin anticipated a million columns to come. Mrs. Dogood claimed to be a pioneer born on shipboard who was taught to read by the clergyman she later married. She wrote briskly of political and religious hypocrisy, women's rights and male chauvinism, the insufferable airs of Harvard, the richness of American vernacular, and the perils of demon rum—staple themes down the ages. Her self-description also is an evergreen:

> I am of an extensive Charity and a great Forgiver of *private* injuries: a hearty Lover of the Clergy and all good Men, and a mortal Enemy to arbitrary Government and unlimited Power . . . I have likewise a natural Inclination to observe and reprove the Faults of others, at which I have an excellent Faculty. I speak this by Way of Warning to all such whose Offenses shall come under my Cognizance, for I never intend to wrap my Talent in a Napkin.

As startling, one encounters a full-blown argument in favor of an unfettered press in letter eight, which Mrs. Dogood quotes from an unnamed London journal. It deserves quotation, because it is the mustard-seed from which was to spring the press's privileged place in the First Amendment:

> Without Freedom of Thought, there can be no such Thing as Wisdom; and no such Thing as publick Liberty, without Freedom of Speech; which is the right of every Man, as far as by it, he does not hurt or control the Right of another. And this is the only Check it ought to suffer, and the only Bounds it ought to know. This sacred privilege is so essential to free Government, that the Security of Property, and the Freedom of Speech always go together; and in those wretched Countries where a Man cannot call his Tongue his own, he can scarce call any Thing else his own. Whoever would overthrow the Liberty of a Nation, must begin by subduing the Freeness of Speech, a Thing terrible to Publick Traytors.

If young Franklin did not identify the author by name, it was because most readers would surely know he was quoting "Cato." And as many in London and Boston were well aware, "Cato" signified John Trenchard and Thomas Gordon, trouble-making Whigs, journalists-about-town, and libertarians in all the senses of that word. Capriciously and unjustly, "Cato" is forgotten, though the essays that appeared under that name in *The Independent Whig* in 1720–21 were to freedom of speech what the *Spectator* essays were to letters: the anchor of a tradition.

The idea that government would benefit from the relentless scrutiny of a free press did not derive from Locke or the other eminent political philosophers said to influence the American Revolution; it derived instead from journalists like Gordon and Trenchard, whose doctrine was taken up with particular ardor by printers like Franklin. When the *New-York Weekly Journal* was brought to the bar for libeling the colonial government in 1733, the paper published the boldest of the "Cato" letters and its editor, James Alexander, added his own thoughts as "An American Cato." In due course, a jury freed the printer, John Peter Zenger, in a case that proved to be a beacon for press freedom.

Be it noted that "Cato" was both libertarian and conservative, believing that security of property and freedom of speech must always go together. This credo was especially congenial to the "Publick Printers" who had set up shop in the colonies after the first printing-press arrived in Boston in 1696. New gazettes proliferated because printshops were reluctant to lock up precious type for producing books that were cheaper to import. Many printers benefited from government patronage, but also needed job work with a quick turnover and assured demand, such as newspapers and almanacs. "The American printer was the servant of literacy rather than of literature," writes Daniel Boorstin, one of the few historians to explore the political economy of the printshop.

The vital result was to give colonial printers a direct commercial interest in diversity of opinions. The point was elaborated with amiable agnosticism by Franklin in "Apology for Printers." He pleaded in 1731 with readers of his *Pennsylvania Gazette* not to be angry at the opinions expressed in his periodicals or pamphlets: "Printers are educated in the Belief, that when Men differ in Opinion, both Sides ought equally to have the Advantage of being heard by the Publick; and that when Truth and Error have fair Play, the former is always an overmatch for the latter. Hence they cheerfully serve all contending Writers that pay them well, without regarding on which side they are of the Question in dispute."

Thus when differences with England became too wide for compromise, the scattered colonies were united by a network of printers whose various journals were busily reprinting tracts from each other's pages. What was wanting was a match for the bonfire. It was supplied by Franklin himself when he wrote from London in 1774 to commend to his family back home "Mr. Thomas Paine . . . an ingenious worthy young man . . . I request you to give him your best advice and countenance, as he is quite a stranger there." Two years later, Paine was well established as a journalist in Philadelphia when he wrote the most infectious of pamphlets, "Common Sense," which demonstrated its thesis by the very speed with which it spread through an America that Paine had dared call a nation.

As arguments turned from independence to constitution-making, the newspapers and their printers kept apace, providing the principal forum for debating the charter urged by the Philadelphia convention. Hamilton, Madison, and Jay collaborated on *The Federalist,* consisting of eighty-five letters signed "Publius" that were published in two New York newspapers in 1787–88. Less familiar, because its authors were on the losing side, were what are now called the Anti-Federalist Papers, essays and letters that also appeared in newspapers under such names as "Centinel," "Brutus," "The Federal Farmer," and the ever serviceable "Cato."

For all its ferocity, the great debate on the Constitution was as remarkable for what the debaters agreed upon. Federalists and anti-Federalists alike were certain that a noble experiment was under way, that an empire of liberty was rising in the New World as a shaming example to the Old. "Degeneracy here is almost a useless word," Tom Paine had written in *Common Sense.* "Those who are conversant with Europe would be tempted to believe that even the air of the Atlantic disagrees with the constitution of foreign vices." It went without saying that American motives were invariably pure, and that a reforming spirit would remedy whatever vices, such as slavery, might remain.

Enter at this moment an Englishman who found the assumed virtue of the new Republic ludicrous and unjustified. This was William Cobbett, freshly arrived in 1796 by way of Canada to teach French in Philadelphia (though he also despised France). Cobbett judged everything American wretched: houses, land, weather. A single English peach was worth a bushel of the American fruit. "There is no spring or autumn . . . The people are worthy of the country—a cheating, sly, roguish gang." Benjamin Franklin was a lecher and a spawner of bastards. George Washington cheated on his expense account. Jefferson was a mountebank. Worst of all, in Cobbett's eyes, were émigré English radicals like Dr. Benjamin Rush and Dr. Joseph Priestley, both fraudulent to the core. As to the vaunted free press:

> These people plead the liberty of the press, in the fullest extent of the word; they claim a right to print and publish whatever they please; they tell you that free discussion must lead to the truth . . . They have calumniated the best of governments and the best of men; they revile all that is good and all that is sacred, and that too in language the most brutal and obscene; and if they are accused of indecency, or called on for proofs of what they advance, they take shelter in their sanctuary, *the liberty of the press,* but on the other hand, if anyone has courage enough to oppose, and is so happy to do it with success; if the mildest of their expressions are retorted, they instantly threaten their opponents with violence and even murder. Their doctrine is, that the press is free for them, and them alone. This is democratic liberty of the press.

To accuse Cobbett of hyperbole is to call the ocean wet. He founded and for four years published *Peter Porcupine's Gazette,* the whole purpose of which was to cause offense. He had no desire to be fair, balanced, objective. His journalism was a series of eruptions meant to disturb the peace. In this Cobbett was successful: writ followed upon writ, timorous printers pleaded with him to moderate his pen, and he was finally forced to leave the country on the heels of a libel trial in which a jury awarded Dr. Benjamin Rush $5,000 in damages and $2,000 in costs, a record for the era. As he returned to England, Peter Porcupine in his fashion took his farewell of his readers in January 1800:

> I congratulate myself on never having, in a single instance, been the sycophant of the Sovereign People, and having persisted in spite of the *savage howlings* of the SANS-CULOTTES and the *soothing serenades* of the FEDERALISTS (for I have heard both under my window)—I congratulate myself on having, in spite of all these things, persisted in openly and unequivocally avowing my attachment to my native country, and my allegiance to my King.

With that, Cobbett closed the first American chapter of an unpredictable career. He also fixed the course for an entire tradition of journalism in which writing becomes a display of personality meant to dazzle and entertain rather than persuade. The question for the coming century was whether the land of the free would still need to look to Britain for models of good prose, and for fearless scolds like William Cobbett. Having proclaimed a devotion to press freedom, would Americans dare use it?

III

> What would this mighty people do,
> If there, alas! were nothing new?
>
> *The New-York Gazette,* April 16, 1770

As the nineteenth century opened, the new nation was already a deeply ink-stained Republic. There were few American books, and the fine arts were not much to speak of, but America boasted upwards of 360 newspapers, more than in the far more populous British Isles. The Reverend Samuel Miller, pastor to a fashionable Presbyterian congregation in New York, saw these newspapers as great moral and political engines, reaching not only the wealthy few but all strata of society: "Never, it may be safely asserted, was the number of political journals so great in proportion to the population of a country as at present in ours . . . And never were they actually perused by so large a minority of all classes since the art of printing was discovered."

Yet one thing troubled the Reverend Dr. Miller. As he wrote in his *Retrospect of the Eighteenth Century,* in former times talents and learning if not virtue were thought necessary in the conductors of political journals. "Towards the close of the century, however," he went on,

> persons of less character, and of humbler qualifications, began, without scruple, to undertake the high task of enlightening the public mind. This remark applies, in some degree, to Europe; but it applies with particular force to our own country, where every judicious observer must perceive, that too many of our gazettes are in the hands of persons destitute at once of the urbanity of gentlemen, the information of scholars, and the principles of virtue. To this source, rather than to any peculiar depravity of national character, we may ascribe the faults of American newspapers, which have been pronounced by travelers the most profligate and scurrilous public prints in the civilized world.

In truth, the worthy divine was deploring what was most interesting about the American experiment. Fine writing in Europe was the product of aristocratic patronage, ancient universities, and an elite readership. That Shakespeare was not an Oxford or Cambridge man was to become *prima facie* evidence to many that he might not be, nay could not be, the author of his own plays. Poor boys did, of course, make their way into the English pantheon, but, revealingly, had to do so through Grub Street, crawling through the transom of journalism, like Defoe, Dr. Johnson, and Dickens. Or in a few rare cases like Blake, they were printers.

How differently they ordered things in the United States! If one lacked a degree, one could matriculate through the printshop without shame, as did Franklin, Walt Whitman, and Samuel Clemens. Indeed, in 1847, as the young Clemens was about to become an apprentice, the *Hannibal Gazette* carried an item captioned "The Poor Boy's College" that observed: "There is something in the very atmosphere of a printing office, calculated to awake the mind and inspire a thirst for knowledge." This is not to ignore the disabling obstacles of race and gender discrimination. But more than anywhere else, literature in America was far more open to the less advantaged.

And more than anywhere else, the writer could hope to address a literate mass audience. An unlikely witness, William Cobbett, affirmed as much. In 1818, Cobbett returned for a year's visit to the United States; no longer an arch-Tory, he was now an ardent democrat, who this time described to his British readers a very different America:

> There are few really ignorant men in America of native growth. Every farmer is more or less a reader. There is no class like that which the French call peasantry . . . They are all well-informed; modest without shyness; always free to communicate what they know, and never ashamed to acknowledge that they have yet to learn . . . They have all been readers from their youth up; and there are few subjects upon which they cannot converse with you, whether of a political or scientific nature.

The consequences were becoming evident in the language. Thanks in part to the precepts of Noah Webster, American spelling was different from that used in England: *defence* had become *defense, labour* was *labor,* and so forth. Moreover, by insensible degrees, the texture of American prose was changing. At first, it is true, written American English was pretty well indistinguishable from written English English. But the spoken language was so different that in the 1840s Sam Slick, a Yankee clockmaker in popular fiction, expressed a commonplace: "I never seed an Englishman yet that spoke good English."

Over the decades, vernacular American English was to transform written American English. The process is set forth by Richard Bridgman in his fascinating study, *The Colloquial Style in America.* Between 1825 and 1925, writes Professor Bridgman, American prose style changed in significant ways, becoming more concrete, simpler in diction, with greater stress on the verbal unit and the use of repetition to bind and unify. He contrasts these passages from Nathaniel Hawthorne and Ernest Hemingway:

> Beyond that darksome verge, the firelight glimmered on the stately trunks and almost black foliage of the pines, intermixed with the lighter verdure of sapling oaks, maples and poplars, while here and there lay the gigantic corpses of dead trees, decaying on the leaf-strewn soil.

> We walked on the road between the thick trunks of the old beeches and the sunlight came through the leaves in light patches on the grass.

Not only is the language and syntax simpler in Hemingway, but the very look of his novels is different. A page of Hawthorne is block-like with dense and tangled interiors whereas, Bridgman observes, "Hemingway crosses his page with thin lines and chips of language, clean and well-lighted."

How did this come about? I think Professor Bridgman has got it right: it was initiated primarily in dialect pieces and fictional dialogue. The demotic conquest came from the bottom. And in this campaign the humble but effective foot soldier was the newspaper columnist and sketch writer.

An early breakthrough occurred in Maine, where Seba Smith, the editor of a Portland daily conjured into his pages a cracker-barrel Yankee sage named Major Jack Downing. In a preface to the collected works of Downing, *My Thirty Years Out of the Senate,* Smith describes how it happened:

> In January, 1830, the first Downing Letter ever written appeared in the *Daily Courier,* published in Portland, Maine. This paper had just been started by the author, and was the first daily paper in the country north or east of Boston. The *Courier* was started as an independent paper, devoted to no political party—a position for a paper in those days likely to command but small support. The Maine Legislature met in Portland on the first of January, and the two political parties were so evenly balanced, and partisan feeling ran so high, that it was six weeks before they got fairly organized and proceeded with the business of legislation . . .

> At this juncture of affairs, the author of these papers, wishing to show the ridiculous position of the Legislature in its true light, and also, by something out of the common track of newspaper writing, to give increased interest and popularity to his little daily paper, bethought himself of a plan to bring a green, unsophisticated lad from the country into town with a load of axe-handles, hoop poles, and other notions for sale, and . . . let him blunder into the halls of the Legislature, and after witnessing for some days their strange doings, sit down and write an account of them to his friends at home in his own plain language.
>
> The plan was successful almost beyond parallel. The first letter made so strong a mark that others had to follow as a matter of course. The whole town read them and laughed; the politicians themselves read them, and their wrathful fire-eating visages relaxed to a broad grin. The Boston papers copied them, and all Boston tittered over them. The series was inaugurated and must go on.

Smith then gave Major Jack Downing a national reputation by sending him to Washington to become confidential adviser of President Andrew Jackson. An entire army of copycat Downings jumped into print as other writers seized the name for their own political purposes. So ubiquitous was the Major that his cartoon likeness, in top hat and vested suit, evolved into the familiar image of Uncle Sam, according to Walter Blair and Hamlin Hill in their splendid book, *America's Humor*.

The rage for dialect humor and cornpone capers flooded newspapers with the works of scores of now forgotten "phunny phellows." In their survey, Blair and Hill list forty of the most prominent, among them Mozis Addums, The Hawkeye Man, Carl Pretzel, Spoopendyke, Julius Caesar Hannibal, Erratic Enrique, Doesticks, Job Shuttle, Bill Arp, Petroleum V. Nasby, Josh Billings and, of course, Artemus Ward. What can be said in defense of this slapstick brigade is that the broadness and badness of their sketches had a liberating effect on Americans intimidated by what James Russell Lowell called the Universal Schoolmaster.

Lowell himself drew upon New England's vernacular for a series of poems that purported to be by Hosea Biglow lampooning the proponents of slavery from the Mexican to the Civil War. Most of the Biglow poems were published first in the *Boston Courier*, and then assembled in a collection to which Lowell added a discursive introduction. "In choosing the Yankee dialect, I did not act without forethought," wrote the poet:

> It had long seemed to me that the great vice of American writing and speaking was a studied want of simplicity, that we were in danger of coming to look on our mother tongue as a dead language, to be sought in the grammar and dictionary rather than in the heart, and that our only chance of escape was by seeking its living sources . . . That we are all made to talk like books is the danger with which we are threatened by the Universal Schoolmaster, who does his best to enslave the minds

> and memories of his victims to what he esteems the best models of English composition, that is to say, to the writers whose style is faultily correct and has no blood-warmth in it. No language after it has faded into *diction*, none that cannot suck up the feeding juices secreted for it in the rich mother-earth of common folk can bring forth a sound and lusty book.

Very properly, Lowell went on to deplore the familiar vices of journalism: banality, prolixity, and pomposity. Only a writer of considerable gifts, it would seem, could take this medium and infuse it with liberating originality.

Such a writer, serendipitously, turned up in California in 1850, drawn not by the gold rush but by the Mexican War. This was George Horatio Derby (1823–1861), born in Dedham, Massachusetts, the sixth generation in what had been a wealthy merchant family in Salem. As a West Point cadet, Derby was remembered for his buffoonery and reckless practical jokes, but nevertheless finished seventh in a class of fifty-nine. Called to Mexico for active service, Derby was wounded in the four-day siege of Vera Cruz while serving under Captain Robert E. Lee, the chief of engineers. Lieutenant Derby's valor was forgotten, however, when he burlesqued the new doorman-style uniforms just introduced by the Secretary of War, Jefferson Davis. One Derby "improvement" was the suggestion that every private be equipped with a stout iron hook sewn securely to the seat of his trousers. When marching, soldiers could hang their spare boots on the useful hook, and when sleeping, suspend themselves on fence rails. All this was illustrated with Derby's meticulous sketches.

As punishment, he was banished to the fringes of civilization: California. There he led a double life for eight years, as an Army engineer mapping crags and as a journalist drawing on and adding to the hilarity of saloon life during the gold rush. In 1850 he began writing in San Francisco for the *Alta California*, the same daily that was to commission Mark Twain's travel letters which became the basis for *Innocents Abroad*. He wrote as "Squibob" until a hated rival borrowed the name; Derby announced the demise of Squibob and reappeared as John Phoenix. Transferred in 1853 to San Diego, Derby/Phoenix found a new venue in *The Herald*, whose editor left him in charge while on a holiday. Derby produced an "illustrated" edition of surreal lunacy, totally reversed the paper's political stance and announced one of Phoenix's many patented inventions: A Blowing Machine "for the manufacture of newspaper puffs." The device was of pure brass for confecting "first-rate notices at a moment's notice" by taking any copy of *The Herald* and turning the crank for fifteen seconds.

His local fame carried to New York, where D. Appleton & Co. published a collection called *Phoenixiana*. The title page was adorned with a sketch of Derby wearing the absurd new Army hat, captioned "In the name of the prophet—FIGS!" Unlike most comic productions of the era, it hasn't dated. At once thoroughly literate and exuberantly populist, Derby salted his parodies with allusions to Virgil, Horace, Shakespeare, Cervantes, Milton, Sterne, Boswell, Byron, Thackeray, Macaulay, Dickens, Pope, Swift, Hood, Emerson, Harriet Beecher Stowe, John Stuart Mill, and Dr. Johnson.

Alas, the Army in its wisdom transferred Derby to Florida in 1857, and there he suffered a sunstroke and went mad. He had been mostly forgotten when he died in New York in 1861 at the age of thirty-eight. *Phoenixiana* is long out of print and impossible to find. Years later, in 1915, Derby was rediscovered by a professor of English, Fred Lewis Pattee, who correctly identified him as "the real father of a new school of humorists" and noted his "grotesque exaggerations . . . euphuistic statements . . . Yankee aphorisms . . . whimsical *non sequiturs*" but incorrectly described him as a Western primitive with "little time for books." (Walter Blair and Hamlin Hill made amends in their survey in 1978.) Derby is the now overlooked link who showed the way to a new generation, most especially to Samuel Clemens.

When Clemens arrived in San Francisco in 1864 to report for *The Morning Call,* he had completed his apprenticeship. He had come from Virginia City, Nevada, where as "Mark Twain" he had mastered the commoner forms of antic journalism on the *Territorial Enterprise.* He was twenty-eight, the Civil War was ending, a new nationalism was waxing. San Francisco's Barbary Coast had already attracted a remarkable group of literary artisans who found their workshop in the proliferating dailies and weeklies. In 1865, Mark Twain wrote his famous Jumping Frog sketch, then sailed for a five-month sojourn in Hawaii, where he concocted artful letters for the *Sacramento Union.* Now well in the forefront, Mark Twain became a roving correspondent for the *Alta California,* embarking by way of Nicaragua for New York to sail on the first American cruise ship, the *Quaker City.* With the publication of *Innocents Abroad* in 1869, Twain became that dubious thing, a literary celebrity. Esteemed as a humorist, he was lionized by his public and patronized by the college-bred establishment. When in 1884 he finally published his masterpiece, *Huckleberry Finn,* this oft-quoted note appeared in the *Boston Transcript:*

> The Concord, Massachusetts, Public Library committee has decided to exclude Mark Twain's latest book . . . One member of the committee says that, while he does not wish to call it immoral, he thinks it contains but little humor, and that of a very coarse type. He regards it as the veriest trash . . . other members entertain similar views, characterizing it as rough, coarse and inelegant . . . the whole book being more suited to the slums than to intelligent, respectable people.

That Mark Twain was in fact liberating American English from an inhibiting corset is plain only in retrospect. Even nowadays there is a certain professorial discomfort in recognizing that in the half-century from

1864 to 1914, much of what was original and interesting in American writing was the product of higher journalism. In Atlanta, Joel Chandler Harris gave a global currency to black English in his Uncle Remus tales that first appeared in *The Constitution*. In San Francisco, Ambrose Bierce relived the Civil War in the unsparing realism of his short fiction. Stephen Crane, Frank Norris, Jack London, Theodore Dreiser, and even William Dean Howells arose from newspaperdom, as did their successor, Ernest Hemingway, who was to write, in his most-quoted literary judgment: "All modern American literature comes from one book by Mark Twain called *Huckleberry Finn* . . . It's the best book we've had. All American writing comes from that. There was nothing before. There has been nothing as good since."

Perhaps not so curiously, the British were quicker to recognize the contagious buoyancy of the new American writing. Bret Harte, Joaquin Miller and Uncle Remus were instantly acclaimed in the old country; Mark Twain was awarded an honorary degree by Oxford. Stephen Crane's genius was heralded by Conrad and the expatriate Henry James. Yet if the English applauded, they did not, with one notable exception, emulate. The exception was interesting: Rudyard Kipling, another vernacular artist. As a young journalist in India, he first read Harte's serialized stories in the newspaper he was editing; "Why buy Bret Harte, I asked, when I was prepared to supply home-grown fiction on the hoof?" He consciously imitated the colloquial style, made heroes of outcasts and adventured riskily into farce. Beyond that, he came to America and sought out Mark Twain. The American preserved in his notebook the text of this letter sent August 16, 1895:

> Dear Kipling: It is reported that you are about to revisit India. This has moved me to journey to that far country in order that I might unload from my conscience a debt long owed to you. Years ago you came from India to Elmira to visit me, as you said at the time. It has always been my purpose to return that visit and that great compliment, some day. I shall arrive next January, and you must be ready. I shall come riding my Ayah, with his tusks adorned with silver bells and ribbons, and escorted by a troop of native Howdahs, richly clad and mounted upon a herd of wild Bungalows, and you must be on hand with a few bottles of Ghee, for I shall be thirsty.

To modern readers who know Kipling only as an idolater of imperialism and nursery entertainer, his high standing among American literary innovators may be a surprise. Yet as Van Wyck Brooks notes, Kipling's influence on his American counterparts "can scarcely be overstated." Kipling made Stephen Crane feel that "I'm just a dry twig on the edge of the bonfire." Finley Peter Dunne's "Dooley" dialogues came out of Kipling's pocket; Vachel Lindsay, Carl Sandburg, and Eugene O'Neill took him as a model for their early verse. "All of us have been Kipling maniacs at some time or another," confessed H. L. Mencken, "I was myself a very ardent one at seventeen." In one of his letters from Japan, Lafcadio Hearn wrote almost in awe: "There is a prodigious compressed force in the man's style. Never in this world will I be able to write one page to compare with a page of his."

Hearn's praise has its special interest. His talent was as exotic as his origins: born in the Ionian Isles, schooled in England and France, Hearn somehow turned up in Cincinnati, where he wrote for the local papers. A few years later, in 1877, he migrated to New Orleans and was making a reputation as the author of iridescent newspaper sketches on Creole life. Short of stature and quirky in manner, he felt himself something of a misfit, and in 1890 sailed for Japan to become the first gifted American interpreter of a still unknown empire. A writer of style and imagination, his judgment carried weight. On September 2, 1881, Hearn had this to say in the *New Orleans Item:*

> A curious feature of journalism has been developed in the United States within comparatively recent years, which has certainly no parallel in the literature of any other civilized country—the humorous sketch . . . sharp, strong, vivid, with something satanic in its flash . . . The American paragraph is the supreme expression of satirical wit. Nothing rivals it. English fun is grotesque compared with it; the brilliancy of French fun becomes a very feeble light in its blaze . . . It is highly probable that a taste will yet be created for literature of a higher order in the daily press . . . This age in the future of our newspaper is as yet remote; but we may feel assured that it will come even from present indications in the great centers of American journalism.

Allowing for self-serving bias, there was much in what Hearn said. A fresh style *was* evolving, and by degrees the Republic *had* found a new voice. The modern column was in its chrysalis, and was about to emerge in Chicago, soon to be the first city of journalism.

IV

> Journalism's a shrew and scold;
>   I like her.
> She makes you sick, she makes you old;
>   I like her.
> She's daily trouble, storm and strife;
> She's love and hate and death and life;
> She ain't no lady—she's my wife;
>   I like her.
>
> FRANKLIN PIERCE ADAMS,
> *The New York World,*
> February 27, 1931 (Its last issue).

Among the new breed of journalists whose work so captivated Lafcadio Hearn was a cheerful young editorialist who called himself "the Good

Knight *sans peur and sans monnaie.*" This was Eugene Field, who more than anyone planted the column on its present-day plinth. Now vaguely remembered as the author of "Wynken, Blynken and Nod" and other children's verse, Field was the envy of a hard-bitten calling in Chicago's boisterous 1890s. His column, "Sharps and Flats," was filled six days a week with whatever came from his wonderfully fertile imagination. The intense bond he formed with his readers is suggested in this passage from Theodore Dreiser's *Book About Myself:*

> During the year 1890 I had been formulating my first dim notion as to what to do in life. For two years and more I had been reading Eugene Field's "Sharps and Flats," a column he wrote daily for the *Chicago Daily News,* and through this, the various phases of life which he suggested in a humorous though at times romantic way, I was beginning to suspect, vaguely at first, that I wanted to write . . . Nothing else that I had so far read—novels, plays, poems, histories—gave me the same feel for constructive thought as did the matter of his daily notes, poems, and aphorisms, which were of Chicago principally, whereas nearly all others dealt with foreign scenes and people.

Field assuredly concentrated on things local, but his approach was the reverse of parochial; his genially bookish columns reflected a solid classical education, his *Wanderjahre* in Europe and an apprenticeship in St. Louis, Kansas City, and mining-boom Denver. He flattered readers with pastiches of Horace's Sabine odes and echoes of Chaucer and Boccaccio, intermingled with colloquial ballads in the manner of Kipling and Bret Harte. He managed to write without condescension for readers eager to learn. As Dreiser wrote, Chicago was young, blithe and new, a city without traditions, "the very thing that everyone seemed to understand and rejoice in . . . It was something wonderful to witness a world metropolis springing up under one's very eyes, and this was what was happening here before me." In a brief thirty years and despite a devastating fire, Chicago had grown from rail depot and frontier village into a conurbation spread over 181 square miles, more than any other American city. Between stockyard and steel plants were endless rows of timber houses, near grandiose lakefront were the earliest skyscrapers, the affluent suburbs were soon to boast the prairie architecture of Frank Lloyd Wright. In 1890, of its two million people, second only to New York, no less than 68 percent were immigrants. "The older of them cry through the twilight in tongues you have not heard since you were last in Budapest or Athens," marveled a later visitor, Rebecca West, "and the younger in a tongue that you will not understand if you come from England, though it is English." It was a new kind of city and it served as the crucible for a new kind of journalism. Chicago produced at its nadir the cutthroat recklessness dramatized in *The Front Page* and at its summit the artistry pioneered by Field and carried on by Finley Peter Dunne, George Ade, and Ring

Lardner. In her study of Dunne's Dooley sketches, Barbara C. Schaaf plausibly reasons that Chicago's melting-pot population gave a potent impetus to the new journalism. Chicago looked to the newspaper as the bearer of gossip and education, she writes, making personal journalism especially popular among readers who "missed the homely communications network they had left behind in their small towns, or where others from abroad wanted to learn the English language and American ways."

To meet the demand there emerged editors with an audacity as boundless as the prairie grasslands. Among them was Melville E. Stone, at twenty-four the managing editor of *The Inter-Ocean,* its masthead carrying the slogan "Republican in everything, independent in nothing." In 1876, Stone joined with Victor Lawson in launching the penny-a-copy *Daily News* to undersell his nickel-a-copy rivals; by 1885, the new paper's daily average sale was 131,992, nearly as much as the combined circulation of seven other English-language dailies. Stone had an eye for talent, one of his unlikely catches being Art Young, a courtroom artist who became the foremost radical cartoonist of his era. Another was Eugene Field, then on the staff of the *Denver Tribune,* where his "Odds and Ends" column was causing a stir. In 1883, while stopping in Denver to hear Emma Abbott sing at Tabor's Opera House, Stone sought out Field and heard the perennial lament of journalists. As the proprietor recalled in his memoirs:

> Field wanted to join me. He was tired of Denver and mistrustful of the limitations upon him there. But if he was to make a change, he must be assured it was for his permanent good. He was a newspaper man not from choice, but because in that field he could earn his daily bread. Behind all he was conscious of great capacity—not vain nor by any means self-sufficient, but certain that by study and endeavor he could take a high rank in the literary world and could win a place of lasting distinction. So he stipulated that he should be given a column of his own, that he might stand or fall by the excellence of his work. Salary was less an object than opportunity.

Stone had just launched *The Morning News,* a two-cent-a-copy offspring of his evening paper, and Field signed up with the new daily for $50 a week the first year, $50.50 the second year and $55 the third year ("I tack on 50 cents for the second year to gratify a desire to say I am earning a little more money each year"). The new columnist rapidly established himself as a civic treasure, writing his daily quota on an architect's drawing board with penmanship as meticulous as a copper-plate engraving. Art Young in his memoirs recalls Field's penchant for practical jokes, such as inviting visitors to sit in a bottomless cane chair, and the curiosa in his den in suburban Buena Park, which included Jefferson Davis's armchair, an inkstand said to belong to Napoleon, and an unexpected album of pornographic pictures. Young writes of him: "A man of reading and discernment, Field was at the same time Wild Western and raw, an odd combination. He chewed tobacco with avidity and swore convincingly, often inventing unique profane phrases which aroused admiration among his less imaginative coworkers."

I set down these homely details because they have been lost in the mist, and because Eugene Field opened a path for other frustrated pilgrims seeking a measure of integrity in a fugitive calling. When he died in 1895 after twelve years of conducting "Sharps and Flats," the column was no longer a novelty; it was an institution. Its new status and influence were soon to be confirmed by one of Field's younger colleagues, Finley Peter Dunne.

If Field was a Westernized New Englander, Dunne was by birth and breeding Chicago Irish-American Democrat. His immigrant parents came by way of Canada to Chicago's West Side, where his father Peter, trained as a carpenter, prospered in real estate. Young Finley (he was born Peter but adopted his mother's family name) was educated on the pavements and in the newspaper trade, beginning as a copy boy and police reporter, becoming at twenty-one the city editor of *The Chicago Times.* In 1892 he wrote his first dialect columns as a newly hired editorialist on *The Evening Post.* He was all of twenty-three, and was awarded a $10 bonus for each column.

Dunne initially experimented with a bartender named Colonel McNeery who favored customers in his downtown saloon with observations like this: "Fame to be a gr-reat thing . . . If it wasn't f'r fame there'd be manny a rayporther that's dancin' around nowadays with his little pincil an his wad iv papers maakin' notes that'd be out be Jackson Park runnin' a push chair." On October 7, 1893, Dunne moved the locale to a tavern in Bridgeport on Archey Road, in the heart of the Gaelic Sixth Ward, where Martin J. Dooley, saloonkeeper, made his debut. The shift freed Dunne to create a convincing milieu for an Irish sage better able to see that the higher a politician moved, the more his bottom showed. The outlook is evident in this exchange. "Which would you rather be," asks the bartender's foil, Mr. Hennessy. "Famous or rich?" Dooley responds, "I'd like to be famous, an' have enough money to buy off all the threatenin' biographers."

Dooley has worn well. His adages are still gratefully pillaged by a thousand editorial writers addressing readers hardly aware of who Dunne was or what a sensation he caused by remarking on the hollowness of numerous celebrities: Theodore Roosevelt (author, according to Dooley, of *Alone in Cuba*), Admiral Dewey, William McKinley, Andrew Carnegie, the elder John D. Rockefeller and assorted quacks, reformers, and popular idols. As Dooley said: No man is a hero to his undertaker . . . The Vice Presidency isn't a crime exactly, but a kind of disgrace . . . I'm strong for any revolution that isn't going to happen in my day . . . We're a grreat people. We are that. And the best of it is, we know we are.

What keeps Dooley alive is a quality not commonly found or prized in American journalism. He is sardonic rather than worldly, a deflater of national bombast. A similar astringency distinguished the best work of Dunne's friend and colleague, George Ade, who dealt with the same failings from a different vantage—that of the migrants arriving in big cities with foolish illusions and dime-store dreams. His world is well described (with Adean capitals) in an admiring essay by another unsentimental Hoosier, Jean Shepherd: "While the South has been drenched with Decadence, the Midwest has been swimming in Futility. It is dotted with cities and towns that have never quite made it: Toledos that want to be Detroit, Detroits that want to be Chicago, and Chicagos that forever want to be New York. And they all know they are running in a race that has been fixed. Between these major metropoli lie countless hamlets whose only ambition is to become incorporated and to beat the County Seat in baseball."

Ade's hamlet was Kentland, Indiana (1980 pop.: 1,234), eighty Greyhound bus miles from Chicago. The red-letter day of his youth was his departure for Lafayette, Indiana, to attend Purdue University. There he met the cartoonist John McCutcheon, his lifelong collaborator, who after graduation led the way to Chicago. And there McCutcheon found a job on Field's old paper, *The Morning News,* and persuaded the editors to give his newly arrived chum a chance. To the dazzled Ade, Chicago was a mining town five stories high, a cosmopolis bustling with wily swindlers and trusting greenhorns, crooked aldermen and visionary settlement workers, porker aristocrats and Division Street bums, yet withal yearning to be Taken Seriously. Hence the new Opera, the Art Institute and an instant Gothic University. Hence, too, the great Columbian Exposition of 1893, which inspired a new series in the *Morning News:* "Stories of the Streets of the Town," unsigned sketches by Ade, illustrated by McCutcheon. Some of Ade's tales were short stories, some were parodies, some essays. Their success encouraged him four years later to attempt his *Fables in Slang,* in which he capitalized cant phrases and vogue words lest his readers write him off as a Big Boob from the Burbs. His first fable told of Luella, a Good Girl who had taken the Prizes at the Mission Sunday School but who was Plain and a Lumpy Dresser. Luella's sister, by contrast, was short on Intellect but long on Shape, and was named Mary, which she changed to Marie and finally Mae. "Although her Grammar was Sad, it made no Odds. Her Picture was on many a Button." So Mae married a Bucket Shop Man who was awful Generous, and got herself an Automobile, a seat at the Vogner Opera and an Irish chambermaid. In her prosperity, did Mae forget Luella? "No indeed. She took Luella away from the Hat Factory, where the Pay was three Dollars a Week, and Gave her a Position as Assistant Cook at five Dollars. Moral: *Industry and Perseverance bring a sure Reward.*"

One Fable led to another, and they were a Solid Hit, impelling a delighted Victor Lawson, now publisher of *The Record*, to assemble a collection, an overnight bestseller in 1899. Ade had an ear for what he called "the sweet vernacular" and an aversion to "long Boston words." His cunning with bromidic phrases anticipated the art of S. J. Perelman, an Ade devotee. But the moral of his fables was as important as the manner; Ade inverted the gospel of boosterism, with its belief in the soul-healing powers of the great Baal, success. Nice guys in his fables, as frequently in life, finish last. Chicago showed Ade that ruthlessness paid.

It certainly helped with newspaper circulation, as attested by the rise of Chicago's remarkable turn-of-the-century editor, the English-born James Keeley. He arrived in 1889, his antecedents obscure, and was hired as a police reporter by the *Chicago Tribune*. Six years later he was city editor, and in 1898 was named managing editor. Only his years (30) were tender; Keeley liked to say, "A good newspaperman has no friends" and "A good newspaperman is always on the job. He never rests." He worked under the awed gaze of Robert M. Patterson, the young son-in-law of the *Tribune*'s proprietor, Joseph Medill (who, in his will, beseeched his paper never to support "that party which sought to destroy the American Union"). To his underlings Keeley was an unsparing martinet; his literary editor, Burton Rascoe, called him a "vicious bulldog" with the egomania "so often present in those who by sheer aggressiveness and limited but concentrated resourcefulness have arisen rapidly from lowly origins."

Notwithstanding, Keeley was a formidable newspaperman who thirsted to make the *Tribune* "the World's Greatest Newspaper," a boast blazoned on page one from August 29, 1911, until it quietly vanished on January 1, 1977. Keeley's round-the-clock alert enabled the *Tribune* to scoop even the White House on Admiral Dewey's destruction in 1898 of the Spanish fleet at Manila. He assembled a superb staff, headed by Walter Howey, the manic original of the city editor in *The Front Page*. ("Don't ever fake a story or anything in a story," Howey admonished Burton Rascoe, "that is, never let me catch you at it.") And though Keeley had little formal education, he saw the news value of learning and the prestige value of good writing. So he energetically devised new genres of journalism.

Keeley recruited Chicago's health commissioner, Dr. William A. Evans, to write a daily "How to Keep Well" column. The bosomy Lillian Russell wrote regularly on charm and beauty care. A "Helping Hand" column was launched; so was "The Friends of the People," which dealt with what Keeley called "the simple annals of a broken sidewalk, street lamps that fail to light . . . the petty burdens of the poor and uninfluential." Laura Jean Libbey, well known as a writer for women's magazines, started an "Advice to the Lovelorn" column that drew 50,000 letters over a two-year period. On the editorial page, Keeley established "A Line O' Type or Two," a potpourri of verse, commentary, and quips, much of it submitted by readers. The idea proved popular, so Keeley applied it to sports in another new column, "In the Wake of the News." Readers welcomed Keeley's efforts to exploit them, with sprightly results in the case of "A Line O' Type."

The column appeared six days a week and measured the full length of the editorial page. Its impresario from 1901 until his death in 1921 was Bert Leston Taylor, "B.L.T.," a transplanted New Englander. Taylor himself set the standard with deft light verse and acerbic paragraphs. His friend and disciple, Franklin P. Adams, "F.P.A.," said Taylor was "full of healthy malice—sometimes the 'sunny malice of a faun,' but the malice that could help destroy the faker and the poser; his column administered hundreds of hard, deserved, and decidedly ungentle hits at pretenders, and shammers." So vigorous was the competition to "make the Line" that some contributors won their reputation there, notably Kurt M. Stein, "K.M.S.," whose poems in Germanized English filled three books. So why did they write, gratis? A contributor, "R. P," attempted an answer on February 7, 1912:

> If B.L.T. should die, of course we'd sob,
> But one of us, perchance, might grab the job.

And yet, allowing for the depreciation of the dollar, the financial rewards of journalism were scant. In 1913, the sports columnist Hugh Keogh died after a decade's production of "In the Wake of the News," seven days a week. As successor, Keeley turned to a twenty-eight-year-old baseball writer, R. W. Lardner, whose excellent work made the editor feel "absolutely sure you will make good." He was hired for a three-month trial, carried on for six years, supported a family of five, and gained a national reputation when he also found time to write *You Know Me Al.* In 1914, his salary was increased to $100 a week, then raised in 1916 to $200 in a three-year contract; by then magazines were paying him the equivalent of a month's wages for a single article.

Ring Lardner thrived on his treadmill, complaining mainly about the seventh day. This prompted a poem, in response to a friend who asked "Why is the Wake so very short on Monday?":

> I told him, and I'll tell you, too—'twas thus my boss did speak:
> "Each week, to rest your brain, we'll give you one day.
> We really can't expect a man to keep a Wake all week,
> So just send in a verse or two for Monday."
> And, yielding to the boss, which is the proper thing to do,
> I set aside each Sunday as a fun day,
> And Tuesday, Wednesday, Thursday, Friday—all the week through—
> I wonder wotinel to write for Monday.

In fact, Lardner's four sons inspired yards of Monday verse, much of it collected in his first book, *Bib Ballads* (1915). And his editors eventually relented, agreeing to a six-day week.

Lardner wrote about everything, not just sports. His first biographer, Donald Elder, found that Lardner had experimented with nearly every form and style: sketches, parodies, short stories, plays, verse. He challenged readers with madcap contests, burlesqued the more pretentious "Line O' Type or Two," confected monologues or letters as they might have been written by baseball bushers, and so forth. His second biographer, Jonathan Yardley, found the overall quality so high "that the work of those years must be counted among the extraordinary accomplishments of American journalism. Virtually everything upon which Ring's later fame was built originated in the Wake, and in some case the Wake version was actually superior to the more celebrated work." And as Yardley notes, an exceptional array of young writers was dazzled by Lardner: Ernest Hemingway, Sherwood Anderson, James T. Farrell, Nathanael West, and James Thurber. So smitten was Hemingway that he wrote imitative columns signed Ring Lardner, Jr., for his Oak Park High School paper.

Yet if Chicago rewarded Lardner with applause, he soon became aware of New York's more material rewards. By 1914 he had won admiring attention there for his baseball sketches. Franklin P. Adams, a transplanted Chicagoan, wrote a fan letter to which he gratefully responded: "It may not sound reasonable, but sometimes I almost prefer appreciation (from real guys) to dough. However, it's dough and the prospect of it that would tempt me to tackle the New York game. I think a gent in this business would be foolish not to go to New York if he had a good chance. From all I can learn, that's where the real money is. I'm not grabbing such a salary from the Trib that I have any trouble carrying it home."

It was the *mene mene tekel upharsin* of Chicago journalism. After his Dooley dialogues caught on, Finley Peter Dunne had moved east, as did Ade and Dreiser, Lardner and Rascoe, Hecht and MacArthur, all drawn to what another émigré, O. Henry, called Baghdad-on-the-Hudson. Paul Dresser, composer of "Moon Over the Wabash," excitedly wrote to his brother, Theodore Dreiser, in 1898: ". . . there was only one place where one might live in a keen and vigorous way, and that was New York. It was *the* city, the only cosmopolitan city, a wonder world in itself. It was great, wonderful, marvelous, the size, the color, the tang, the beauty."

However, the outback had its revenge. Though it provided the money, New York crucially relied on invaders for ideas. There was thus a Midwest Mafia, a Southern Mafia, a California Mafia, and a Texas Mafia, subcultures whose members knew—and helped—each other. Inventive new periodicals tended to be established by Middle Americans like Henry Luce, Harold Ross, and DeWitt Wallace. And the most successful tabloid, the *New York Daily News,* was founded in 1925 by Joseph Medill Patterson, nephew of Colonel Robert R. McCormick, proprietor of *The Tribune.* The millions who read *The News* were/are unaware that Captain Patterson was responsible for its splashy photojournalism and its staple comics, Dick

Tracy, Orphan Annie and Gasoline Alley—just as more upscale readers were/are unaware that another Chicagoan, Marshall Field, underwrote the innovative, adless, designer packaged daily newspaper *PM*, which in the 1940s was the first to experiment with what is now called lifestyle journalism: consumer reporting, trendy recipes, tips on remodeling houses, and so forth.

An early transplant, Franklin P. Adams, showed New York editors how they could also exploit readers in the fashion of Bert Liston Taylor's "Line O' Type or Two." Born in Chicago in 1881 of German-Jewish stock, young Adams was a college dropout; his reading was voracious, his memory for verse and popular songs prodigious, his pen silver. He became a regular contributor to B.L.T.'s *Tribune* column with aphoristic verse and ingenious limericks like the following (its feeble second line redeemed by a bravura finale):

> A sculptor there was named Praxiteles
> Whose critics would give him blamed little ease;
> They claimed that his pieces
> Were not up to Greece's
> High standard, but way down to Italy's.

In 1903, Adams had already won a berth as a columnist on the *Chicago Journal*. A year later, he turned up in New York and wangled a job on the *Evening Mail*. He started a column called "A Line or Two in Jest," which built a following with clever reader contributions and entered folklore with his line in July 1910, about the great Chicago Cub infield:

> These are the saddest of possible words:
> "Tinker to Evers to Chance."
> Trio of bear cubs, and fleeter than birds,
> Tinker and Evers and Chance.
> Ruthlessly pricking our gonfalon bubble,
> Making a Giant hit into a double—
> Words that are heavy with nothing but trouble:
> "Tinker to Evers to Chance."

Four years later, F. P. A. accepted a better offer from the *New York Tribune*, and devised a catchier, double-edged name, "The Conning Tower."

Adams was a disciplined planner who went far with a modest talent. This was not the case with his rival Don Marquis, whose gifts were larger and origins humbler. Marquis recalled his boyhood in Walnut, Illinois, in a fine *Saturday Evening Post* essay, "Confessions of a Columnist":

> I was inoculated in early youth; when I was a kid I read, every day, Eugene Field's column in a Chicago paper; and later, George Ade's sketches, and I decided I wanted to do something like that. After teach-

> ing at a country school—an occupation into which I naturally drifted because I had very little education—clerking in a drug store and other makeshift jobs, I finally went into a country printing office. The owner and editor was good natured, and before I had learned to be a really competent printer I was helping to edit a weekly paper . . . [put in charge of a weekly] I had to collect all the news, write all the editorials, solicit the advertisements and write them, set about half the type myself, saw boiler plate to fit holes in the columns, make up the paper, run off part of the edition myself on the old flat bed press, fold and wrap the papers and take them to the post office. But this was incidental, in my mind, to the main thing, which was writing and printing a column; the other work was really the price I paid for the privilege of seeing my verse and sketches and paragraphs and fables in print.

In 1902, in pursuit of that grail, Marquis had moved on to Atlanta, first to a newspaper job, then to *Uncle Remus's Magazine,* newly launched by Joel Chandler Harris. But though he regarded Harris as a "great man," being No. 2 on a children's monthly was not what Marquis wanted. So in 1909, he pressed on to New York with $7.50 in cash; after three lean years, he was rewarded with an offer to start a new column, "The Sun Dial," on the editorial page of the *New York Evening Sun.* "When I first got the signed column, I was ready for it," Marquis later recalled, "I had written for it and saved up for it some of the best general stuff I could do; and the day I got it, I began slamming into it the stuff I had saved." The column's success was instantaneous. Marquis tried to explain why:

> Showmanship figures in everything. The difference between failure and success is frequently the difference between nonpareil type and brevier. Nobody had noticed the "Notes and Comment," but "The Sun Dial"—with no better stuff in it, really—got across at once. A column must have plenty of white space, a challenging make-up, constant variation in typographical style; not only must it catch the eye but it must have points and corners and barbs that prick and stimulate the vision, a surface and a texture that intrigue and cling to and pull at the sight. Franklin P. Adams is the master hand at this sort of thing . . . I tried to get as much variety in the stuff as there was in its typographical presentation. So besides the verse, paragraphs, sketches, fables and occasional serious expression of opinion, I began to create characters . . .

Marquis worked in a crammed, untidy cubicle, typing on a mammoth Remington that looked (and sounded) like a printing-press. On March 29, 1916, as the columnist matter-of-factly informed his readers, something peculiar happened: he discovered "a gigantic cockroach jumping about on the keys" who cast himself with all his force on the keys, but lacked the strength to work the capital letters. The first lines of archy's first poem was a timeless sigh: "expression is the need of my soul/i was once a vers libre bard/ but i died and my soul went into the body of a cockroach/it has given me a new outlook on life." Very like his insect pal, Marquis threw all his energy into what he came to call a "twenty-three inch grave." But though he created other comic characters he accurately foresaw that he would be recalled only as "the creator of a goddam cockroach."

In time, Marquis moved from the *Sun* to the *New York Tribune,* where he wrote a new column, "The Lantern," until 1925, when he quit to write for magazines and stage. By this time, the newspaper column was well established as the "most sophisticated" of the minor arts in America, according to Gilbert Seldes in his *Seven Lively Arts.* And the editor most responsible for that was Herbert Bayard Swope of the *New York Evening World,* who in 1921 invented the Op-Ed page.

Its prototype was "the highbrow page" introduced *circa* 1912 in the *Chicago Tribune's* Sunday edition by the new head of the Sunday department, Joseph M. Patterson. He used the page facing the editorials as a showcase for provocative authors like Bernard Shaw, H. G. Wells, G. K. Chesterton and, nearer home, H. L. Mencken. The page was edited by the quarrelsome Burton Rascoe, who in his memoirs says Swope "imitated" the Chicago innovation. This is unfair. Swope's idea was more ambitious and fifty years later its rediscovery was to reshape American newspapers.

In 1920, when Swope took over, *The World* had already changed. Founder Joseph Pulitzer had died a decade earlier and was succeeded as publisher by his eldest son, Ralph. The paper had crusaded vainly for Woodrow Wilson's visionary League of Nations under Frank Cobb, who was succeeded as editorial page editor by a chastened skeptic, Walter Lippmann. H. L. Mencken was again writing for the *Baltimore Sun,* after a wartime exile owing to his pro-Germanism and his scorn for Woodrow Wilson. Now came Harding, upon whose pompous language Mencken fell with a whoop of delight, calling it "Gamaliese." The mood had switched from uplift to mockery, especially among the young who were now asked to observe a Prohibition law openly flouted by their elders. The mood of the hour was irreverence, and the resourceful Swope turned it to the *World*'s advantage.

Swope noticed that a prize page, facing the editorials, was a catchall for book reviews, society boilerplate, and obituaries. As he later explained in a letter to Gene Fowler: "It occurred to me that nothing is more interesting than opinion when opinion is interesting, so I devised a method of cleaning off the page opposite the editorial, which became the most important in America . . . and thereon I decided to print opinions, ignoring facts." Opportunely, he had just been approached for a job by Heywood Broun, a sports writer turned social critic then chafing under the restraints he felt at the starchier *New York Tribune.* Swope hired Broun at $15,000 a year, and made Op-Ed space for a daily column named by the editor "It Seems to Me." In due course, F.P.A. departed from the same competitor, bringing "The Conning Tower" with him.

This was undoubtedly Adams's decade. As founder along with Broun of the fabled Round Table at the Algonquin Hotel, F.P.A. was able to make *The World* resound with the unpaid contributions of a stellar group: George S. Kaufman, Marc Connolly, Ring Lardner, E. B. White, Edna Ferber, Gellett Burgess and best of all, Dorothy Parker, who later quipped that Adams had "raised me from a couplet." But what gave his column its special eclat were the talented writers adorning the spaces around it. Swope made Op-Ed his special concern, recruiting for it during the 1920s Alexander Woollcott (drama critic), Deems Taylor and Samuel Chotzinoff (music), Lawrence Stallings and Harry Hansen (books), with cameo appearances by Robert Benchley, Frank Sullivan, and William Bolitho (the South African-born foreign correspondent now remembered as a Hemingway mentor). Though *The World* could not compete with *The Times* in the scope of its coverage, its talented, temperamental Op-Ed writers were the talk of the town—an effect amplified by shameless back-scratching.

Thus when there was a family quarrel, everybody took notice. The most memorable flared over the Sacco-Vanzetti case after a special committee headed by the president of Harvard, A. Lawrence Lowell, upheld the death sentence handed down against the two radicals by another Harvardman, Judge Thayer. Heywood Broun ('10) erupted in the spirit of one of his classmates (John Reed) but to the displeasure of his publisher ('00) and the editorial page editor, Walter Lippmann ('10). On the same day a *World* editorial said it still faulted the trial of the two anarchists but the law had to be obeyed, Broun wrote: "It is not every person who has a President of Harvard throw the switch for him. If this is lynching, at least the fish peddler and his friend, the factory hand, may take unction to their souls that they will die at the hands of men in dinner jackets and academic gown. . . ." Next day, a *World* editorial again asked for clemency, while Broun pushed a different switch: "Shall the institution of learning at Cambridge, which we once called Harvard, be known as Hangman's House?" This was too much for an editorialist at the *Times*, which saw in Broun's "educated sneer" at Lowell's "great civic duty" proof of "the wild and irresponsible spirit" in the land.

It was also too much for the publisher, Ralph Pulitzer, who refused to publish a third such column on Sacco-Vanzetti, announcing his decision in Broun's regular space. Broun stood his ground, saying the difference was not on substance but on his vehemence: "This could have been a duet with the editorial page carrying the air in sweet and tenor tones while in my compartment bass rumblings were added." Pulitzer was unmoved, and editor Swope, who might have made peace, was off at the races in Saratoga. So Broun left the paper on "permanent strike" though to Pulitzer he was on "witch's sabbath." What to the columnist was an issue of censorship was to the proprietor a question of whether Broun "may direct the *World* to publish a column against its conscience."

The *World* had lost more than Broun; the fizz also went out of Op-Ed. The absence of "It Seems To Me" spoke all too loudly about the realities of authority. For his part, Broun found a new berth on the Scripps-Howard flagship, the *Telegram,* and was now pressing his campaign for what was to become the American Newspaper Guild. In 1930, as the Great Depression lengthened breadlines, Broun announced himself a Socialist and ran for Congress in the hopeless silk-stocking district of Manhattan's Upper East Side. And over at the *World,* publisher Pulitzer gloomily contemplated a first year of serious losses. In February 1931 the unthinkable occurred. The Pulitzer heirs won court approval to revoke the founder's will and sold the *World,* over the objections of Swope and the entire staff, to the Scripps-Howard chain. In time, the *World-Telegram* absorbed the *Sun,* the *Herald Tribune,* and the *Journal-American,* before the surviving paper with its agglutinized masthead, also sank in 1965, as if pulled down by the ghosts of journalism past.

Still, the same drift to concentrated ownership and the worrisome rise of the one-newspaper town had an interesting and useful side-effect: it recalled Op-Ed to life. The surviving dominant newspapers in bigger cities found it both equitable and expedient to adopt a more ecumenical policy on opinion features. Conservative papers like the *Chicago Tribune* and the *Los Angeles Times* sought greater balance, as did less conservative survivors like the *New York Times* and *Boston Globe.* When the *Washington Post* absorbed its morning rival, the very conservative *Times Herald,* in 1954, the new owners kept not just popular comics but also rightwing columnists like George Sokolsky in the combined paper. At least part of the page opposite editorials was devoted to columnists. But the *New York Times* continued to publish obituaries in the same choice space. In 1970, the publisher, Arthur O. Sulzberger, cut through jurisdictional disputes and acted on an idea long pressed by the editorial page editor, John B. Oakes. So Op-Ed was reborn, and similar pages soon became the rule elsewhere.

This brings us to our own era. Whether the new breed of columnists is better or worse than the old is an interesting but subjective question. It is always the best of times and worst of times in the world of journalism, where golden ages tend to be embalmed in memory and unconsulted archives. Thus E. B. White, ruminating on the excellence of Don Marquis, declared that the newspaper column had grievously degenerated:

> In 1916 to hold a job on a daily paper, a columnist was expected to be something of a scholar and poet, or if not a poet at least to harbor the soul of a transmigrated poet. Nowadays, to get a columning job a man need only have the soul of a Peep Tom or a third-rate prophet. There are plenty of loud clowns and bad poets at work on papers today, but

> there are not many columnists adding to *belles lettres,* and certainly there is no Don Marquis . . . Mr. Marquis's cockroach was more than the natural issue of a creative and humorous mind. Archy was the child of compulsion, the stern compulsion of journalism. The compulsion is as great today as it ever was, but is met in a different spirit. Archy used to come back from the golden companionship of the tavern with the poet's report of life as seen from the under side. Today's columnist returns from the platinum companionship of the night club with a dozen pieces of watered gossip and a few bottomless anecdotes. Archy returned carrying a heavy load of wine and dreams. These later cockroaches come sober from their taverns, carrying a basket of fluff.

So it seemed to White in 1950, yet one may reasonably wonder if he did not put his thumb on the scales. For every Don Marquis, there were a hundred third-rate poets and insipid paragraphers. There were no political columnists with the grace of a James Reston, the subtlety of a Murray Kempton, or the craft of a George Will. Even as the number of newspapers shrank, the boundaries of discourse widened, so that a Miss Manners or an Anna Quindlen have been able to write unblushingly about hitherto unmentionable matters, in and out of wedlock. And an age rich in B's—Baker, Buchwald, Breslin, Bombeck, and Buckley—may in hindsight be seen as not entirely contemptible, perhaps even as a silver age.

Besides, as the columnist has escaped the six-day treadmill, successful practitioners have learned new devices for building and holding an audience. When A. M. Rosenthal moved from the executive editorship to the role of columnist on the *New York Times,* he was favored with astute advice from his columnist colleague William Safire. Writing in the *Times* house publication, Safire tendered these trade secrets:

> *Never put the story in the lead.* Forget all that journalism-school stuff you drilled into your reporters; this new life is not straight-line newspapering, it is hot-shot philosophizing. You do not have to pander to your readers habits; let them get into your habit, which may include reading your column warily, wondering what your message really is. Let 'em have a hot shot of ambiguity right between the eyes.
>
> *As you cultivate the garden of controversy, burn the bridges of objectivity.* Show me an evenhanded columnist and I'll show you an odds-on favorite soporific. What is required in a great editor and expected of a great reporter is death to a provocative pundit . . . Keep 'em guessing, muttering, off balance. Even a jerk who knees is better than a knee that jerks. (That is one of those meaningless turnaround sentences that copy editors are trained to cut out as cutesy; Op-Ed copy editors are forced to leave them in, for fear of stifling a profundity.)

Strategies like these enable a columnist to cross a vital threshold and form a bond with his or her audience. In an anthology of essays compiled long ago by J. B. Priestley, the late novelist and playwright offered an important truth:

> To anyone with any knowledge of literary history, there is nothing more amusing than the not infrequent complaint of critics and reviewers, who imagine that they are standing for the dignity of letters, against the practice of collecting contributions to the Press, essays or critical articles, and making books of them.
>
> We are always led to infer that this is a new and reprehensible practice, a mark of a degenerate age. The truth is, of course, that all the best essays in the language have first seen the light in the periodical press . . . The economic influence is easy to understand. What is not so easy to understand is the influence this periodical work has upon the essayist's attitude of mind . . . When a man is writing regularly in one place for one set of readers (and nearly all the essayists were regular contributors to the Press, appearing in the same periodical at regular intervals), he tends to lose a certain stiffness, formality, self-consciousness, that would inevitably make its appearance if he were writing a whole book alone.
>
> He comes to feel that he is among friends and can afford as it were to let himself go, and the secret of writing a good essay is to let oneself go. . . . it also encourages him to focus his attention upon passing little things that he might have disdained were he not writing for the next week's paper. He is a snapper-up of unconsidered trifles, and it is his pleasure and privilege to glimpse the significance of such trifles, so that for a second we see them, surprisingly, against the background of the Eternities.

What Priestley expressed was itself a glimpse of a not-so-obvious truth about the fruitful communion between writer and audience. There's nothing shameful about journalism as a venue for letters. Without overstating the claims for what is at best an uneven popular art, one can say that in the voice of the columnists one can hear, if at times discordantly, the joyful noise of a free people.

Pundits, Poets, and Wits

Benjamin Franklin

(Silence Dogood)

Benjamin Franklin (1706–90) led the way as the first recognizable columnist in the American colonies. The place was Boston, and the annus mirabilis *1722, when the sixteen-year-old Benjamin surreptitiously contributed fourteen essays signed "Silence Dogood" to the* New-England Courant *published by his half-brother James. Pretending in these letters to be a sharp-tongued young widow, Franklin borrowed his manner from Addison and Steele but his matter was local, persuading his readers that he was "a young Genius that had a Turn for Libelling & Satyr."*

Letters 1, 2, and 12 are reproduced here. I have resisted selecting Poor Richard's aphorisms from Franklin's Almanacks (1733–58) because they all can be readily found in the Library of America edition of his Writings *(New York, 1987), along with the rest of the Dogood letters and much else. Of special interest are Franklin's pseudonymous contributions to the London press when he was colonial agent, 1758–75, traced down and collected in* Benjamin Franklin's Letters to the Press *(Chapel Hill, 1950), edited by Verner W. Crane.*

SILENCE DOGOOD INTRODUCES HERSELF

New-England Courant *April 2, 1722*

Sir,

It may not be improper in the first place in inform your Readers, that I intend once a Fortnight to present them, by the Help of this Paper, with a short Epistle, which I presume will add somewhat to their Entertainment.

And since it is observed, that the Generality of People, now a days, are unwilling either to commend or dispraise what they read, until they are in some measure informed who or what the Author of it is, whether he be *poor* or *rich, old* or *young,* a *Schollar* or a *Leather Apron Man, &c.* and give their Opinion of the Performance, according to the Knowledge which they have of the Author's Circumstances, it may not be amiss to begin with a short Account of my past Life and present Condition, that the Reader may not be at a Loss to judge whether or no my Lucubrations are worth his reading.*

At the time of my Birth, my Parents were on Ship-board in their Way from London to N. England. My Entrance into this troublesome World was attended with the Death of my Father, a Misfortune, which tho' I was not then capable of knowing, I shall never be able to forget; for as he, poor Man, stood upon the Deck rejoycing at my Birth, a merciless Wave entred the Ship, and in one Moment carry'd him beyond Reprieve. Thus, was the *first Day* which I saw, the *last* that was seen by my Father; and thus was my disconsolate Mother at once made both a *Parent* and a *Widow.*

When we arrived at Boston (which was not long after) I was put to Nurse in a Country Place, at a small Distance from the Town, where I went to School, and past my Infancy and Childhood in Vanity and Idleness, until I was bound out Apprentice, that I might no longer be a Charge to my Indigent Mother, who was put to hard Shifts for a Living.

My Master was a Country Minister, a pious good-natur'd young Man, and a Batchelor: He labour'd with all his Might to instil vertuous and godly Principles into my tender Soul, well knowing that it was the most suitable Time to make deep and lasting Impressions on the Mind, while it was yet untainted with Vice, free and unbiass'd. He endeavour'd that I might be instructed in all that Knowledge and Learning which is necessary for our Sex, and deny'd me no Accomplishment that could possibly be attained in a Country Place; such as all Sorts of Needle-Work, Writing, Arithmetick, &c. and observing that I took a more than ordinary Delight in reading ingenious Books, he gave me the free Use of his Library, which tho' it was but small, yet it was well chose, to inform the Understanding rightly, and enable the Mind to frame great and noble Ideas.

Before I had liv'd quite two Years with this Reverend Gentleman, my indulgent Mother departed this Life, leaving me as it were by my self, having no Relation on Earth within my Knowledge.

I will not abuse your Patience with a tedious Recital of all the frivolous Accidents of my Life, that happened from this Time until I arrived

*Franklin's inspiration was the opening paragraph of the first *Spectator* essay by Addison (March 1, 1710): "I have observed, that a Reader seldom peruses a Book with Pleasure, 'till he knows whether the Writer of it be a black or a fair Man, of a mild or cholerick Disposition, Married or a Batchelor, with other Particulars of a like nature . . ." But as Carl van Doren notes, it remained for Franklin to add "or a Leather Apron Man."

to Years of Discretion, only inform you that I liv'd a chearful Country Life, spending my leisure Time either in some innocent Diversion with the neighbouring Females, or in some shady Retirement, with the best of Company, *Books*. Thus I past away the Time with a Mixture of Profit and Pleasure, having no affliction but what was imaginary, and created in my own Fancy; as nothing is more common with us Women, than to be grieving for nothing, when we have nothing else to grieve for.

As I would not engross too much of your Paper at once, I will defer the Remainder of my Story until my next Letter; in the mean time desiring your Readers to exercise their Patience, and bear with my Humours now and then, because I shall trouble them but seldom. I am not insensible of the Impossibility of pleasing all, but I would not willingly displease any; and for those who will take Offence where none is intended, they are beneath the Notice of Your Humble Servant,

Silence Dogood

SHE TELLS MORE OF HER LIFE

New-England Courant *April 16, 1722*

Sir,

Histories of Lives are seldom entertaining, unless they contain something either admirable or exemplar: And since there is little or nothing of this Nature in my own Adventures, I will not tire your Readers with tedious Particulars of no Consequence, but will briefly, and in as few Words as possible, relate the most material Occurrences of my Life, and according to my Promise, confine all to this Letter.

My Reverend Master who had hitherto remained a Batchelor, (after much Meditation on the Eighteenth verse of the Second Chapter of Genesis,) took up a Resolution to marry; and having made several unsuccessful fruitless Attempts on the more topping Sort of our Sex, and being tir'd with making troublesome Journeys and Visits to no Purpose, he began unexpectedly to cast a loving Eye upon Me, whom he had brought up cleverly to his Hand.

There is certainly scarce any Part of a Man's Life in which he appears more silly and ridiculous, than when he makes his first Onset in Courtship. The aukward Manner in which my Masster first discover'd his Intentions, made me, in spite of my Reverence to his Person, burst out into an unmannerly Laughter: However, having ask'd his Pardon, and with much ado compos'd my Countenance, I promis'd him I would take his Proposal into serious Consideration, and speedily give him an Answer.

As he had been a great Benefactor (and in a Manner a Father to me) I could not well deny his Request, when I once perceived he was in earnest. Whether it was Love, or Gratitude, or Pride, or all Three that made me consent, I know not; but it is certain, he found it no hard Matter, by the Help of his Rhetorick, to conquer my Heart, and perswade me to marry him.

This unexpected Match was very astonishing to all the Country round about, and served to furnish them with Discourse for a long Time after; some approving it, others disliking it, as they were led by their various Fancies and Inclinations.

We lived happily together in the Heighth of conjugal Love and mutual Endearments, for near Seven Years, in which Time we added Two likely Girls and a Boy to the Family of the Dogoods: But alas! When my Sun was in its meridian Altitude, inexorable unrelenting Death, as if he had envy'd my Happiness and Tranquility, and resolv'd to make me entirely miserable by the Loss of so good an Husband, hastened his Flight to the Heavenly World, by a sudden unexpected Departure from this.

I have now remained in a State of Widowhood for several Years, but it is a State I never much admir'd, and I am apt to fancy that I could be easily perswaded to marry again, provided I was sure of a good-humour'd, sober, agreeable Companion: But one, even with these few good Qualities, being hard to find, I have lately relinquish'd all Thoughts of that Nature.

At present I pass away my leisure Hours in Conversation, either with my honest Neighbour Rusticus and his Family, or with the ingenious Minister of our Town, who now lodges at my House, and by whose Assistance I intend now and then to beautify my Writings with a Sentence or two in the learned Languages, which will not only be fashionable, and pleasing to those who do not understand it, but will likewise be very ornamental.

I shall conclude this with my own Character, which (one would think) I should be best able to give. *Know then,* That I am an Enemy to Vice, and a Friend to Vertue. I am one of an extensive Charity, and a great Forgiver of *private* Injuries: A hearty Lover of the Clergy and all good Men, and a mortal Enemy to arbitrary Government and unlimited Power. I am naturally very jealous for the Rights and Liberties of my Country; and the least appearance of an Incroachment on those invaluable Priviledges, is apt to make my Blood boil exceedingly. I have likewise a natural Inclination to observe and reprove the Faults of others, at which I have an excellent Faculty. I speak this by Way of Warning to all such whose Offences shall come under my Cognizance, for I never intend to wrap my Talent in a Napkin. To be brief; I am courteous and affable, good humour'd (unless I am first provok'd,) and handsome, and sometimes witty, but always, Sir, Your Friend and Humble Servant,

Silence Dogood

SHE EXAMINES INEBRIATION

New-England Courant *September 10, 1722*

Sir,

It is no unprofitable tho' unpleasant Pursuit, diligently to inspect and consider the Manners and Conversation of Men, who, insensible of the greatest Enjoyments of humane Life, abandon themselves to Vice from a false Notion of *Pleasure* and *good Fellowship.* A true and natural Representation of any Enormity, is often the best Argument against it and Means of removing it, when the most severe Reprehensions alone, are found ineffectual.

I would in this letter improve the little Observation I have made on the Vice of *Drunkeness,* the better to reclaim the *good Fellows* who usually pay the Devotions of the Evening to Bacchus.

I doubt not but *moderate Drinking* has been improv'd for the Diffusion of Knowledge among the ingenious Part of Mankind, who want the Talent of a ready Utterance, in order to discover the Conceptions of their Minds in an entertaining and intelligible Manner. 'Tis true, drinking does not *improve* our Faculties, but it enables us to *use* them; and therefore I conclude, that much Study and Experience, and a little Liquor, are of absolute Necessity for some Tempers, in order to make them accomplish'd Orators. Dic. Ponder discovers an excellent Judgment when he is inspir'd with a Glass or two of *Claret,* but he passes for a Fool among those of small Observation, who never saw him the better for Drink. And here it will not be improper to observe, That the moderate Use of Liquor, and a well plac'd and well regulated Anger, often produce this same Effect; and some who cannot ordinarily talk but in broken Sentences and false Grammar, do in the Heat of Passion express themselves with as much Eloquence as Warmth. Hence it is that my own Sex are generally the most eloquent, because the most passionate. "It has been said in the Praise of some Men, (says an ingenious Author,) that they could talk whole Hours together upon any thing; but it must be owned to the Honour of the other Sex, that there are many among them who can talk whole Hours together upon Nothing. I have known a Woman branch out into a long extempore Dissertation on the Edging of a Petticoat, and chide her Servant for breaking a China Cup, in all the Figures of Rhetorick."

But after all it must be consider'd, that no Pleasure can give Satisfaction or prove advantageous to a *reasonable Mind,* which is not attended with the *Restraints of Reason.* Enjoyment is not to be found by Excess in any sensual Gratification; but on the contrary, the immoderate Cravings of the Voluptuary, are always succeeded with Loathing and a palled Appetite. What Pleasure can the Drunkard have in the Reflection, that, while in his Cups, he retain'd only the Shape of a Man, and acted the Part of a Beast; or that from reasonable Discourse a few Minutes before, he descended to Impertinence and Nonsense?

I cannot pretend to account for the different Effects of Liquor on Persons of different Dispositions, who are guilty of Excess in the Use of it. 'Tis strange to see Men of a regular Conversation become rakish and profane when intoxicated with Drink, and yet more surprizing to observe, that some who appear to be the most profligate Wretches when sober, become mighty religious in their Cups, and will then, and at no other Time address their Maker, but when they are destitute of Reason, and actually affronting him. Some shrink in the Wetting, and others swell to such an unusual Bulk in their Imaginations, that they can in an Instant understand all Arts and Sciences, by the liberal Education of a little vivifying *Punch,* or a sufficient Quantity of other exhilerating Liquor.

And as the Effects of Liquor are various, so are the Characters given to its Devourers. It argues some Shame in the Drunkards themselves, in that they have invented numberless Words and Phrases to cover their Folly, whose proper Significations are harmless, or have no Signification at all. They are seldom known to be *drunk,* tho' they are very often *boozey, cogey, tipsey, fox'd, merry, mellow, fuddl'd, groatable, Confoundedly cut, See two Moons,* are *Among the Philistines, In a very good Humour, See the Sun,* or, *The Sun has shone upon them;* they *Clip the King's English,* are *Almost froze, Feavourish, In their Altitudes, Pretty well enter'd,* &c. In short, every Day produces some new Word or Phrase which might be added to the Vocabulary of the *Tiplers:* But I have chose to mention these few, because if at any Time a Man of Sobriety and Temperance happens to *cut himself confoundedly,* or is *almost froze,* or *feavourish,* or accidentally *sees the Sun,* &c. he may escape the Imputation of being *drunk,* when his Misfortune comes to be related. I am Sir, Your Humble Servant,

Silence Dogood

James Alexander

The first colonist to formulate a doctrine of press freedom was James Alexander (1691–1756), writing as an American "Cato." He edited the New-York Weekly Journal, *whose printer, John Peter Zenger, became a byword for free speech. Alexander was Scots-born, a supporter of the Stuart cause in the failed Rebellion of 1715, who had fled to America. His gifts in mathematics and law won him the posts of Surveyor General for New York and New Jersey, and of Attorney General of New Jersey. He later fell out with New York's high-handed Governor, William Cosby. Opponents of Cosby launched the* Journal, *enlivened with articles by Alexander that the Governor deemed "scandelous, virulent, false and seditious." Zenger was jailed as the responsible printer, but the paper continued to publish, reprinting the famous "Cato" essays by the English Whigs, John Trenchard and Thomas Gordon, the boldest proponents of press freedom. As legal strategist in Zenger's trial, Alexander was disbarred by the presiding judge, a Cosby crony, and the defense turned to the eminent colonial lawyer, Andrew Hamilton of Philadelphia. Ignoring precedent, he implored the jury to consider the truth of the alleged libels: "It is not the cause of a poor printer, nor of New York alone, which you are trying. It is the best cause. It is the cause of liberty." Though Zenger was acquitted, the verdict had little practical effect. In fact, Leonard Levy notes, the worst offenders against colonial press freedom were not governors or judges, but intolerant provincial assemblies. As symbol, however, the Zenger case was to become, in the words of Gouverneur Morris, "the morning star of that liberty which subsequently revolutionized America." Alexander's appeal as an American "Cato" is in Levy's* Freedom of the Press: From Zenger to Jefferson *(Indianapolis, 1966). For a prime source and modern reconsideration, see James Alexander,* A Brief Narrative of the Case and Trial of John Peter Zenger *(Cambridge, Mass., 1972), edited and introduced by Stanley N. Katz.*

AN AMERICAN "CATO" DEFENDS CRITICISM OF THE GOVERNMENT

New-York Weekly Journal *November 12, 1733*

Mr. Zenger.

Incert the following in your next, and you'll oblige your Friend,

CATO.

The Liberty of the Press is a Subject of the greatest Importance, and in which every Individual is as much concern'd as he is in any other Part of Liberty: Therefore it will not be improper to communicate to the publick the Sentiments of a late excellent Writer upon this Point. Such is the Elegance and Perspicuity of his Writings, such the inimitable Force of his Reasoning, that it will be difficult to say any Thing new that he has not said, or not to say that much worse which he has said.

There are two Sorts of Monarchies, an absolute and a limited one. In the first, the Liberty of the Press can never be maintained, it is inconsistent with it; for what absolute Monarch would suffer any Subject to animadvert on his Actions when it is in his Power to declare the Crime and to nominate the Punishment? This would make it very dangerous to exercise such a Liberty. Besides the Object against which those Pens must be directed is their Sovereign, the sole Supreme Magistrate; for there being no Law in those Monarchies but the Will of the Prince, it makes it necessary for his Ministers to consult his Pleasure before any Thing can be undertaken: He is therefore properly chargeable with the Grievances of his Subjects, and what the Minister there acts being in Obedience to the Prince, he ought not to incur the Hatred of the People; for it would be hard to impute that to him for a Crime which is the Fruit of his Allegiance, and for refusing which he might incur the Penalties of Treason. Besides, in an absolute Monarchy, the Will of the Prince being the Law, a Liberty of the Press to complain of Grievances would be complaining against the Law and the Constitution, to which they have submitted or have been obliged to submit; and therefore, in one Sense, may be said to deserve Punishment; so that under an absolute Monarchy, I say, such a Liberty is inconsistent with the Constitution, having no proper Subject to Politics on which it might be exercis'd, and if exercis'd would incur a certain Penalty.

But in a limited Monarchy, as *England* is, our Laws are known, fixed, and established. They are the streight Rule and sure Guide to direct the King, the Ministers, and other his Subjects: And therefore an Offense against the Laws is such an Offense against the Constitution as ought to receive a proper adequate Punishment; the several Constituents of the

Government, the Ministry, and all subordinate Magistrates, having their certain, known, and limited Sphere in which they move; one part may certainly err, misbehave, and become criminal, without involving the rest or any of them in the Crime or Punishment.

But some of these may be criminal, yet above Punishment, which surely cannot be denied, since most Reigns have furnished us with too many instances of powerful and wicked Ministers, some of whom by their Power have absolutely escaped Punishment, and the Rest, who met their Fate, are likewise Instances of this Power as much to the Purpose; for it was manifest in them that their Power had long protected them, their Crimes having often long preceded their much desired annd deserved Punishment and Reward.

That *Might over comes Right,* or which is the same Thing, that Might preserves and defends Men from Punishment, is a Proverb established and confirmed by Time and Experience, the surest Discoverers of Truth and Certainty. It is this therefore which makes the Liberty of the Press in a limited Monarchy and in all its Colonies and Plantations proper, convenient, and necessary, or indeed it is rather incorporated and interwoven with our very Constitution; for if such an over grown Criminal, or an impudent Monster in Iniquity, cannot immediately be come at by ordinary Justice, let him yet receive the Lash of satyr, let the glaring Truths of his ill Administration, if possible, awaken his Conscience, and if he has no Conscience, rouse his Fear by Shewing him his Deserts, sting him with the Dread of Punishment, cover him with Shame, and render his Actions odious to all honest Minds. These Methods may in Time, and by watching and exposing his Actions, make him at least more Cautious, and perhaps at last bring down the great haughty and secure Criminal within the Reach and Grasp of ordinary Justice. This Advantage therefore of Exposing the exorbitant Crimes of wicked Ministers under a limited Monarchy makes the Liberty of the Press not only consistent with, but a necessary Part of, the Constitution itself.

It is indeed urged, that the Liberty of the Press ought to be restrained, because not only the Actions of evil Ministers may be exposed, but the Character of good ones traduced. Admit it in the strongest Light that Calumny and Lyes would prevail, and blast the Character of a great and good Minister; yet that is a less Evil than the Advantages we reap from the Liberty of the Press, as it is a Curb, a Bridle, a Terror, a Shame, and Restraint to evil Ministers; and it may be the only punishment, especially for a Time. But when did Calumnies and Lyes ever destroy the Character of one good Minister? Their benign Influences are known, tasted, and felt by every body: Or if their Characters have been clouded for a Time, yet they have generally shined forth in greater Luster: Truth will always prevail over Falsehood.

The Facts exposed are not to be believed, because said or published;

but it draws People's Attention, directs their View, and fixes the Eye in a proper Position that everyone may judge for himself whether those Facts are true or not. People will recollect, enquire and search, before they condemn; and therefore very few good Ministers can be hurt by Falsehood, but many wicked Ones by seasonable Truth: But however the Mischief that a few many possibly, but improbably, suffer by the Freedom of the Press is not to be put in Competition with the Danger which the KING and the *people* may suffer by a shameful, cowardly Silence under the Tyranny of an insolent, rapacious, infamous Minister.

Thomas Paine

When Thomas Paine (1737–1809) arrived in Philadelphia in November 1774, the rudiments of a national press existed. The Massachusetts Spy *carried on its front page the names of the colonies on a segmented snake with the words "Join or Die" under the slogan: "Do thou Great LIBERTY inspire our Souls." And the* Spy*—"rabid, yellow and very successful," writes the historian Esther Forbes—was representative of dozens of papers eager for lively copy with a patriotic savor; moreover, something like a syndication network also existed. Paine more than anyone gave the discordant colonies a unifying voice. Arriving with a letter from Benjamin Franklin to his printer son-in-law, Richard Bache, Paine was soon at work on the latter's just-founded* Pennsylvania Magazine, *filling its pages with topical pieces signed "Atlanticus," "Aesop," or "Vox Populi." His radical temper was evident in his attacks on imperial misrule that anticipated "Common Sense." Published in January 1776, his pamphlet urging independence rapidly sold a half-million copies. A horrified Philadelphia loyalist sought to reply, resorting to the (by now) hackneyed pen name "Cato." This provoked four vigorous letters from Paine, who signed himself "Forester." But Paine's plea in the third,* "Forget not the hapless African," *was itself forgotten when the Signers made separation a fact. The rest is folklore: Paine at Valley Forge, writing the first* Crisis *paper ("These are the times that . . ."). After the Revolution, Paine was cursed or forgotten in the nation he helped invent. The fullest biography is by Moncure Conway (New York, 1892); his works were last collected in the* Complete Writings, *edited by Philip S. Foner (New York, 1945), from which these selections are taken.*

REFLECTIONS ON TITLES

Pennsylvania Magazine *May, 1775*

> Ask me what's honor? I'll the truth impart:
> Know, honor then, is *Honesty of Heart.*
> WHITEHEAD

When I reflect on the pompous titles bestowed on unworthy men, I feel an indignity that instructs me to despise the absurdity. The *Honorable* plunderer of his country, or the *Right Honorable* murderer of mankind, create such a contrast of ideas as exhibit a monster rather than a man. Virtue is inflamed at the violation, and sober reason calls it nonsense.

Dignities and high sounding names have different effects on different beholders. The lustre of the *Star* and the title of *My Lord,* overawe the superstitious vulgar, and forbid them to inquire into the character of the possessor: Nay more, they are, as it were, bewitched to admire in the great, the vices they would honestly condemn in themselves. This sacrifice of common sense is the certain badge which distinguishes slavery from freedom; for when men yield up the privilege of thinkinng, the last shadow of liberty quits the horizon.

But the reasonable freeman sees through the magic of a title, and examines the man before he approves him. To him the honors of the worthless serve to write their masters' vices in capitals, and their stars shine to no other end than to read them by. The possessors of undue honors are themselves sensible of this; for when their repeated guilt renders their persons unsafe, they disown their rank, and, like glowworms, extinguish themselves into common reptiles, to avoid discovery. Thus Jeffries sunk into a fisherman, and his master escaped in the habit of a peasant.

Modesty forbids men, separately or collectively, to assume titles. But as all honors, even that of kings, originated from the public, the public may justly be called the fountain of true honor. And it is with much pleasure I have heard the title of *Honorable* applied to a body of men, who nobly disregarding private ease and interest for public welfare, have justly merited the address of The Honorable Continental Congress.

Vox Populi.

REFLECTIONS ON SLAVERY

Pennsylvania Journal *October 18, 1775*

When I reflect on the horrid cruelties exercised by Britain in the East Indies—How thousands perished by artificial famine—How religion and every manly principle of honor and honesty were sacrificed to luxury and pride—When I read of the wretched natives being blown away, for no other crime than because, sickened with the miserable scene, they refused to fight—When I reflect on these and a thousand instances of similar barbarity, I firmly believe that the Almighty, in compassion to mankind, will curtail the power of Britain. And when I reflect on the use she has made of the discovery of this new world—that the little paltry dignity of early kings has been set up in preference to the great cause of the King of kings—That instead of Christian examples to the Indians, she has basely tampered with their passions, imposed on their ignorance, and made them tools of treachery and murder—And when to these and many other melancholy reflections I add this sad remark, that ever since the discovery of America she has employed herself in the most horrid of all traffics, that of human flesh, unknown to the most savage nations, has yearly (without provocation and in cold blood) ravaged the hapless shores of Africa, robbing it of its unoffending inhabitants to cultivate her stolen dominions in the West—When I reflect on these, I hesitate not for a moment to believe that the Almighty will finally separate America from Britain. Call it independence or what you will, if it is the cause of God and humanity it will go on.

And when the Almighty shall have blest us, and made us a people *dependent only upon Him,* then may our first gratitude be shown by an act of continental legislation, which shall put a stop to the importation of Negroes for sale, soften the hard fate of those already here, and in time procure their freedom.

Humanus.

THE MAGAZINE: NURSERY OF GENIUS

Pennsylvania Magazine *January 24, 1775*

In a country whose reigning character is the love of science, it is somewhat strange that the channels of communication should continue so narrow and limited. The weekly papers are at present the only vehicles of public information. Convenience and necessity prove that the opportunities of acquiring and communicating knowledge ought always to enlarge with the circle of population. America has now outgrown the state of infancy; her strength and commerce make large advances to manhood; and science in all its branches has not only blossomed, but even ripened on the soil. The cottages as it were of yesterday have grown to villages, and the villages to cities; and while proud antiquity, like a skeleton in rags, parades the streets of other nations, their genius, as if sickened and disgusted with the phantom, comes hither for recovery.

The present enlarged and improved state of things gives every encouragement which the editor of a new magazine can reasonably hope for. The failure of former ones cannot be drawn as a parallel now. Change of times adds propriety to new measures. In the early days of colonization, when a whisper was almost sufficient to have negotiated all our internal concerns, the publishing even of a newspaper would have been premature. Those times are past; and population has established both their use and their credit. But their plan being almost wholly devoted to news and commerce, affords but a scanty residence to the Muses. Their path lies wide of the field of science, and has left a rich and unexplored region for new adventurers.

It has always been the opinion of the learned and curious, that a magazine, when properly conducted, is the nursery of genius; and by constantly accumulating new matter, becomes a kind of market for wit and utility. The opportunities which it affords to men of abilities to communicate their studies, kindle up a spirit of invention and emulation. An unexercised genius soon contracts a kind of mossiness, which not only checks its growth, but abates its natural vigor. Like an untenanted house it falls into decay, and frequently ruins the possessor.

The British magazines, at their commencement, were the repositories of ingenuity. They are now the retailers of tale and nonsense. From elegance they sunk to simplicity, from simplicity to folly, and from folly to voluptuousness. *The Gentleman's*, the *London*, and the *Universal, Magazines*, bear yet some marks of their originality; but the *Town* and *Country*, the *Covent-Garden*, and the *Westminster*, are no better than incentives to profligacy and dissipation. They have added to the dissolution of manners, and supported Venus against the Muses.

America yet inherits a large portion of her first-imported virtue. Degeneracy is here almost a useless word. Those who are conversant with Europe would be tempted to believe that even the air of the Atlantic disagrees with the constitution of foreign vices; if they survive the voyage, they either expire on their arrival, or linger away in an incurable consumption. There is a happy something in the climate of America, which disarms them of all their power both of infection and attraction.

But while we give no encouragement to the importation of foreign vices, we ought to be equally as careful not to create any. A vice begotten might be worse than a vice imported. The latter, depending on favor, would be a sycophant; the other, by pride of birth, would be a tyrant: to the one we should be dupes, to the other slaves.

There is nothing which obtains so general an influence over the manners and morals of a people as the Press; from *that,* as from a fountain, the streams of vice or virtue are poured forth over a country. And of all publications, none are more calculated to improve or infect than a periodical one. All others have their rise and their exit; but *this* renews the pursuit. If it has an evil tendency, it debauches by the power of repetition; if a good one, it obtains favor by the gracefulness of soliciting it. Like a lover, it woos its mistress with unabated ardor, nor gives up the pursuit without a conquest.

The two capital supports of a magazine are utility and entertainment. The first is a boundless path, the other an endless spring. To suppose that arts and sciences are exhausted subjects, is doing them a kind of dishonor. The divine mechanism of creation reproves such folly, and shows us by comparison, the imperfection of our most refined inventions. I cannot believe that this species of vanity is peculiar to the present age only. I have no doubt but that it existed before the flood, and even in the wildest ages of antiquity. 'Tis folly we have inherited, not created; and the discoveries which every day produces, have greatly contributed to dispossess us of it. Improvement and the world will expire together: And till that period arrives, we may plunder the mine, but can never exhaust it! That *"We have found out everything,"* has been the motto of every age. Let our ideas travel a little into antiquity, and we shall find larger portions of it than now; and so unwilling were our ancestors to descend from this mountain of perfection, that when any new discovery exceeded the common standard, the discoverer was believed to be in alliance with the devil. It was not the ignorance of the age only, but the vanity of it, which rendered it dangerous to be ingenious. The man who first planned and erected a tenable hut, with a hole for the smoke to pass, and the light to enter, was perhaps called an able architect, but he who first improved it with a chimney, could be no less than a prodigy; yet had the same man been so unfortunate as to have embellished it with glass windows, he might probably have been burnt for a magician. Our fancies would be highly diverted could we look back, and behold a circle or original Indians haranguing on the sublime perfection of the age: Yet 'tis not impossible but future times may exceed us almost as much as we have exceeded them.

I would wish to extirpate the least remains of this impolitic vanity. It has a direct tendency to unbrace the nerves of invention, and is peculiarly hurtful to young colonies. A magazine can never want matter in America, if the inhabitants will do justice to their own abilities. Agriculture and manufactures owe much of their improvement in England, to hints first thrown out in some of their magazines. Gentlemen whose abilities enabled them to make experiments, frequently chose that method of communication, on account of its convenience. And why should not the same spirit operate in America? I have no doubt of seeing, in a little time, an

American magazine full of more useful matter than I ever saw an English one: Because we are not exceeded in abilities, have a more extensive field for enquiry; and, whatever may be our political state, *Our happiness will always depend upon ourselves.*

Something useful will always arise from exercising the invention, though perhaps, like the witch of Endor, we shall raise up a being we did not expect. We owe many of our noblest discoveries more to accident than wisdom. In quest of a pebble we have found a diamond, and returned enriched with the treasure. Such happy accidents give additional encouragement to the making experiments; and the convenience which a magazine affords of collecting and conveying them to the public, enhances their utility. Where this opportunity is wanting, many little inventions, the forerunners of improvement, are suffered to expire on the spot that produced them; and, as an elegant writer beautifully expresses on another occasion,

> "They waste their sweetness on the desert air."
> *Gray.*

In matters of humor and entertainment there can be no reason to apprehend a deficiency. Wit is naturally a volunteer, delights in action, and under proper discipline is capable of great execution. 'Tis a perfect master in the art of bush-fighting; and though it attacks with more subtility than science, has often defeated a whole regiment of heavy artillery. Though I have rather exceeded the line of gravity in this description of wit, I am unwilling to dismiss it without being a little more serious. 'Tis a qualification which, like the passions, has a natural wildness that requires governing. Left to itself, it soon overflows its banks, mixes with common filth, and brings disrepute on the fountain. We have many valuable springs of it in America, which at present run purer streams, than the generality of it in other countries. In France and Italy, 'tis froth highly fomented; in England it has much of the same spirit, but rather a browner complexion. European wit is one of the worst articles we can import. It has an intoxicating power with it, which debauches the very vitals of chastity, and gives a false coloring to every thing it censures or defends. We soon grow fatigued with the excess, and withdraw like gluttons sickened with intemperance. On the contrary, how happily are the sallies of innocent humor calculated to amuse and sweeten the vacancy of business! We enjoy the harmless luxury without surfeiting, and strengthen the spirits by relaxing them.

The Press has not only a great influence over our manners and morals, but contributes largely to our pleasures; and a magazine when properly enriched, is very conveniently calculated for this purpose. Voluminous works weary the patience, but here we are invited by conciseness and variety. As I have formerly received much pleasure from perusing these kind of publications, I wish the *present* success; and have no doubt of seeing a proper diversity blended so agreeably together, as to furnish out an *Olio* worthy of the company for whom it is designed.

I consider a magazine as a kind of bee-hive, which both allures the swarm, and provides room to store their sweets. Its division into cells, gives every bee a province of its own; and though they all produce honey, yet perhaps they differ in their taste for flowers, and extract with greater dexterity from one than from another. Thus, we are not all Philosophers, all Artists, nor all Poets.

James Madison

(Publius)

Having helped invent the United States, newspapers then became vitally implicated in explaining the new Republic and the Constitution drafted in 1787. The critical ratification vote was in New York, so urgent that three of the charter's Framers, signing themselves ''Publius,'' wrote eighty-four essays defending their work. What came to be called The Federalist *originated as columns appearing in four New York newspapers, written by Alexander Hamilton, James Madison, and John Jay.*

It can be reasonably maintained that no more influential collection of newspaper articles has appeared. At various times and to different people, The Federalist *has been used to judge the intent of the Framers, to expose their class bias and/or economic realism, to argue the need for Federal Union with or in Europe, to illuminate the anti-majoritarian flaws of American democracy, and to illustrate theories of faction. Debating authorship of specific numbers has become an academic cottage industry. But Madison's authorship of Number 10, the most celebrated, is not in dispute.*

The elegance of Madison's argument was equal to the difficulty of his task: to convince Americans that a liberty would be safeguarded by a Federal system through which the evils of faction inherent in human nature could be checked. A Niagara of words has spilled over Number 10. Charles Beard saw in it the key to his Economic Interpretation of the Constitution *(New York, 1913); Robert E. Brown begged to differ in* Charles Beard and the Constitution *(Princeton, 1956); so did Morton White in* Philosophy, The Federalist and the Constitution *(New York, 1987). Garry Wills, a columnist as well as academic, sees in Number 10 the very chemistry of representation in* Explaining America: The Federalist *(New York, 1981).*

"LIBERTY IS TO FACTION WHAT AIR IS TO FIRE"

Federalist No. 10

New York Packet *November 23, 1787*

Among the numerous advantages promised by a well-constructed Union, none deserves to be more accurately developed than its tendency to break and control the violence of faction. The friend of popular governments never finds himself so much alarmed for their character and fate, as when he contemplates their propensity to this dangerous vice. He will not fail, therefore, to set a due value on any plan which, without violating the principles to which he is attached, provides a proper cure for it. The instability, injustice, and confusion introduced into the public councils, have, in truth, been the mortal diseases under which popular governments have everywhere perished; as they continue to be the favorite and fruitful topics from which the adversaries to liberty derive their most specious declamations. The valuable improvements made by the American constitutions on the popular models, both ancient and modern, cannot certainly be too much admired; but it would be an unwarrantable partiality, to contend that they have as effectually obviated the danger on this side, as was wished and expected.

Party Conflicts Make for Disorder. Complaints are everywhere heard from our most considerate and virtuous citizens, equally the friends of public and private faith, and of public and personal liberty, that our governments are too unstable, that the public good is disregarded in the conflicts of rival parties, and that measures are too often decided, not according to the rules of justice and the rights of the minor party, but by the superior force of an interested and overbearing majority. However anxiously we may wish that these complaints had no foundation, the evidence of known facts will not permit us to deny that they are in some degree true. It will be found, indeed, on a candid review of our situation, that some of the distresses under which we labor have been erroneously charged on the operation of our governments; but it will be found, at the same time, that other causes will not alone account for many of our heaviest misfortunes; and, particularly, for that prevailing and increasing distrust of public engagements, and alarm for private rights, which are echoed from one end of the continent to the other. These must be chiefly, if not wholly, effects of the unsteadiness and injustice with which a factious spirit has tainted our public administrations.

By a faction, I understand a number of citizens, whether amounting to a majority or minority of the whole, who are united and actuated by some common impulse of passion, or of interest, adverse to the rights of other citizens, or to the permanent and aggregate interests of the community.

There are two methods of curing the mischiefs of faction: the one, by removing its causes; the other, by controlling its effects.

There are again two methods of removing the causes of faction: the one, by destroying the liberty which is essential to its existence; the other, by giving to every citizen the same opinions, the same passions, and the same interests.

It could never be more truly said than of the first remedy, that it was worse than the disease. Liberty is to faction what air is to fire, an aliment without which it instantly expires. But it could not be less folly to abolish liberty, which is essential to political life, because it nourishes faction, than it would be to wish the annihilation of air, which is essential to animal life, because it imparts to fire its destructive agency.

Liberty Fosters Diversity of Opinion. The second expedient is as impracticable as the first would be unwise. As long as the reason of man continues fallible, and he is at liberty to exercise it, different opinions will be formed. As long as the connection subsists between his reason and his self-love, his opinions and his passions will have a reciprocal influence on each other; and the former will be objects to which the latter will attach themselves. The diversity in the faculties of men, from which the rights of property originate, is not less an insuperable obstacle to a uniformity of interests. The protection of these faculties is the first object of government. From the protection of different and unequal faculties of acquiring property, the possession of different degrees and kinds of property immediately results; and from the influence of these on the sentiments and views of the respective proprietors, ensues a division of the society into different interests and parties.

Tenacity of the Combative Spirit. The latent causes of faction are thus sown in the nature of man; and we see them everywhere brought into different degrees of activity, according to the different circumstances of civil society. A zeal for different opinions concerning religion, concerning government, and many other points, as well of speculation as of practice; an attachment to different leaders ambitiously contending for pre-eminence and power; or to persons of other descriptions whose fortunes have been interesting to the human passions, have, in turn, divided mankind into parties, inflamed them with mutual animosity, and rendered them much more disposed to vex and oppress each other than to cooperate for their common good. So strong is this propensity of mankind to fall into mutual animosities, that where no substantial occasion presents itself, the most frivolous and fanciful distinctions have been sufficient to kindle their unfriendly passions and excite their most violent conflicts.

Economic Roots of Conflicting Interests. But the most common and durable source of factions has been the various and unequal distribution of property. Those who hold and those who are without property have ever formed distinct interests in society. Those who are creditors, and those who are debtors, fall under a like discrimination. A landed interest, a manufacturing interest, a mercantile interest, a moneyed interest, with many lesser interests, grow up of necessity in civilized nations, and divide them into different classes, actuated by different sentiments and views. The regulation of these various and interfering interests forms the principal task of modern legislation, and involves the spirit of party and faction in the necessary and ordinary operations of the government.

Legislation by Special Interests. No man is allowed to be a judge in his own cause, because his interests would certainly bias his judgment, and, not improbably, corrupt his integrity. With equal, nay with greater reason, a body of men are unfit to be both judges and parties at the same time; yet what are many of the most important acts of legislation, but so many judicial determinations, not indeed concerning the rights of single persons, but concerning the rights of large bodies of citizens? And what are the different classes of legislators but advocates and parties to the causes which they determine? Is a law proposed concerning private debts? It is a question to which the creditors are parties on one side and the debtors on the other. Justice ought to hold the balance between them. Yet the parties are, and must be, themselves the judges; and the most numerous party, or, in other words, the most powerful faction must be expected to prevail. Shall domestic manufactures be encouraged, and in what degree, by restrictions on foreign manufactures? are questions which would be differently decided by the landed and the manufacturing classes, and probably by neither with a sole regard to justice and the public good. The apportionment of taxes on the various descriptions of property is an act which seems to require the most exact impartiality; yet there is, perhaps, no legislative act in which greater opportunity and temptation are given to a predominant party to trample on the rules of justice. Every shilling with which they overburden the inferior number, is a shilling saved to their own pockets.

It is in vain to say that enlightened statesmen will be able to adjust these clashing interests, and render them all subservient to the public good. Enlightened statesmen will not always be at the helm. Nor, in many cases, can such an adjustment be made at all without taking into view indirect and remote considerations, which will rarely prevail over the immediate interest which one party may find in disregarding the rights of another or the good of the whole.

The inference to which we are brought is, that the *causes* of faction cannot be removed, and that relief is only to be sought in the means of controlling its *effects.*

Problem of How to Control Powers of Majorities. If a faction consists of less than a majority, relief is supplied by the republican principle, which enables the majority to defeat its sinister views by regular vote. It may clog the administration, it may convulse the society; but it will be unable to execute and mask its violence under the forms of the Constitution. When a majority is included in a faction, the form of popular government, on the other hand, enables it to sacrifice to its ruling passion or interest both the public good and the rights of other citizens. To secure the public good and private rights against the danger of such a faction, and at the same time to preserve the spirit and the form of popular government, is then the great object to which our inquiries are directed. Let me add that it is the great desideratum by which this form of government can be rescued from the opprobrium under which it has so long labored, and be recommended to the esteem and adoption of mankind.

By what means is this object attainable? Evidently by one of two only. Either the existence of the same passion or interest in a majority at the same time must be prevented, or the majority, having such coexistent passion or interest, must be rendered, by their number and local situation, unable to concert and carry into effect schemes of oppression. If the impulse and the opportunity be suffered to coincide, we well know that neither moral nor religious motives can be relied on as an adequate control. They are not found to be such on the injustice and violence of individuals, and lose their efficacy in proportion to the number combined together, that is, in proportion as their efficacy becomes needful.

Pure Democracy No Cure for Factionalism. From this view of the subject it may be concluded that a pure democracy, by which I mean a society consisting of a small number of citizens, who assemble and administer the government in person, can admit of no cure for the mischiefs of faction. A common passion or interest will, in almost every case, be felt by a majority of the whole; a communication and concert result from the form of government itself; and there is nothing to check the inducements to sacrifice the weaker party or an obnoxious individual. Hence it is that such democracies have ever been spectacles of turbulence and contention; have ever been found incompatible with personal security or the rights of property; and have in general been as short in their lives as they have been violent in their deaths. Theoretic politicians, who have patronized this species of government, have erroneously supposed that by reducing mankind to a perfect equality in their political rights, they would, at the same time, be perfectly equalized and assimilated in their possessions, their opinions, and their passions.

Republican Government Refines Popular Passions. A republic, by which I mean a government in which the scheme of representation takes place, opens a different prospect, and promises the cure for which we are seeking. Let us examine the points in which it varies from pure democracy, and we shall comprehend both the nature of the cure and the efficacy which it must derive from the Union.

The two great points of difference between a democracy and a republic are: first, the delegation of the government, in the latter, to a small number of citizens elected by the rest; secondly, the greater number of citizens, and greater sphere of country, over which the latter may be extended.

The effect of the first difference is, on the one hand, to refine and enlarge the public views, by passing them through the medium of a chosen body of citizens, whose wisdom may best discern the true interest of their country, and whose patriotism and love of justice will be least likely to sacrifice it to temporary or partial considerations. Under such a regulation, it may well happen that the public voice, pronounced by the representatives of the people, will be more consonant to the public good than if pronounced by the people themselves, convened for the purpose. On the other hand, the effect may be inverted. Men of factious tempers, of local prejudices, or of sinister designs, may, by intrigue, by corruption, or by other means, first obtain the suffrages, and then betray the interests, of the people. The question resulting is, whether small or extensive republics are more favorable to the election of proper guardians of the public weal; and it is clearly decided in favor of the latter by two obvious considerations:

Larger Republic Better than Small. In the first place, it is to be remarked that, however small the repbulic may be, the representatives must be raised to a certain number, in order to guard against the cabals of a few; and that, however large it may be, they must be limited to a certain number, in order to guard against the confusion of a multitude. Hence, the number of representatives in the two cases not being in proportion to that of the two constituents, and being proportionally greater in the small republic, it follows that, if the proportion of fit characters be not less in the large than in the small republic, the former will present a greater option, and consequently a greater probability of a fit choice.

In the next place, as each representative will be chosen by a greater number of citizens in the large than in the small republic, it will be more difficult for unworthy candidates to practise with success the vicious arts by which elections are too often carried; and the suffrages of the people being more free, will be more likely to centre in men who possess the most attractive merit and the most diffusive and established characters.

It must be confessed that in this, as in most other cases, there is a mean, on both sides of which inconveniences will be found to lie. By enlarging too much the number of electors, you render the representative too little acquainted with all their local circumstances and lesser interests; as by reducing it too much, you render him unduly attached to these, and too little fit to comprehend and pursue great and national objects. The federal Constitution forms a happy combination in this respect; the great and aggregate interests being referred to the national, the local and particular to the State legislatures.

Republican Government Applicable to Large Territories. The other point of difference is, the greater number of citizens and extent of territory which may be brought within the compass of republican than of democratic government; and it is this circumstance principally which renders factious combinations less to be dreaded in the former than in the latter. The smaller the society, the fewer probably will be the distinct parties and interests composing it; the fewer the distinct parties and interests, the more frequently will a majority be found of the same party; and the smaller the number of individuals composing a majority, and the smaller the compass within which they are placed, the more easily will they concert and execute their plans of oppression. Extend the sphere, and you take in a greater variety of parties and interests; you make it less probable that a majority of the whole will have a common motive to invade the rights of other citizens; or if such a common motive exists, it will be more difficult for all who feel it to discover their own strength, and to act in unison with each other. Besides other impediments, it may be remarked that, where there is a consciousness of unjust or dishonorable purposes, communication is always checked by distrust in proportion to the number whose concurrence is necessary.

Large Unions Include a Greater Diversity of Interests. Hence, it clearly appears, that the same advantage which a republic has over a democracy, in controlling the effects of faction, is enjoyed by a large over a small republic,—is enjoyed by the Union over the States composing it. Does the advantage consist in the substitution of representatives whose enlightened views and virtuous sentiments render them superior to local prejudices and to schemes of injustice? It will not be denied that the representation of the Union will be most likely to possess these requisite endowments. Does it consist in the greater security afforded by a greater variety of parties, against the event of any one party being able to outnumber and oppress the rest? In an equal degree does the increased variety of parties comprised within the Union, increase this security. Does it, in fine, consist in the greater obstacles opposed to the concert and accomplishment of the secret wishes of an unjust and interested majority? Here, again, the extent of the Union gives it the most palpable advantage.

Local Factions Held in Bounds by the Extent and Structure of the Government. The influence of factious leaders may kindle a flame within their particular States, but will be unable to spread a general conflagration through the other States. A religious sect may degenerate into a political faction in a part of the Confederacy; but the variety of sects dispersed over the entire face of it must secure the national councils against any danger from that source. A rage for paper money, for an abolition of debts, for an equal division of property, or for any other improper or wicked project, will be less apt to pervade the whole body of the Union than a particular member of it; in the same proportion as such a malady is more likely to taint a particular county or district, than an entire State.

In the extent and proper structure of the Union, therefore, we behold a republican remedy for the diseases most incident to republican government. And according to the degree of pleasure and pride we feel in being republicans, ought to be our zeal in cherishing the spirit and supporting the character of Federalists.

Publius [Madison]

William Cobbett

(Peter Porcupine)

If Paine was England's radical gift to America, William Cobbett (1763–1835) was the stalwart opposite: a despiser of democratic humbug who while in Philadelphia assumed the bristling persona of Peter Porcupine. An odd thing happened after Cobbett's return to England in 1800. He emerged as an agrarian radical in his "Rural Rides" and in his celebrated Political Register. *In 1819, Cobbett came back to America and in New Rochelle exhumed Tom Paine's bones, in the unrealized hope of suitably reburying them in England. After Cobbett's death, the bones passed to an aged workman, then to a furniture dealer, and finally vanished. Cobbett's memory has been almost as ill-served. The twelve volumes of* Porcupine's Works *(London, 1801) are to be found in rare book vaults. A Folio Society collection, edited by J. E. Morpurgo,* Cobbett's America *(London, 1985), is inadequate. I have drawn on the former in the selections that follow, with two items added from* American Press Opinion *(Boston, 1927), edited by Allan Nevins.*

PORCUPINE'S FIRST BOW

Introduction to *Porcupine's Works (London, 1801)*

In the Spring of the year 1796, I took a house in Second Street, Philadelphia, for the purpose of carrying on the bookselling business, which I looked upon as being at once a means of getting money, and of propagating writings against the French. I went into my house in May, but the shop could not be gotten ready for some time; and, from one delay and another, I was prevented from opening till the second week in July.

Till I took this house, I had remained almost entirely unknown, as a writer. A few persons did, indeed, know that I was the person, who had assumed the name of PETER PORCUPINE; but the fact was by no means a matter of notoriety. The moment, however, that I had taken a lease of a large house, the transaction became a topic of public conversation, and the eyes of the Democrats and the French, who still lorded it over the city, and who owed me a mutual grudge, were fixed upon me.

I thought my situation somewhat perilous. Such truths as I had published, no man had dared to utter, in the United States, since the rebellion. I knew that these truths had mortally offended the leading men amongst the Democrats, who could, at any time, muster a mob quite sufficient to destroy my house, and to murder me. I had not a friend, to whom I could look with any reasonable hope of receiving efficient support; and, as to the *law,* I had seen too much of republican justice, to expect anything but persecution from that quarter. In short, there were, in Philadelphia, about ten thousand persons, all of whom would have rejoiced to see me murdered; and there might, probably, be two thousand, who would have been very sorry for it; but not above fifty of whom would have stirred an inch to save me.

As the time approached for opening my shop, my friends grew more anxious for my safety. It was recommended to me, to be cautious how I exposed, at my window, any thing that might provoke the people; and, above all, not to put up any *aristocratical portraits,* which would certainly cause my windows to be demolished.

I saw the danger; but also saw, that I must, at once, set all danger at defiance, or live in everlasting subjection to the prejudices and caprice of the democratical mob. I resolved on the former; and, as my shop was to open on a Monday morning, I employed myself all day on Sunday, in preparing an exhibition, that I thought would put the courage and the power of my enemies to the test. I put up in my windows, which were very large, all the portraits that I had in my possession of *kings, queens, princes,* and *nobles.* I had all the English Ministry; several of the Bishops and Judges; the most famous Admirals; and, in short, every picture that I thought likely to excite rage in the enemies of Great Britain.

Early on the Monday morning, I took down my shutters. Such a sight had not been seen in Philadelphia for twenty years. Never since the beginning of the rebellion, had any one dared to hoist at his window the portrait of George the Third.

In order to make the test as perfect as possible, I had put up some of the *''worthies of the Revolution,''* and had found out fit companions for them. I had coupled *Franklin* and *Marat* together; and, in another place, *McKean* and *Ankerstrom*—The following tract records some amongst the consequences. . . .

PORCUPINE'S WAY WITH HIS CRITICS

Porcupine's Gazette *July 1796*

To Mr. John Olden Merchant, Chesnut Street.

Sir,

A certain William Cobbett alias Peter Porcupine, I am informed is your tenant. This daring *scoundrell,* not satisfied with having repeatedly traduced the people of this country, vilified the most eminent and patriotic characters among us and *grosly* abused our allies the French, in his detestable productions, has now the astonishing effrontery to expose those very publications at his window for sale, as well as certain prints indicative of the prowess of our enemies the British and the disgrace of the French. Calculating largely upon the moderation or rather *pucellanimity* of our citizens, this puppy supposes he may even *insults* us with impunity. But he will e'er long find himself dreadfully mistaken. *'Tho* his miserable publications have not been hiterto considered worthy of notice, the late *manifestations* of his impudence and enmity to this country will not be passed over. With a view therefore of preventing your feeling the blow designed for him, I now address you. When the time of retribution arrives, it may not be convenient to discriminate between the innocent and the guilty. Your property therefore may suffer. For depend upon it brick walls will not skreen the rascal from punishment when once the business is undertaken. As a friend therefore I advice you to save your property by either compelling Mr. Porcupine to leave your house or at all events oblige him to cease exposing his abominable production or any of his courtley prints at his window for sale. In this way only you may avoid danger to your house and perhaps save the rotten *carcase* of your tenant for the present.

A Hint

July 16th, 1796.

I have copied this loving epistle, word for word, and letter for letter, preserving the false orthography, as the manner of spelling may probably lead some of my readers to a discovery of the writer.

When Mr. Vicessimus Knox (who is a sort of a Democrat), publishes his next edition of Elegant Epistles, he will do well to give this a place amongst them; for, it is certainly a master-piece in its way. It will be a good pattern for the use of future ruffians, who wish to awe a man into silence, when they are incapable of resisting him in print. But, the worst of it will be, the compiler will not have it in his power to say, that this was attended with success.

If I am right in my guess, the family of the author of this powder blunderbuss, makes a considerable figure in the Tyburn Chronicle. His grandfather was hanged for house-breaking, and his *papa* came to the southern part of these States on his travels, by the direction of a righteous judge, and twelve honest men.

AMERICAN LOTTERIES: ADVICE TO THOSE WHO NEED IT

Pocupine's Gazette *August 25, 1797*

Have you an itching propensity to use your wits to advantage? Make a lottery. A splendid scheme is a bait that cannot fail to catch the gulls. Be sure to spangle it with rich prizes: the fewer blanks—on paper—the better; for on winding up the business, you know, it is easy to make as many blanks as you please. Witness a late lottery on the Potowmack. The *winding up,* however, is not absolutely necessary: you know what a noise the winding up of a certain clock once made. The better way is to delay the drawing; or should it *ever begin,* there is no hurry about the *end,* or rather, let it have no end at all. If, in either case, a set of discontended adventurers should happen to say hard things of you, show them that you despise their unmannerly insinuations, by humming the tune of *Yankee Doodle.* This may dumbfound them; but should they persist, there is a mode left that cannot fail to stop their mouths. The scheme of the lottery is your contract with the purchasers of tickets: produce this, and defy them to point out any breach of it on your part. *Entre nous;* I am supposing you discreet enough to avoid in your scheme anything that might look like a promise to commence the drawing on this or that particular day; or to finish at any given period. It would be enough to promise a beginning *when a sufficient number of tickets shall be sold;* of the sufficiency you would be the sole judge. Now they ought to know as well as you that, like Peter Pindar's razors, your tickets were "made to sell"; so that if but one ticket remained unsold, you are under no obligation to draw, a *sufficient number* not having been disposed of.

THE BENEFITS OF WAR WITH FRANCE

Porcupine's Gazette *January 1798*

Let us see what would be gained by war. The immediate effect would be, a free passage over the ocean, without the hazard of seizure, or even of examination.

The commerce of America would immediately raise its drooping head; the confidence of commercial men would be reestablished, and the spirit of trade and enterprise renewed. American seamen would no longer be shot at, and flogged, within sight of their own shores; nor would the red-headed ruffians add to the twenty millions they have already seized: no peace should be made with them till they refund their plunder, which would amply discharge all the debts incurred by the war.

Louisiana they might be compelled to relinquish; and thus would these States be completely rid of the most alarming danger than ever menaced them; and which, if not soon removed, must and will, in a few years, effect their disunion and destruction.

But above all, the alliance with Great Britain would cut up the French faction here. It is my sincere opinion, that they have formed the diabolical plan of *revolutionizing* (to use one of their execrable terms) the whole continent of America. They have their agents and partizans without number, and very often where we do not imagine. Their immoral and blasphemous principles have made a most alarming progress. They have explored the community to its utmost boundaries and its inmost recesses, and have left a partizan on every spot, ready to preach up *the holy right of insurrection.*

They have no intention of invading these States with the fair and avowed purpose of *subjugating them.* No; they will come as they went to the Brabanters and the Dutch, as "friends and deliverers." A single spark of their fraternity would set all the Southern States in a flame, the progress of which, as far as Connecticut, would be as rapid as the chariot of Apollo. This dreadful scourge nothing can prevent but a war. That would naturally disarm and discredit their adherents; would expel their intriguing agents, who are now in our streets, in our houses, and at our tables. It would cut off the cankering, poisonous, *sans-culottes* connexion, and leave the country once more sound and really independent.

PORCUPINE TO THE PUBLIC

The Rushlight *February 24, 1800*

When I determined to discontinue the publication of *Porcupine's Gazette,* I intended to remain for the future if not an unconcerned, at least a silent spectator of public transactions and political events; but the unexpected and sweeping result of a lawsuit, since decided against me, has induced me to abandon my lounging intention. The suit to which I allude was an action of slander, commenced against me in the autumn of 1797 by Dr. Benjamin Rush, the noted bleeding physician of Philadelphia. I was tried on the 14th of December last, when "the upright, enlightened, and impartial Republican jury" assessed, as damages, *five thousand dollars;* a sum surpassing the aggregate amount of all the damages assessed for all the torts of this kind, ever sued for in these States, from their first settlement to the present day. To the five thousand dollars must be added the costs of suit, the loss incurred by the interruption in collecting debts in Pennsylvania, and by the sacrifice of property taken in execution, and sold by the sheriff at public auction in Philadelphia, where a great number of books in sheets (among which was a part of the new edition of Porcupine's Works) were sold, or rather given away, as waste paper; so that the total of what has been, or will be, wrested away from me by Rush, will fall little short of eight thousand dollars.

To say that I do not feel this stroke, and very sensibly too, would be great affectation; but to repine at it would be folly, and to sink under it would be cowardice. I knew an Englishman in the Royal Province of New Brunswick, who had a very valuable house, which was, I believe, at that time nearly his all, burnt to the ground. He was out of town when the fire broke out, and happened to come just after it had exhausted itself. Everyone, knowing how hard he had earned the property, expected to see him bitterly bewail its loss. He came very leisurely up to the spot, stood about five minutes, looking earnestly at the rubbish, and then, stripping off his coat, *"here goes,"* said he, *"to earn another!"* and immediately went to work, raking the spikes and bits of iron out of the ashes. This noble-spirited man I have the honor to call my friend, and if ever this page should meet his eye, he will have the satisfaction to see that, should it be impossible for me to follow, I at least remember his example.

In the future exertions of my industry, however, pecuniary emolument will be, as it always has been with me, an object of only secondary consideration. Recent incidents, amongst which I reckon the unprecedented proceedings against me at Philadelphia, have imposed on me the discharge of a duty which I owe to my own country as well as this, and the sooner I begin the sooner I shall have done.

Seba Smith

(Jack Downing)

The linguistic break with England took longer than the political. A pivotal figure was Seba Smith (1792–1868), who introduced dialect sketches in the pages of the Portland *(Maine)* Courier. *His great creation was Major Jack Downing, a hayseed whose letters home in Down East dialect were instantly reprinted and plagiarized. Smith collected his sketches in* The Life and Writings of Major Jack Downing *(Boston, 1834) and* My Thirty Years Out of the Senate, *by Major Jack Downing (New York, 1859). Of imitators, the most influential was Charles Augustus Davis, who hijacked the character for partisan attacks on Jackson Democrats. Milton and Patricia Rickels have written an informative monograph,* Seba Smith *(Boston, 1977).*

THE GENESIS OF DOWNING

From the Author's Preface to *My Thirty Years Out of the Senate (1859)*

The name of Downing was entirely original with the author, who had never heard or seen the name before, and did not then even know that there was a Downing street in London, or an oyster dealer by that name in New York. In a year or two the letters became national in their character, and young Mr. Downing repaired to Washington, where he became the right hand man and confidential adviser of President Jackson. The author continued the letters in the *Portland Courier* for seven years, when he sold that paper and removed to New York. After an interval of a few years he resumed the series again, publishing the letters in the *National Intelligencer* at Washington, and continuing them till near the close of the administration of President Pierce.

Thus these papers, begun and continued partly for emolument, partly for amusement, and partly from a desire to exert a salutary influence upon public affairs and the politics of the country, have grown up to their present condition. In presenting them in this collected form, with original illustrations, to render them more attractive, the author could not let them go out into the world to make new acquaintances, and possibly down to posterity to help furnish political lessons to "Young America" for generations yet to come, without a careful retrospection to consider their whole moral and political character and influence. For should they contain

"One line which, dying, he could wish to blot,"

he would certainly wish to blot it now. But, believing the work will be harmless, and, he hopes, salutary, he leaves it to his countrymen, praying for Heaven's blessing on our whole common country.

Seba Smith.
New York, February, 1859

"SO HERE I AM . . ."

Portland Courier *January 18, 1830*

To Cousin Ephraim Downing, up in Downingville

Dear Cousin Ephraim:

I now take my pen in hand to let you know that I am well, hoping these few lines will find you enjoying the same blessing. When I come down to Portland I didn't think o' staying more than three or four days, if I could sell my load of ax handles, and mother's cheese, and cousin Nabby's bundle of footings; but when I got here I found Uncle Nat was gone a freighting down to Quoddy, and aunt Sally said as how I shouldn't stir a step home till he come back agin, which won't be this month. So here I am, loitering about this great town, as lazy as an ox. Ax handles don't fetch nothing; I couldn't hardly give 'em away. Tell Cousin Nabby I sold her footings for nine-pence a pair, and took it all in cotton cloth. Mother's cheese come to seven-and-sixpence; I got her half a pound of shushon, and two ounces of snuff, and the rest in sugar. When Uncle Nat

*Smith's Note.—The political struggle in the Legislature of Maine in the winter of 1830 will long be remembered. The preceding electioneering campaign had been carried on with a bitterness and personality unprecedented in the State, and so nearly were the parties divided, that before the meeting of the Legislature to count the votes for Governor, both sides confidently claimed the victory. Hence the members came together with feelings highly excited, prepared to dispute every inch of ground, and ready to take fire at the first spark which collision might produce.

comes home I shall put my ax handles aboard of him, and let him take 'em to Boston next time he goes; I saw a feller tother day, that told me they'd fetch a good price there. I've been here now a whole fortnight, and if I could tell ye one half I've seen, I guess you'd stare worse than if you'd seen a catamount. I've been to meeting, and to the museum, and to both Legislaters, the one they call the House, and the one they call the Sinnet. I spose Uncle Joshua is in a great hurry to hear something about these Legislaters; for you know he's always reading newspapers, and talking politics, when he can get anybody to talk with him. I've seen him when he had five tons of hay in the field well made, and a heavy shower coming up, stand two hours disputing with Squire W. about Adams and Jackson—one calling Adams a tory and a fed, and the other saying Jackson was a murderer and a fool; so they kept it up, till the rain began to pour down, and about spoilt all his hay.

Uncle Joshua may set his heart at rest about the bushel of corn that he bet 'long with the postmaster, that Mr. Ruggles would be Speaker of that Legislater they call the House; for he's lost it, slick as a whistle. . . . Well, now everybody says it has turned out jest as that queer little paper, called the *Daily Courier,* said 'twould. That paper said it was such a close rub it couldn't hardly tell which side would beat. And it's jest so, for they've been here now most a fortnight acting jest like two boys playin see-saw on a rail. First one goes up, and then 'tother; but I reckon one of the boys is rather heaviest, for once in a while he comes down chuck, and throws the other up into the air as though he would pitch him head over heels. Your loving cousin till death.

Jack Downing.

THE JOYS OF LAND SPECULATION

Portland Courier *March, 1833*

To Major Jack Downing, at President Jackson's house, in Washington City.

Dear Cousin Jack:

The Legislater folks have all cleared out to-day, one arter t'other, jest like a flock of sheep; and some of 'em have left me in the lurch tu, for they cleared out without paying me for my apples. Some of 'em went off in my debt as much as twenty cents, and some ninepence, and a shilling, and so on. They all kept telling me when they got paid off they'd settle up with me. And so I waited with patience till they adjourned, and thought I was as sure of my money as though it was in the bank.

But, my patience, when they did adjourn, such a hubbub I guess you never see. They were flying about from one room to another, like so many pigeons shot in the head. They run into Mr. Harris' room, and clawed the money off his table, hand over fist. I brustled up to some of 'em, and tried to settle. I come to one man that owed me twelve cents, and he had a ninepence in change; but he wouldn't let me have that, because he should lose half a cent. So, while we were bothering about it, trying to get it changed, the first I knew the rest of 'em had got their money in their pockets, and were off lik a shot—some of 'em in stages, and some in sleighs, and some footing it. I out and followed arter 'em, but 't was no use; I couldn't catch one of 'em. And as for my money, and apples tu, I guess I shall have to whistle for 'em now. . . .

I don't care so much about the apple business after all, for I've found out a way to get rich forty times as fast as I can by retailing apples, or as you can by hunting after an office. And I advise you to come right home, as quick as you can come. Here's a business going on here that you can get rich by ten times as quick as you can in any office, even if you should get to be President. The President don't have but twenty-five thousand dollars a year; but in this 'ere business that's going on here, a man can make twenty-five thousand dollars in a week if he's a mind to, and not work hard neither.

I s'pose by this time you begin to feel rather in a pucker to know what this business is. I'll tell you; but you must keep it to yourself, for if all them are Washington folks and Congress folks should come on here and go dipping into it, I'm afraid they'd cut us all out. But between you and me, it's only jest buying and selling land. Why, Jack, it's forty times more profitable than money digging, or any other business that you ever see. I knew a man here t'other day from Bangor, that made ten thousand dollars, and I guess he wan't more than an hour about it. Most all the folks here, and down to Portland and Bangor, have got their fortunes made, and now we are beginning to take hold of it up in the country.

They've got a slice up in Downingville, and I missed it by being down here selling apples, or I should had a finger in the pie. Uncle Joshua Downing—you know he's an old fox, and always knows where to jump; well, he see how everybody was getting rich, so he went and bought a piece of township up back of Downginville, and give his note for a thousand dollars for it. And then he sold it to Uncle Jacob, and took his note for two thousand dollars; and Uncle Jacob sold it to Uncle Zackary, and took his note for three thousand dollars; and Uncle Zackary sold it to Uncle Jim, and took his note for four thousand dollars; and Uncle Jim sold it to Cousin San, and took his note for five thousand dollars; and Cousin Sam sold it to Bill Johnson, and took his note for six thousand dollars. So you see there's five of 'em, that wan't worth ninepence apiece, (except Uncle Joshua,) have now got a thousand dollars apiece clear, when their notes are paid. And Bill Johnson's going to logging off of it, and they say he'll make more than any of 'em.

Come home, Jack; come home by all means, if you want to get rich. Give up your commission, and think no more about being President, or anything else, but come home and buy land before it's all gone.

Your loving cousin,
Ephraim Downing.

William Lloyd Garrison

Looming still like an Old Testament prophet in the journalism of rage and commitment is William Lloyd Garrison (1805–79), editor of the Abolitionist weekly The Liberator. *Moderation and gradualism were anathema to Garrison, who viewed a Constitution that tolerated slavery as a "Covenant with Death and an Agreement with Hell." This noble agitator learned his trade from Benjamin Lundy, editor of an anti-slavery journal in Baltimore,* The Genius of Universal Emancipation. *Having decided that Lundy was too cautious, Garrison returned to his native Massachusetts in 1831 to found, on a pittance, his percussive weekly. The selections that follow telescope his life and cause; they are drawn from* Documents of Upheaval *(New York, 1966), edited by Truman Nelson.*

GARRISON INDICTS MODERATION

The Liberator *January 1, 1831*

In the month of August, I issued proposals for publishing *The Liberator* in Washington City; but the enterprise, though hailed in different sections of the country, was palsied by public indifference. Since that time, the removal of the *Genius of Universal Emancipation* to the Seat of Government has rendered less imperious the establishment of a similar periodical in that quarter.

During my recent tour for the purpose of exciting the minds of the people by a series of discourses on the subject of slavery, every place that I visited gave fresh evidence of the fact, that a greater revolution in public sentiment was to be effected in the free states—*and particularly in New*

England—than at the south. I found contempt more bitter, opposition more active, detraction more relentless, prejudice more stubborn, and apathy more frozen, than among slave owners themselves. Of course, there were individual exceptions to the contrary. This state of things afflicted, but did not dishearten me. I determined, at every hazard, to lift up the standard of emancipation in the eyes of the nation, *within sight of Bunker Hill and in the birth place of liberty.* That standard is now unfurled; and long may it float, unhurt by the spoliations of time or the missiles of a desperate foe—yea, till every chain be broken, and every bondman set free! Let Southern oppressors tremble—let their secret abettors tremble—let their Northern apologists tremble—let all the enemies of the persecuted blacks tremble.

I deem the publication of my original Prospectus unnecessary, as it has obtained a wide circulation. The principles therein inculcated will be steadily pursued in this paper, excepting that I shall not array myself as the political partisan of any man. In defending the great cause of human rights, I wish to derive the assistance of all religions and of all parties.

Assenting to the "self evident truth" maintained in the American Declaration of Independence, "that all men are created equal, and endowed by their Creator with certain inalienable rights—among which are life, liberty and the pursuit of happiness," I shall strenuously contend for the immediate enfranchisement of our slave population. In Park-Street Church, on the Fourth of July, 1829, in an address on slavery, I unreflectingly assented to the popular but pernicious doctrine of *gradual* abolition. I seize this opportunity to make a full and unequivocal recantation, and thus publicly to ask pardon of my God, of my country, and of my brethren the poor slaves, for having uttered a sentiment so full of timidity, injustice and absurdity. A similar recantation, from my pen, was published in the *Genius of Universal Emancipation* at Baltimiore, in September, 1829. My conscience is now satisfied.

I am aware, that many object to the severity of my language; but is there not cause for severity? I *will be* as harsh as truth, and as uncompromising as justice. On this subject, I do not wish to think, or speak, or write, with moderation. No! No! Tell a man whose house is on fire, to give a moderate alarm; tell him to moderately rescue his wife from the hands of the ravisher; tell the mother to gradually extricate her babe from the fire into which it has fallen;—but urge me not to use moderation in a cause like the present. I am in earnest—I will not equivocate—I will not excuse—I will not retreat a single inch—*AND I WILL BE HEARD.* The apathy of the people is enough to make every statue leap from its pedestal, and to hasten the resurrection of the dead.

It is pretended, that I am retarding the cause of emancipation by the coarseness of my invective, and the precipitancy of my measures. *The charge is not true.* On this question my influence,—humble as it is,—is felt at this moment to a considerable extent, and shall be felt in coming years—not perniciously, but beneficially—not as a curse, but as a blessing; and posterity will bear testimony that I was right. I desire to thank God, that he enables me to disregard "the fear of man which bringeth a snare," and to speak his truth in its simplicity and power.

A SHORT CATECHISM ON RACISM

The Liberator *November 17, 1837*

1. Why is American slaveholding in all cases not sinful?
   Because its victims are ***black.***
2. Why is gradual emancipation right?
   Because the slaves are ***black.***
3. Why is immediate emancipation wrong and dangerous?
   Because the slaves are ***black.***
4. Why ought one-sixth portion of the American population to be exiled from their native soil?
   Because they are ***black.***
5. Why would the slaves if emancipated, cut the throats of their masters?
   Because they are ***black.***
6. Why are our slaves not fit for freedom?
   Because they are ***black.***
7. Why are American slaveholders not thieves, tyrants and men-stealers?
   Because their victims are ***black.***
8. Why does the Bible justify American slavery?
   Because its victims are ***black.***
9. Why ought not the Priest and the Levite, 'passing by on the other side,' to be sternly rebuked?
   Because the man who has fallen among thieves, and lies welterinng in his blood, is ***black.***
10. Why are abolitionists fanatics, madmen and incendiaries?
    Because those for whom they plead are ***black.***
11. Why are they wrong in their principles and measures?
    Because the slaves are ***black.***
12. Why is all the prudence, moderation, judiciousness, philanthropy and piety on the side of their opponents?
    Because the slaves are ***black.***
13. Why ought not the free discussion of slavery to be tolerated?
    Because its victims are ***black.***

14. Why is Lynch law, as applied to abolitionists, better than common law?
Because the slaves, whom they seek to emancipate, are ***black.***
15. Why are the slaves contented and happy?
Because they are ***black!***
16. Why don't they want to be free?
Because they are ***black!***
17. Why are they not created in the image of God?
Because their skin is ***black.***
18. Why are they not cruelly treated, but enjoy unusual comforts and privileges?
Because they are ***black!***
19. Why are they not our brethren and countrymen?
Because they are ***black.***
20. Why is it unconstitutional to pity and defend them?
Because they are ***black.***
21. Why is it a violation of the national compact to rebuke their masters?
Because they are ***black.***
22. Why will they be lazy, improvident, and worthless, if set free?
Because their skin is ***black.***
23. Why will the whites not wish to amalgamate with them in a state of freedom?
Because they are ***black!!***
24. Why must the Union be dissolved, should Congress abolish slavery in the District of Columbia?
Because the slaves in that District are ***black.***
25. Why are abolitionists justly treated as outlaws in one half of the Union?
Because those whose cause they espouse are ***black.***
26. Why is slavery 'the corner-stone of our republican edifice?'
Because its victims are ***black.***

We have thus given twenty-six replies to those who assail our principles and measures—that is, one reply, unanswerable and all-comprehensive, to all the cavils, complaints, criticisms, objections and difficulties which swarm in each State in the Union, against out holy enterprize. The victims are BLACK! 'That alters the case!' There is not an individual in all this country, who is not conscious before God, that if the slaves at the South should be to-day miraculously transformed into men of white complexions, to-morrow the abolitionists would be recognised and cheered as the best friends of their race; their principles would be eulogised as sound and incontrovertible, and their measures as rational and indispensable! Then, indeed, immediate emancipation would be the right of the slaves, and the duty of the masters! . . .

GARRISON'S VALEDICTORY

The Liberator *December 29, 1865. (The last number.)*

> "The last! the last! the last!
> O, by that little word
> How many thoughts are stirred—
> That sister of The Past!"

The present number of the *Liberator* is the completion of its thirty-fifth volume, and the termination of its existence.

Commencing my editorial career when only twenty years of age, I have followed it continuously till I have attained my sixtieth year—first, in connection with *The Free Press,* in Newburyport, in the spring of 1826; next, with *The National Philanthropist,* in Boston, in 1827; next, with *The Journal of the Times,* in Bennington, Vt., in 1828–9; next, with *The Genius of Universal Emancipation,* in Baltimore, in 1829–30; and, finally, with the *Liberator,* in Boston, from the 1st of January, 1831, to the 1st of January, 1866;—at the start, probably the youngest member of the editorial fraternity in the land, now, perhaps, the oldest, not in years, but in continuous service,—unless Mr. Bryant, of the New York *Evening Post,* be an exception.

Whether I shall again be connected with the press, in a similar capacity, is quite problematical; but, at my period of life, I feel no prompting to start a new journal at my own risk, and with the certainty of struggling against wind and tide, as I have done in the past.

I began the publication of the *Liberator* without a subscriber, and I end it—it gives me unalloyed satisfaction to say—without a farthing as the pecuniary result of the patronage extended to it during thirty-five years of unremitted labors.

From the immense change wrought in the national feeling and sentiment on the subject of slavery, the *Liberator* derived no advantage at any time in regard to its circulation. The original "disturber of the peace," nothing was left undone at the beginning, and up to the hour of the late rebellion, by Southern slaveholding villainy on the one hand, and Northern pro-slavery malice on the other, to represent it as too vile a sheet to be countenanced by any claiming to be [Christian or patriotic]; and it always required rare moral courage or singular personal independence to be among its patrons. Never had a journal to look such opposition in the face—never was one so constantly belied and caricatured. . . .

The object for which the *Liberator* was commenced—the extermination of chattel slavery—having been gloriously consummated, it seems to me specially appropriate to let its existence cover the historic period of the great struggle; leaving what remains to be done to complete the work of emancipation to other instrumentalities, (of which I hope to avail myself,) under new auspices, with more abundant means, and with millions instead of hundreds for allies.

Most happy am I to be no longer in conflict with the mass of my fellow-countrymen on the subject of slavery. For no man of any refinement or sensibility can be indifferent to the approbation of his fellow-men, if it be rightly earned. But to obtain it by going with the multitude to do evil—by pandering to despotic power or a corrupt public sentiment—is self-degradation and personal dishonor:

> "For more true joy Marcellus exiled feels,
> Than Caesar with a senate at his heels."

Better to be always in a minority of one with God—branded as madman, incendiary, fanatic, heretic, infidel—frowned upon by "the powers that be," and mobbed by the populace—or consigned ignominiously to the gallows, like him whose "soul is marching on," though his "body lies mouldering in the grave," or burnt to ashes at the stake like Wickliffe, or nailed to the cross liek him who "gave himself for the world,"—in defence of the RIGHT, than like Herod, having the shouts of a multitude, crying, "It is the voice of a god, and not of a man!"

Farewell, tried and faithful patrons! Farewell, generous benefactors, without whose voluntary but essential pecuniary contributions the *Liberator* must have long since been discontinued! Farewell, noble men and women who have wrought so long and so successfully, under God, to break every yoke! Hail, ye ransomed millions! Hail, year of jubilee! With a grateful heart and a fresh baptism of the soul, my last invocation shall be—

> "Spirit of Freedom. On—
>   On! pause not in thy flight
> Till every clime is won
>   To worship in thy light:
> Speed on thy glorious way,
>   And wake the sleeping lands!
> Millions are watching for the ray,
>   And lift to thee their hands.
> Still 'Onward!' be thy cry—
>   Thy banner on the blast;
> And, like a tempest, as thou rushest by,
>   Despots shall shrink aghast.
> On! till thy name is known
>   Throughout the peopled earth;
> On! till thou reign'st alone,
>   Man's heritage by birth;
> On! till from every vale, and where the mountains rise,
> The beacon lights of Liberty shall kindle to the skies!"

WM. LLOYD GARRISON

Margaret Fuller

To describe Margaret Fuller (1810–50) as a feminist author is to diminish her. She was editor (with Emerson) of the Transcendental journal The Dial; *author of the path-opening* Woman in the Nineteenth Century; *translator of Goethe; critic, travel writer, foreign correspondent; fighter (with Mazzini) for Italy's unification. Her death at forty in a shipwreck just beyond Long Island beaches, along with her daughter and Italian husband, Marquis Ossoli, was a poignant calamity. In 1844, Fuller was lured from Boston to New York by Horace Greeley, editor of the* Tribune. *For twenty months she contributed book reviews twice a week, and essays on whatever moved her. On her departure for Europe she continued as a contributor, describing her encounters with Carlyle and George Sand, and the siege of Rome. What follows is from* Margaret Fuller: American Romantic *(New York, 1963), edited by Perry Miller; and* The Writings of Margaret Fuller *(New York, 1941), edited by Mason Wade.*

MISS FULLER REVIEWS POE

New York Daily Tribune *July 11, 1845*

Mr. Poe's tales need no aid of newspaper comment to give them popularity; they have secured it. We are glad to see them given to the public in this neat form, so that thousands more may be entertained by them without injury to their eyesight.*

No form of literary activity has so terribly degenerated among us as the tale. Now that everybody who wants a new hat or bonnet takes this way to earn one from the magazines or annuals, we are inundated with

Edgar A. Poe, *Tales* (New York: Wiley and Putnam, 1845).

the very flimsiest fabrics ever spun by mortal brain. Almost every person of feeling or fancy could supply a few agreeable and natural narratives, but when instead of using their materials spontaneously they set to work with geography in hand to find unexplored nooks of wild scenery in which to locate their Indians or interesting farmers' daughters, or with some abridgment of history to hunt monarchs or heroes yet unused to become the subjects of their crude coloring, the sale-work produced is a sad affair indeed and "gluts the market" to the sorrow both of buyers and lookers-on.

In such a state of things the writings of Mr. Poe are a refreshment, for they are the fruit of genuine observations and experience, combined with an invention which is not "making up," as children call their way of contriving stories, but a penetration into the causes of things which leads to original but credible results. His narrative proceeds with vigor, his colors are applied with discrimination, and where the effects are fantastic they are not unmeaningly so.

The "Murders in the Rue Morgue" especially made a great impression upon those who did not know its author and were not familiar with his mode of treatment. Several of his stories make us wish he would enter the higher walk of the metaphysical novel and, taking a mind of the self-possessed and deeply marked sort that suits him, give us a deeper and longer acquaintance with its life and the springs of its life than is possible in the compass of these tales.

As Mr. Poe is a professed critic and of all the band the most unsparing to others, we are surprised to find some inaccuracies in the use of words, such as these: "he had with him many books, but rarely *employed* them."—"His results have, in truth, the *whole air* of intuition."

The degree of skill shown in the management of revolting or terrible circumstances makes the pieces that have such subjects more interesting than the others. Even the failures are those of an intellect of strong fiber and well-chosen aim.

SHE BIDS FAREWELL TO NEW YORK

New York Daily Tribune *August 1, 1846*

Farewell to New York city, where twenty months have presented me with a richer and more varied exercise for thought and life, than twenty years could in any other part of these United States.

It is the common remark about New York, that it has at least nothing petty or provincial in its methods and habits. The place is large enough: there is room enough, and occupation enough, for men to have no need or excuse for small cavils or scrutinies. A person who is independent, and knows what he wants, may lead his proper life here, unimpeded by others.

Vice and crime, if flagrant and frequent, are less thickly coated by hypocrisy than elsewhere. The air comes sometimes to the most infected subjects.

New York is the focus, the point where American and European interests converge. There is no topic of general interest to men, that will not betimes be brought before the thinker by the quick turning of the wheel.

Too quick that revolution,—some object. Life rushes wide and free, but *too fast.* Yet it is in the power of every one to avert from himself the evil that accompanies the good. he must build for his study, as did the German poet, a house beneath the bridge; and then all that passes above and by him will be heard and seen, but he will not be carried away with it.

Earlier views have been confirmed, and many new ones opened. On two great leadings, the superlative importance of promoting national education by heightening and deepening the cultivation of individual minds, and the part which is assigned to woman in the next stage of human progress in this country, where most important achievements are to be effected, I have received much encouragement, much instruction, and the fairest hopes of more.

On various subjects of minor importance, no less than these, I hope for good results, from observation, with my own eyes, of life in the old world, and to bring home some packages of seed for life in the new.

These words I address to my friends, for I feel that I have some. The degree of sympathetic response to the thoughts and suggestions I have offered, through the columns of the *Tribune,* has indeed surprised me, conscious as I am of a natural and acquired aloofness from many, if not the most popular tendencies of my time and place. It has greatly encouraged me, for none can sympathize with thoughts like mine, who are permanently insnared in the meshes of sect or party; none who prefer the formation and advancement of mere opinions to the free pursuit of truth. I see, surely, that the topmost bubble or sparkle of the cup is no voucher for the nature of its contents throughout, and shall, in future, feel that in our age, nobler in that respect than most of the preceding ages, each sincere and fervent act or word is secure, not only of a final, but of a speedy response.

I go to behold the wonders of art, and the temples of old religion. But I shall see no forms of beauty and majesty beyond what my country is capable of producing in myriad variety, if she has but the soul to will it; no temple to compare with what she might erect in the ages, if the catchword of the time, a sense of *divine order,* should become no more a mere word of form, but a deeply-rooted and pregnant idea in her life. Beneath the light of a hope that this may be, I say to my friends once more a kind farewell!

James Russell Lowell

(Hosea Biglow)

James Russell Lowell (1819–91), who dealt somewhat snidely with Margaret Fuller in his A Fable for Critics, *was also intrigued by the popular press. Half as* jeu d'esprit, *half for political reasons, Lowell began in June 1846 to contribute dialect poetry to the* Boston Courier. *He later gathered two years' production in* The Biglow Papers *(1848), saying of his political purposes: "I believed and still believe that slavery is the Achilles heel of our polity: that it is a temporary and false supremacy of the white races." After fifteen letters in a first series purporting to be by Hosea Biglow came a second collection in 1867 under the poet's own name, comprising eleven more letters. Lowell introduced the second series with an essay on American English. Reproduced here is a Biglow poem provoked by a Senate debate on slavery and an excerpt from Lowell's essay. The text is from* Lowell's Poetical Works *(Boston, 1894). The story of the Biglow papers, listing place and date of publication, can be found in* James Russell Lowell *(Boston, 1901) by Horace Elisha Scudder.*

TO MR. BUCKENAM

Boston Courier *May 3, 1848*

Mr. Editer,

As i wuz kinder prunin round, in a little nussry sot out a year or 2 a go, the Dbait in the sennit cum inter my mine An so i took & Sot it to wut I call a nussry rime. I hev made sum onnable Gentlemun speak thut dident speak in a Kind uv Poetikul lie sense the seeson is dreffle backered up This way

ewers as ushul
Hosea Biglow.

"Here we stan' on the Constitution, by thunder!
  It's a fact o' wich ther's bushils o' proofs;
Fer how could we trample on 't so, I wonder,
  Ef't worn't thet it's ollers under our hoofs?"
    Sez John C. Calhoun, sez he;
      "Human rights haint no more
      Right to come on this floor,
      No more 'n the man in the moon," sez he.

"The North haint no kind o' bisness with nothin',
  An' you've no idee how much bother it saves;
We aint none riled by their frettin' an' frothin',
  We're *used* to layin' the string on our slaves,"
    Sez John C. Calhoun, sez he;—
      Sez Mister Foote,
      "I should like to shoot
    The holl gang, by the gret horn spoon!" sez he.

"Freedom's Keystone is Slavery, thet ther's no doubt on,
  It's sutthin' thet's—wha' d'ye call it?—divine,—
An' the slaves thet we ollers *make* the most out on
  Air them north o' Mason an' Dixon's line,"
    Sez John C. Calhoun, sez he;—
      "Fer all thet," sez Mangum,
      " 'T would be better to hang 'em,
    An' so git red on 'em soon," sez he.

"The mass ough' to labor an' we lay on soffies,
  Thet's the reason I want to spread Freedom's aree;
It puts all the cunninest on us in office,
  An' reelises our Maker's orig'nal idee,"
    Sez John C. Calhoun, sez he;—
      "Thet's ez plain," sez Cass,
      "Ez thet some one's an ass,
    It's ez clear ez the sun is at noon," sez he.

"Now don't go to say I'm the friend of oppression,
  But keep all your spare breath fer coolin' your broth,
Fer I ollers hev strove (at least thet's my impression)
  To make cussed free with the rights o' the North,"
    Sez John C. Calhoun, sez he;—
      "Yes," sez Davis o' Miss.,
      "The perfection o' bliss
    Is in skinnin' thet same old coon," sez he.

"Slavery's a thing thet depends on complexion,
It's God's law thet fetters on black skins don't chafe;
Ef brains wuz to settle it (horrid reflection!)
"Wich of our onnable body 'd be safe?"
Sez John C. Calhoun, sez he;—
Sez Mister Hannegan,
Afore he began agin,
"Thet exception is quite oppertoon," sez he.

"Gen'nle Cass, Sir, you need n't be twitchin' your collar,
Your merit's quite clear by the dut on your knees,
At the North we don't make no distinctions o' color;
You can all take a lick at our shoes wen you please,"
Sez John C. Calhoun, sez he;—
Sez Mister Jarnagin,
"They wun't hev to larn agin,
They all on 'em know the old toon," sez he.

"The slavery question aint no ways bewilderin',
North an' South hev one int'rest, it's plain to a glance;
No'thern men, like us patriarchs, don't sell their childrin,
But they *du* sell themselves, ef they git a good chance,"
Sez John C. Calhoun, sez he;—
Sez Atherton here,
"This is gittin' severe,
I wish I could dive like a loon," sez he.

"It'll break up the Union, this talk about freedom,
An' your fact'ry gals (soon ez we split) 'll make head,
An' gittin' some Miss chief or other to lead 'em,
'll go to work raisin' permiscoous Ned,"
Sez John C. Calhoun, sez he;—
"Yes, the North," sez Colquitt,
"Ef we Southerners all quit,
Would go down like a busted balloon," sez he.

"Jest look wut is doin', wut annyky's brewin'
In the beautiful clime o' the olive an' vine,
All the wise aristoxy's a tumblin' to ruin,
An' the sankylots drorin' an' drinkin' their wine,"
Sez John C. Calhoun, sez he;—
"Yes," sez Johnson, "in France

They're beginnin' to dance
Beëlzebub's own rigadoon," sez he.

"The South 's safe enough, it don't feel a mite skeery,
Our slaves in their darkness an' dut air tu blest
Not to welcome with proud hallylugers they ery
Wen our eagle kicks yourn from the naytional nest,"
Sez John C. Calhoun, sez he;—
"Oh," sez Westcott o' Florida,
"Wut treason is horrider
Then our priv'leges tryin' to proon?" sez he.

"It's 'coz they're so happy, thet, wen crazy sarpints
Stick their nose in our bizness, we git so darned riled;
We think it's our dooty to give pooty sharp hints,
Thet the last crumb of Edin on airth sha' n't be spiled,"
Sez John C. Calhoun, sez he;—
"Ah," sez Dixon H. Lewis,
"It perfectly true is
Thet slavery's airth's grettest boon," sez he.

ON AMERICAN VERNACULAR

From introduction to The Biglow Papers, Second Series (1867)

In choosing the Yankee dialect, I did not act without forethought. It had long seemed to me that the great vice of American writing and speaking was a studied want of simplicity, that we were in danger of coming to look on our mother-tongue as a dead language, to be sought in the grammar and dictionary rather than in the heart, and that our only chance of escape was by seeking it at its living sources among those who were, as Scottowe says of Major-General Gibbons, "divinely illiterate." President Lincoln, the only really great public man whom these latter days have seen, was great also in this, that he was master—witness his speech at Gettysburg—of a truly masculine English, classic because it was of no special period, and level at once to the highest and lowest of his countrymen. I learn from the highest authority that his favorite reading was in Shakespeare and Milton, to which, of course, the Bible should be added. But whoever should read the debates in Congress might fancy himself present at a meeting of the city council of some city of Southern Gaul in the decline of the Empire, where barbarians with a Latin varnish emulated each other in being more than Ciceronian. Whether it be want of culture, for the highest outcome of that is simplicity, or for whatever reason, it is certain that very few American writers or speakers wield their native language with the directness, precision, and force that are common as the day in the mother country. We use it like Scotsmen, not as if it belonged to us, but as if we wished to prove that we belonged to it, by showing our intimacy with its written rather than with its spoken dialect. And yet all the while our popular idiom is racy with life and vigor and originality, bucksome (as Milton used the word) to our new occasions, and proves itself no mere graft by sending up new suckers from the old root in spite of us. It is only from its roots in the living generations of men that a language can be reinforced with fresh vigor for its needs; what may be called a literate dialect grows ever more and more pedantic and foreign, till it becomes at last a unfitting a vehicle for living thought as monkish Latin. That we should all be made to talk like books is the danger with which we are threatened by the Universal Schoolmaster, who does his best to enslave the minds and memories of his victims to what he esteems the best models of English composition, that is to say, to the writers whose style is faultily correct and has no blood-warmth in it. No language after it has faded into *diction,* none that cannot suck up the feeding juices secreted for it in the rich mother-earth of common folk, can bring forth a sound and lusty book. True vigor and heartiness of phrase do not pass from page to page, but from man to man, where the brain is kindled and the lips suppled by downright living interests and by passion in its very throe. Language is the soil of thought, and our own especially is a rich leaf-mould, the slow deposit of ages, the shed foliage of feeling, fancy, and imagination, which has suffered an earth-change, that the vocal forest, as Howell called it, may clothe itself anew with living green. There is death in the dictionary; and, where language is too strictly limited by convention, the ground for expression to grow in is limited also; and we get a *potted* literature, Chinese dwarfs instead of healthy trees.

But while the schoolmaster has been busy starching our language and smoothing it flat with the mangle of a supposed classical authority, the newspaper reporter has been doing even more harm by stretching and swelling it to suit his occasions. A dozen years ago I began a list, which I have added to from time to time, of some of the changes which may be fairly laid at his door. I give a few of them as showing their tendency, all the more dangerous that their effect, like that of some poisons, is insensibly cumulative, and that they are sure at last of effect among a people whose chief reading is the daily paper. I give in two columns the old style and its modern equivalent.

Old Style	*New Style*
Was hanged.	Was launched into eternity.
When the halter was put round his neck.	When the fatal noose was adjusted about the neck of the unfortunate victim of his own unbridled passions.
A great crowd came to see.	A vast concourse was assembled to witness.
Great fire.	Disastrous conflagration.
The fire spread.	The conflagration extended its devastating career.
House burned.	Edifice consumed.
The fire was got under.	The progress of the devouring element was arrested.
Man fell.	Individual was precipitated.
Began his answer.	Commenced his rejoinder.
Asked him to dine.	Tendered him a banquet.

Frederick Douglass

Born a slave in Talbot County, Eastern Shore, Maryland, Frederick Douglass (1817–95) escaped to Massachusetts (1838), and there became an Abolitionist speaker. Fearing recapture, he sailed to Britain, returning after several years, to found a weekly antislavery newspaper in Rochester, New York. His authorship questioned, Douglass began to initial his contributions to the North Star *(later* Frederick Douglass' Paper*). Hostility was unremitting. He recalled in his memoirs: "The* New York Herald, *true to the spirit of the times, counseled the people of the place to throw my printing press into Lake Ontario and to banish me to Canada." Less fiery than Garrison's* Liberator, *the paper continued until 1864. Douglass recruited two Negro regiments during the Civil War, and later served in Federal posts. These selections are from* The Life and Writings of Frederick Douglass *(New York, 1946), edited by Philip S. Foner.*

TO MY OPPRESSED COUNTRYMEN

North Star *December 3, 1847*

We solemnly dedicate the *North Star* to the cause of our long oppressed and plundered fellow countrymen. May God bless the offering to your good! It shall fearlessly assert your rights, faithfully proclaim your wrongs, and earnestly demand for you instant and even-handed justice. Giving no quarter to slavery at the South, it will hold no truce with oppressors at the North. While it shall boldly advocate emancipation for our enslaved brethren, it will omit no opportunity to gain for the nominally free, complete enfranchisement. Every effort to injure or degrade you or your cause

—originating wheresoever, or with whomsoever—shall find in it a constant, unswerving and inflexible foe.

We shall energetically assail the ramparts of Slavery and Prejudice, be they composed of church or state, and seek the destruction of every refuge of lies, under which tyranny may aim to conceal and protect itself.

Among the multitude of plans proposed and opinions held, with reference to our cause and condition, we shall try to have a mind of our own, harmonizing with all as far as we can, and differing from any and all where we must, but always discriminating between men and measures. We shall cordially approve every measure and effort calculated to advance your sacred cause, and strenuously oppose any which in our opinion may tend to retard its progress. In regard to our position, on questions that have unhappily divided the friends of freedom in this country, we shall stand in our paper where we have ever stood on the platform. Our views written shall accord with our views spoken, earnestly seeking peace with all men, when it can be secured without injuring the integrity of our movement, and never shrinking from conflict or division when summoned to vindicate truth and justice.

While our paper shall be mainly Anti-Slavery, its columns shall be freely opened to the candid and decorous discussion of all measures and topics of a moral and humane character, which may serve to enlighten, improve, and elevate mankind. Temperance, Peace, Capital Punishment, Education,—all subjects claiming the attention of the public mind may be freely and fully discussed here.

While advocating your rights, the *North Star* will strive to throw light on your duties: while it will not fail to make known your virtues, it will not shun to discover your faults. To be faithful to our foes it must be faithful to ourselves, in all things.

Remember that we are one, that our cause is one, and that we must help each other, if we would succeed. We have drunk to the dregs the bitter cup of slavery; we have worn the heavy yoke; we have sighed beneath our bonds, and writhed beneath the bloody lash;—cruel mementoes of our oneness are indelibly marked in our living flesh. We are one with you under the ban of prejudice and proscription—one with you under the slander of inferiority—one with you in social and political disfranchisement. What you suffer, we suffer; what you endure, we endure. We are indissolubly united, and must fall or flourish together.

We feel deeply the solemn responsibility which we have now assumed. We have seriously considered the importance of the enterprise, and have now entered upon it with full purpose of heart. We have nothing to offer in the way of literary ability to induce you to encourage us in our laudable undertaking. You will not expect or require this at our hands. The most that you can reasonably expect, or that we can safely promise, is a paper of which you need not be ashamed. Twenty-one years of severe bondage at the South, and nine years of active life at the North, while it has afforded us the best possible opportunity for storing our mind with much practical and important information, has left us little time for literary pursuits or attainments. We have yet to receive the advantage of the first day's schooling. In point of education, birth and rank, we are one with yourselves, and of yourselves. What we are, we are not only without help, but against trying opposition. Your knowledge of our history for the last seven years makes it unnecessary for us to say more on this point. What we have been in your cause, we shall continue to be; and not being too old to learn, we may improve in many ways. Patience and Perseverance shall be our motto.

We shall be the advocates of learning, from the very want of it, and shall most readily yield the deference due to men of education among us; but shall always bear in mind to accord most merit to those who have labored hardest, and overcome most, in the praiseworthy pursuit of knowledge, remembering "that the whole need not a physician, but they that are sick," and that "the strong ought to bear the infirmities of the weak."

Brethren, the first number of the paper is before you. It is dedicated to your cause. Through the kindness of our friends in England, we are in possession of an excellent printing press, types, and all other materials necessary for printing a paper. Shall this gift be blest to our good, or shall it result in our injury? It is for you to say. With your aid, co-operation and assistance, our enterprise will be entirely successful. We pledge ourselves that no effort on our part shall be wanting, and that no subscriber shall lose his subscription—"The *North Star* Shall Live."

THE DESTINY OF COLORED AMERICANS

North Star *November 16, 1849*

It is impossible to settle, by the light of the present, and by the experience of the past, any thing, definitely and absolutely, as to the future condition of the colored people of this country; but, so far as present indications determine, it is clear that this land must continue to be the home of the colored man so long as it remains the abode of civilization and religion. For more than two hundred years we have been identified with its soil, its products, and its institutions; under the sternest and bitterest circumstances of slavery and oppression—under the lash of Slavery at the South—under the sting of prejudice and malice at the North—and under hardships the most unfavorable to existence and population, we have lived, and continue to live and increase. The persecuted red man of the forest, the original owner of the soil, has, step by step, retreated from the Atlantic lakes and rivers; escaping, as it were, before the footsteps of the white man, and gradually disappearing from the face of the country. He looks upon the steamboats, the railroads, and canals, cutting and crossing his former hunting grounds; and upon the ploughshare, throwing up the bones of his venerable ancestors, and beholds his glory departing—and his heart sickens at the desolation. He spurns the civilization—he hates the race which has despoiled him, and unable to measure arms with his superior foe, he dies.

Not so with the black man. More unlike the European in form, feature and color—called to endure greater hardships, injuries and insults than those to which the Indians have been subjected, he yet lives and prospers under every disadvantage. Long have his enemies sought to expatriate him, and to teach his children that this is not their home, but in spite of all their cunning schemes, and subtle contrivances, his footprints yet mark the soil of his birth, and he gives every indication that America will, for ever, remain the home of his posterity. We deem it a settled point that the destiny of the colored man is bound up with that of the white people of this country; be the destiny of the latter what it may.

It is idle—worse than idle, ever to think of our expatriation, or removal. The history of the colonization society must extinguish all such speculations. We are rapidly filling up the number of four millions; and all the gold of California combined, would be insufficient to defray the expenses attending our colonization. We are, as laborers, too essential to the interests of our white fellow-countrymen, to make a very grand effort to drive us from this country among probable events. While labor is needed, the laborer cannot fail to be valued; and although passion and prejudice may sometimes vociferate against us, and demand our expulsion, such efforts will only be spasmodic, and can never prevail against the sober second thought of self-interest. *We are here,* and here we are likely to be. To imagine that we shall ever be eradicated is absurd and ridiculous. We can be remodified, changed, and assimilated, but never extinguished. We repeat, therefore, that *we are here;* and that this is *our* country; and the question for the philosophers and statesmen of the land ought to be, What principles should dictate the policy of the action towards us? We shall neither die out, nor be driven out; but shall go with this people, either as a testimony against them, or as an evidence in their favor throughout their generations. We are clearly on their hands, and must remain there for ever. All this we say for the benefit of those who hate the Negro more than they love their country. In an article, under the caption of "Government and its Subjects," (published in our last week's paper,) we called attention to the unwise, as well as the unjust policy usually adopted, by our Government, towards its colored citizens. We would continue to direct attention to that policy, and in our humble way, we would remonstrate against it, as fraught with evil to the white man, as well as to his victim.

The white man's happiness cannot be purchased by the black man's misery. Virtue cannot prevail among the white people, by its destruction among the black people, who form a part of the whole community. It is evident that white and black "must fall or flourish together." In the light of this great truth, laws ought to be enacted, and institutions established —all distinctions, founded on complexion, ought to be repealed, repudiated, and for ever abolished—and every right, privilege, and immunity, now enjoyed by the white man, ought to be as freely granted to the man of color.

Where "knowledge is power," that nation is the most powerful which has the largest population of intelligent men; for a nation to cramp, and circumscribe the mental faculties of a class of its inhabitants, is as unwise as it is cruel, since it, in the same proportion, sacrifices its power and happiness. The American people, in the light of this reasoning, are, at this moment, in obedience to their pride and folly, (we say nothing of the wickedness of the act,) wasting one sixth part of the energies of the entire nation by transforming three millions of its men into beasts of burden.— What a loss to industry, skill, invention, (to say nothing of its foul and corrupting influence,) is *Slavery!* How it ties the hand, cramps the mind, darkens the understanding, and paralyses the whole man! Nothing is more evident to a man who reasons at all, than that America is acting an irrational part in continuing the slave system at the South, and in oppressing its free colored citizens at the North. Regarding the nation as an individual, the act of enslaving and oppressing thus, is as wild and senseless as it would be for Nicholas to order the amputation of the right arm of every Russian soldier before engaging in a war with France. We again repeat that Slavery is the peculiar weakness of America, as well as its peculiar crime; and the day may yet come when this visionary and oft repeated declaration will be found to contain a great truth.—F. D.

Frederick Law Olmsted

(Yeoman)

The author of a classic account of the Old South, The Cotton Kingdom, *is better known as the designer of New York's Central Park. Yet Frederick Law Olmsted (1822–1903) was also an outstanding shoeleather journalist. His account was based on three extended journeys as a correspondent for New York newspapers. Each trip produced a book, which Olmsted then wove into a two-volume work published simultaneously in England and the United States in 1861, just as the Civil War was beginning. His journeys arose from a recurring argument with an Abolitionist friend; to settle it, Olmsted agreed to judge conditions first-hand. His friend mentioned the project to Henry J. Raymond, who had recently launched the* New York Times. *In a five-minute encounter, the editor engaged Olmsted. "The only intimation I received of his expectations as to the matter I should write," Olmsted was to recall, "was a request that it should be confined to personal observations, and the expression of a wish that I would not feel myself at all restricted or constrained by regard to consistency with the general position of the paper."*

The first letter appeared February 16, 1853, under what came to be a running title, "The South." All forty-nine letters were signed "Yeoman." A trip to the Southwest a year later resulted in fifteen reports, and a final journey was described in ten articles for the New York Tribune. *Olmsted's reportage was distinguished by its meticulous details and verbatim conversations, as in the excerpts that follow. A summary essay for the* Times *that he omitted from his books is also appended, reproduced from a reissue of* The Cotton Kingdom *(New York, 1953), edited by Arthur M. Schlesinger, Sr.*

CONVERSATIONS IN DIXIE

New York Times *circa March 1854*

On the Chockolate [in Louisiana]—"Which way did you come?" asked some one of the old man.

"From ——."

"See anything of a runaway nigger over there, anywhar?"

"No, sir. What kind of a nigger was it?"

"A small, black, screwed-up-faced nigger."

"How long has he been out?"

"Nigh two weeks."

"Whose is he?"

"Judge——'s, up here. And he cut the judge right bad. Like to have killed the judge. Cut his young master, too."

"Reckon, if they caught him 'twould go rather hard with him."

"Reckon 'twould. We caught him once, but he got away from us again. We was just tying his feet together, and he give me a kick in the face, and broke. I had my six-shooter handy, and I tried to shoot him, but every barrel missed fire. Been loaded a week. We shot at him three times with rifles, but he'd got too far off, and we didn't hit, but we must have shaved him close. We chased him, and my dog got close to him once. If he'd grip'd him, we should have got him; but he had a dog himself, and just as my dog got within about a yard of him, his dog turned and fit my dog, and he hurt him so bad we couldn't get him to run him again. We run him close, though, I tell you. Run him out of his coat, and his boots, and a pistol he'd got. But 'twas getting towards dark, and he got into them bayous, and kept swimming from one side to another."

"How long ago was that?"

"Ten days."

"If he's got across the river, he'd get to the Mexicans in two days, and there he'd be safe. The Mexicans'd take care of him."

"What made him run?"

"The judge gave him a week at Christmas, and when the week was up, I s'pose he didn't want to go to work again. He got unruly, and they was a goin' to whip him."

"Now, how much happier that fellow'd 'a' been, if he'd just stayed and done his duty. He might have just worked and done his duty, and his master'd 'a' taken care of him, and given him another week when Christmas come again, and he'd 'a' had nothing to do but enjoy himself again. These niggers, none of 'em, knows how much happier off they are than if they was free. Now, very likely, he'll starve to death, or get shot."

"Oh, the judge treats his niggers too kind. If he was stricter with them, they'd have more respect for him, and be more contented, too."

"Never do to be too slack with niggers."

* * *

Houston. We were sitting on the gallery of the hotel. A tall, jet black negro came up, leading by a rope a downcast mulatto, whose hands were lashed by a cord to his waist, and whose face was horribly cut, and dripping with blood. The wounded man crouched and leaned for support against one of the columns of the gallery—faint and sick.

"What's the matter with that boy?" asked a smoking lounger.

"I run a fork into his face," answered the negro.

"What are his hands tied for?"

"He's a runaway, sir."

"Did you catch him?"

"Yes, sir. He was hiding in the hay-loft, and when I went up to throw some hay to the horses, I pushed the fork down into the mow and it struck something hard. I didn't know what it was, and I pushed hard, and gave it a turn, and then he hollered, and I took it out."

"What do you bring him here for?"

"Come for the key of the jail, sir, to lock him up."

"What!" said another, "one darkey catch another darkey? Don't believe that story."

"Oh yes, mass'r, I tell for true. He was down in our hay-loft, and so you see when I stab him, I *have* to catch him."

"Why, he's hurt bad, isn't he?"

"Yes, he says I pushed through the bones."

"Whose nigger is he?"

"He says he belong to Mass'r Frost, sir, on the Brazos."

The key was soon brought, and the negro led the mulatto away to jail. He walked away limping, crouching, and writhing, as if he had received other injuries than those on his face. The bystanders remarked that the negro had not probably told the whole story.

We afterwards happened to see a gentleman on horseback, and smoking, leading by a long rope through the deep mud, out into the country, the poor mulatto, still limping and crouching, his hands manacled, and his arms pinioned.

There is a prominent slave-mart in town, which holds a large lot of likely-looking negroes, waiting purchasers. In the windows of shops, and on the doors and columns of the hotel, are many written advertisements, headed "A likely negro girl for sale." "Two negroes for sale." "Twenty negro boys for sale," etc.

* * *

June 2nd. I met a ragged old negro, of whom I asked the way, and at what house within twelve miles I had better stop. He advised me to go to one more than twelve miles distant.

"I suppose," said I, "I can stop at any house along the road here, can't I? They'll all take in travellers?"

"Yes, sir, if you'll take rough fare, such as travellers has to, sometimes. They're all damn'd rascals along dis road, for ten or twelve miles, and you'll get nothin' but rough fare. But I say, massa, rough fare's good enough for dis world; ain't it, massa? Dis world ain't nothin; dis is hell, dis is, I calls it; hell to what's a comin' arter, ha! ha! Ef you's prepared? you says. I don't look much's if I was prepared, does I? nor talk like it, nuther. De Lord he cum to me in my cabin in de night time, in de year '45."

"What?"

"De Lord! massa, de bressed Lord! He cum to me in de night time, in de year '45, and he says to me, says he, "I'll spare you yet five year longer, old boy!' So when '50 cum round I thought my time had cum, sure; but as I didn't die, I reckon de Lord has 'cepted of me, and I 'specs I shall be saved, dough I don't look much like it, ha! ha! ho! ho! de Lord am my rock, and he shall nor prewail over me. I will lie down in green pastures and take up my bed in hell, yet will not His mercy circumwent me. Got some baccy, master?"

A little after sunset I came to an unusually promising plantation, the dwelling being within a large enclosure, in which there was a well-kept southern sward shaded by fine trees. The house, of the usual form, was painted white, and the large number of neat out-buildings seemed to indicate opulence, and, I thought, unusual good taste in its owner. A lad of sixteen received me, and said I could stay; I might fasten my horse, and when the negroes came up he would have him taken care of. When I had done so, and had brought the saddle to the verandah, he offered me a chair, and at once commenced a conversation in the character of entertainer. Nothing in his tone or manner would have indicated that he was not the father of the family, and proprietor of the establishment. No prince royal could have had more assured and nonchalant dignity. Yet a Northern stable-boy, or apprentice, of his age, would seldom be found as ignorant.

"Where do you live, sir, when you are at home?" he asked.

"At New York."

"New York is a big place, sir, I expect?"

"Yes, very big."

"Big as New Orleans, is it, sir?"

"Yes, much bigger."

"Bigger 'n New Orleans? It must be a bully city."

"Yes; the largest in America."

"Sickly there now, sir?"

"No, not now; it is sometimes."

"Like New Orleans, I suppose?"

"No, never so bad as New Orleans sometimes is."

"Right healthy place, I expect, sir?"

"Yes, I believe so, for a place of its size."

"What diseases do you have there, sir?"

"All sorts of diseases—not so much fever, however, as you have hereabouts."

"Measles and whooping-cough, sometimes, I reckon?"

"Yes, 'most all the time, I dare say."

"All the time! People must die there right smart. Some is dyin' 'most every day, I expect, sir?"

"More than a hundred every day, I suppose."

"Gosh! a hundred every day! Almighty sickly place 't must be?"

"It is such a large place, you see—seven hundred thousand people."

"Seven hundred thousand—expect that's a heap of people, ain't it?"

His father, a portly, well-dressed man, soon came in, and learning that I had been in Mexico, said, "I suppose there's a heap of Americans flocking in and settling up that country along on the line, ain't there, sir?"

"No, sir, very few. I saw none, in fact—only a few Irishmen and Frenchmen, who called themselves Americans. Those were the only foreigners I saw, except negroes."

"Niggers! Where were they from?"

"They were runaways from Texas."

"But their masters go there and get them again, don't they?"

"No, sir, they can't."

"Why not?"

"The Mexicans are friendly to the niggers, and protect them."

"But why not go to the Government?"

"The Government considers them as free, and will not let them be taken back."

"But that's stealing, sir. Why don't our Government make them deliver them up? What good is the Government to us if it don't preserve the rights of property, sir? Niggers are property, ain't they? and if a man steals my property, ain't the Government bound to get it for me? Niggers are property, sir, the same as horses and cattle, and nobody's any more right to help a nigger that's run away than he has to steal a horse."

He spoke very angrily, and was excited. Perhaps he was indirectly addressing me, as a Northern man, on the general subject of fugitive slaves. I said that it was necessary to have special treaty stipulations about such matters. The Mexicans lost the *peóns*—bounden servants; they ran away to our side, but the United States Government never took any measures to restore them, nor did the Mexicans ask it. "But," he answered, in a tone of indignation, "those are not niggers, are they? They are white people, sir, just as white as the Mexicans themselves, and just as much right to be free."

A SLAVERY'S EFFECTS ON THE MASTER CLASS

New York Times *January 12, 1854*

The wealthy and educated, and especially the fashionable people of all civilized countries, are now so nearly alike in their ordinary manners and customs, that the observations of a passing traveler upon them, must commonly be of much too superficial a character to warrant him in deducing from them, with confidence, any important conclusions. I have spent an evening at the plantation residence of a gentleman in Louisiana, in which there was very little in the conversation or customs and manners of the family to distinguish them from others whom I have visited in Massachusetts, England and Germany. I shall, therefore, undertake with diffidence to describe certain apparently general and fundamental peculiarities of character in the people, which it is a part of my duty to notice, from their importance with reference to the condition and prospects of the Slave States and their institution.

Slavery exerts an immense quiet influence upon the character of the master, and the condition of the slave is greatly affected by the modifications of character thus effected. I do not believe there are any other people in the world with whom the negro would be as contented, and, if contentment is happiness, so happy, as with those who are now his masters. The hopeless perpetuation of such an intolerable nuisance as this labor-system, it is, however, also apparent, depends mainly upon the careless, temporizing, *shiftless* disposition, to which the negro is indebted for this mitigation of the natural wretchedness of Slavery.

The calculating, indefatigable New-Englander, the go-ahead Western man, the exact and stern Englishman, the active Frenchman, the studious, observing, economical German would all and each lose patience with the frequent disobedience and the constant indolence, forgetfulness and carelessness, and the blundering, awkward, brute-like manner of work of the plantation-slave. The Southerner, if he sees anything of it, generally disregards it and neglects to punish it. Although he is naturally excitable and passionate, he is less subject to impatience and passionate anger with the slave, than is, I believe, generally supposed, because he is habituated to regard him so completely as his inferior, dependent and subject. For the same reason, his anger, when aroused, is usually easily and quickly appeased, and he forgives him readily and entirely, as we do a child or a dog who has annoyed us. And, in general, the relation of master and slave on small farms, and the relations of the family and its household servants everywhere, may be considered a happy one, developing, at the expense of decision, energy, self-reliance and self-control, some of the most beautiful traits of human nature. But it is a great error,—although one nearly universal with Southerners themselves,—to judge of Slavery by the light alone of the master's fireside.

The direct influence of Slavery is, I think, to make the Southerner indifferent to small things; in some relations, we would say rightly, *superior* to small things; prodigal, improvident, and ostentatiously generous. His ordinarily uncontrolled authority, (and from infancy the Southerner is more free from control, in all respects, I should judge, than any other person in the world) leads him to be habitually impulsive, impetuous, and enthusiastic: gives him self-respect and dignity of character, and makes him bold, confident, and true. Yet is has not appeared to me that the Southerner was frank as he is, I believe, commonly thought to be. He seems to me to be very secretive, or at least reserved, on topics which most nearly concern himself. He minds his own business, and lets alone that of others; not in the English way, but in a way peculiarly his own; resulting partly, perhaps, from want of curiosity, in part from habits formed by such constant intercourse as he has with his inferiors, (negroes,) and partly from the caution in conversation which the "rules of honor" are calculated to give. Not, I said, in the English way, because he meets a stranger easily, and without timidity, or thought of how he is himself appearing, and is ready and usually accomplished in conversation. He is much given to vague and careless generalization, and greatly disinclined to exact and careful reasoning. He follows his natural impulses nobly, has nothing to be ashamed of, and is, therefore, habitually truthful; but his carelessness, impulsiveness, vagueness, and want of exactness in everything, make him speak from his mouth that which is in point of fact untrue, rather oftener than any one else.

From early intimacy with the negro, (an association fruitful in other respects of evil,) he has acquired much of his ready, artless and superficial benevolence, good nature and geniality. The comparatively solitary nature and somewhat monotonous duties of plantation life, make guests usually exceedingly welcome, while the abundance of servants at command, and other circumstances, make the ordinary duties of hospitality very light. The Southerner, however, is greatly wanting in hospitality of mind, closing his doors to all opinions and schemes to which he has been bred a stranger, with a contempt and bigotry which sometimes seems incompatible with his character as a gentleman. He has a large but unexpansive mind.

The Southerner has no pleasure in labor except with reference to a result. He enjoys life itself. He is content with being. Here is the grand distinction between him and the Northerner; for the Northerner enjoys progress in itself. He finds his happiness in doing. Rest, in itself, is irksome and offensive to him, and however graceful or beatific that rest may be, he values it only with reference to the power of future progress it will bring him. Heaven itself will be dull and stupid to him, if there is no work to be done in it—nothing to struggle for—if he reaches perfection at a jump, and has no chance to make an improvement.

The Southerner cares for the end only; he is impatient of the means.

He is passionate, and labors passionately, fitfully, with the energy and strength of anger, rather than of resolute will. He fights rather than works to carry his purpose. He has the intensity of character which belongs to Americans in general, and therefore enjoys excitement and is fond of novelty. But he has much less curiosity than the Northerner; less originating genius, less inventive talent, less patient and persevering energy. And I think this all comes from his want of aptitude for close observation and his dislike for application to small details. And this, I think, may be reasonably supposed to be mainly the result of habitually leaving all matters not either of grand and exciting importance, or of immediate consequence to his comfort, to his slaves, and of being accustomed to see them slighted or neglected as much as he will, in his indolence, allow them to be by them.

Walt Whitman

Walt Whitman (1819–92), journeyman printer and aspiring national poet, wrote extensively for no less than fifteen newspapers. Having left school at eleven, Whitman learned his trade as editor of the Long Island Democrat, *moved on to bustling Brooklyn, then a city of its own, and later wrote regularly for papers as different as the* New Orleans Crescent *and the* New York Times. *The sources for the following are* The Uncollected Poetry and Prose of Walt Whitman *(New York, 1921), edited by Emery Holloway; and* The Gathering of the Forces *(New York, 1920), edited by Cleveland Rodgers and John Black.*

IN PRAISE OF LOAFERS

Sun–Down Papers No. 9

Long Island Democrat *November 28, 1840*

How I do love a loafer! Of all human beings, none equals your genuine, inbred, unvarying loafer. Now when I say loafer, I *mean* loafer; not a fellow who is lazy by fits and starts—who today will work his twelve or fourteen hours, and to-morrow doze and idle. I stand up for no such half-way business. Give me your calm, steady, philosophick son of indolence; one that doesn't swerve from the beaten track; a man who goes the undivided beast. To such an one will I doff my beaver. No matter whether he be a street loafer or a dock loafer—whether his hat be rimless, and his boots slouched, and his coat out at the elbows: he belongs to the ancient and honourable fraternity, whom I venerate above all your upstarts, your dandies, and your political oracles.

All the old philosophers were loafers. Take Diogenes for instance. He lived in a tub, and demeaned himself like a true child of the great loafer family. Or go back farther, if you like, even to the very beginning. What was Adam, I should like to know, but a loafer? Did he do anything but loaf? Who is foolish enough to say that Adam was a working man? Who dare aver that he dealt in stocks, or was busy in the sugar line?

I hope you will not so far expose yourself as to ask, who was the founder of loafers. Know you not, ignorance, that there never was such a thing as the *origin* of loaferism? We don't acknowledge any founder. There have always been loafers as they were in the beginning, are now, and ever shall be—having no material difference. Without any doubt, when Chaos has his acquaintance cut, and the morning stars sang together, and the little rivers danced a cotillion for pure fun—there were loafers somewhere about, enjoying the scene in all their accustomed philosophick quietude.

When I have been in a dreamy, musing mood, I have sometimes amused myself with picturing out a nation of loafers. Only think of it! an entire loafer kingdom! How sweet it sounds! Repose,—quietude,—roast duck,—loafer. Smooth and soft are the terms to our jarred tympanums.

Imagine some distant isle inhabited altogether by loafers. Of course there is a good deal of sunshine, for sunshine is the loafer's natural element. All breathes peace and harmony. No hurry, or bustle, or banging, or clanging. Your ears ache no more with the din of carts; the noisy politician offends you not, no wrangling, no quarreling, no loco focos, no British whigs.

Talk about your commercial countries, and your national industry, indeed! Give us the facilities of loafing, and you are welcome to all the benefits of your tariff system, your manufacturing privileges, and your cotton trade. For my part, I have had serious thoughts of getting up a regular ticket for President and Congress and Governor and so on, for the loafer community in general. I think we loafers should organize. We want somebody to carry out "our principles." It is my impression, too, that we should poll a pretty strong vote. We number largely in the land. At all events our strength would enable us to hold the balance of power, and we should be courted and coaxed by all the rival factions. And there is not telling but what we might elect our men. Stranger things than that have come to pass.

These last hints I throw out darkly, as it were. I by no means assert that we positively *will* get up and vote for, a regular ticket to support the "great measures of our party." I am only telling what *may* be done, in case we are provoked. Mysterious intimations have been thrown out—dark sayings uttered, by those high in society, that the grand institution of loaferism was to be abolished. People have talked of us sneeringly and frowningly. Cold eyes have been turned upon us. Overbearing men have spoken in derogatory terms about our rights and our dignity. You had better be careful, gentlemen. You had better look out how you irritate us. It would make you look sneaking enough, if we were to come out at the next election, and carry away the palm before both your political parties.

CONTRA ALBION

Brooklyn Eagle *February 10, 1847*

Several of the papers—among the rest *Yankee Doodle*—took us to task, not long since, for some strictures on an incident which occurred at the anniversary dinner of the Hamilton Association, in this city. One of the guests at that dinner proposed a toast involving an assertion of American literary independence—which was partially hissed, *not* by any member of the Hamilton Society, however, but by some other guest or guests. (In our former brief notice, certain words were used which we are now convinced did injustice to this really talented band of young men; they are, many of them gentlemen of much literary taste, and true perception.) *Yankee Doodle* thinks not only that the toast was very properly hissed, but that instead of mere hissing, the hearers should have hooted its author from the room. This horrible toast, was in the following words: "The United States of America—an independent country, and not a mere suburb of London." Truly a frightful and audacious sentiment! a most treasonable, rebellious toast! *Are* we not a "mere suburb of London?" We trow yes, as long as such sentiments as that of the hapless toast are condemned by a periodical whose very foundation starts in the idea of nationality—as long as we copy with a servile imitation, the very cast-off literary fashions of London—as long as we wait for English critics to stamp our books and our authors, before *we* presume to say they are very good or very bad—as long as the floods of British manufactured books are poured over the land, and give their color to all the departments of taste and opinion—as long as an American society, meeting at the social board, starts with wonder to hear any of its national names, or any national sentiment, mentioned in the same hour with foreign authors or foreign greatness. . . .

LETTER FROM CIVIL WAR WASHINGTON

New York Times *October 4, 1863*

It is doubtful whether justice has been done to Washington, D.C.; or rather, I should say, it is certain there are layers of originality, attraction, and even local grandeur and beauty here, quite unwritten, and even to the inhabitants unsuspected and unknown. Some are in the spot, soil, air and the magnificent amplitude in the laying out of the City. I continually enjoy these streets, planned on such a generous scale, stretching far, without stop or turn, giving the eye vistas. I feel freer, larger in them. Not the squeezed limits of Boston, New-York, or even Philadelphia; but royal plenty and nature's own bounty—American, prairie-like. It is worth writing a book about, this point alone. I often find it silently, curiously making up to me the absence of the ocean tumult of humanity I always enjoyed in New-York. Here, too, is largeness, in another more impalpable form; and I never walk Washington, day or night, without feeling its satisfaction.

Like all our cities, so far, this also, in its inner and outer channels, gives obedient reflex of European customs, standards, costumes, &c. There is the immortal black broadcloth coat, and there is the waiter standing behind the chair. But inside the costume, America can be traced in glimpses. Item, here an indolent largeness of spirit, quite native. No man minds his exact change. The vices here, the extravagancies, (and worse), are not without something redeeming; there is such a flowing hem, such a margin.

We all know the chorus: Washington, dusty, muddy, tiresome Washington is the most awful place, political and other; it is the rendezvous of the national universal axe-grinding, caucusing, and of our never-ending ballot-chosen shysters, and perennial smouchers, and windy bawlers from every quarter far and near. We learn, also, that there is no society, no art, in Washington; nothing of the elaborated high-life attractions of the charming capitals (for rich and morbid idlers) over sea. Truly this particular sort of charm is not in full blossom here; *n'importe.* Let those miss it who miss it (we have a sad set among our rich young men,) and, if they will, go voyage over sea to find it. But there are man's studies, objects here, never more exhilirating ones. What themes, what fields this national city affords, this hour, for eyes of live heads, and for souls fit to feed upon them!

This city, this hour, in its material sights, and what they and it stand for, the point of the physical and moral America, the visible fact of this war, (how at last, after sleeping long as it may, one finds war ever-dearest fact to man, though most terrible, and only arbiter, after all said about the pen being mightier, &c.) This city, concentre to-day of the innauguration of the new adjustment of the civilized world's political power and geography, with vastest consequences of Presidential and Congressional action; things done here, these days, bearing on the status of man, long centuries; the spot and the hour here making history's basic materials and widest ramifications; the city of the armies of the good old cause, full of significant signs, surrounded with weapons and armaments on every hill as I look forth, and THE FLAG flying over all. The city that launches the direct laws, the imperial laws of American Union and Democracy, to be henceforth compelled, when needed, at the point of the bayonet and the muzzle of cannon—launched over continental areas, three millions of square miles, an empire large as Europe. The city of wounded and sick, city of hospitals, full of the sweetest, bravest children of time or lands; tens of thousands, wounded, bloody, amputated, burning with fever, blue with diarrhœa. The city of the wide Potomac, the queenly river, lined with softest, greenest hills and uplands. The city of Congress, with debates, agitations, (petty, if you please, but full of future fruit,) of chaotic formings; of Congress knowing not itself, as it sits there in its rooms of gold, knowing not the depths of consequence belonging to it, that lie below the scum and eructations of its surface.

But where am I running to? I meant to make a few observations of Washington on the surface.

The Dome and The Genius. We are soon to see a thing accomplished here which I have often exercised my mind about, namely, the putting of the Genius of America away up there on the top of the dome of the Capitol. A few days ago, poking about there, eastern side, I found the Genius, all dismembered, scattered on the ground, by the basement front—I suppose preparatory to being hoisted. This, however, cannot be done forthwith, as I know that an immense pedestal surmounting the dome, has yet to be finished—about eighty feet high—on which the Genius is to stand, (with her back to the city).

But I must say something about the dome. All the great effects of the Capitol reside in it. The effects of the Capitol are worth study, frequent and varied I find they grow upon one. I shall always identify Washington with that huge and delicate towering bulge of pure white, where it emerges calm and lofty from the hill, out of a dense mass of trees. There is no place in the city, or for miles and miles off, or down or up the river, but what you see this tiara-like dome quietly rising out of the foliage (one of the effects of first-class architecture is its serenity, its *aplomb*).

A vast eggshell, built of iron and glass, this dome—a beauteous bubble, caught and put in permanent form. I say a beauty and genuine success. I have to say the same, upon the whole, (after some qualms, may be,) with respect to the entire edifice. I mean the entire Capitol is a sufficient success, if we accept what is called architecture [of] the orthodox styles, (a little mixed here,) and indulge them for our purposes until further notice.

The dome I praise with the aforesaid Genius, (when she gets up, which she probably will by the time next Congress meets,) will then aspire about three hundred feet above the surface. And then, remember that our National House is set upon a hill. I have stood over on the Virginia hills, west of the Potomac, or on the Maryland hills, east, and viewed the structure from all positions and distances; but I find myself, after all, very fond of getting somewhere near, somewhere within fifty or a hundred rods, and gazing long and long at the dome rising out of the mass of green umbrage, as aforementioned.

The dome is tiara or triple. The lower division is surrounded with a ring of columns, pretty close together. There is much ornament everywhere, but it is kept down by the uniform white; then lots of slender oval-topt windows. Ever as I look, especially when near, (I repeat it,) the dome is a beauty, large and bold. From the east side it shows immensely. I hear folks say it is too large. Not at all, to my eye. Some say, too, the columns front and rear of the Old Capitol part, there in the centre, are now so disproportionably slender by the enlargement, that they must be removed. I say no; let them stand. They have a pleasant beauty as they are; the eye will get accustomed to them, and approve them.

Of our Genius of America, a sort of compound of handsome Choctaw squaw with the well-known Liberty of Rome, (and the French revolution,) and a touch perhaps of Athenia Pallas, (but very faint,) it is to be further described as an extensive female, cast in bronze, with much drapery, especially ruffles. The Genius has for a year or two past been standing in the mud, west of the Capitol; I saw her there all Winter, looking very harmless and innocent, although holding a huge sword. For pictorial representation of the Genius, see any five-dollar United States greenback; for there she is at the left hand. But the artist has made her twenty times brighter in expression, &c., than the bronze Genius is.

I have curiosity to know the effect of this figure crowning the dome. The pieces, as I have said, are at present all separated, ready to be hoisted to their place. On the Capitol generally, much work remains to be done. I nearly forgot to say that I have grown so used to the sight, over the Capitol, of a certain huge derrick which has long surmounted the dome, swinging its huge one-arm now south, now north, &c., that I believe I shall have a sneaking sorrow when they remove it and substitute the Genius. . . .

Then the trees and their dark and glistening verdure play their part. Washington, being full of great white architecture, takes through the Summer a prevailing color-effect of white and green. I find this everywhere, and very pleasing to my sight. So, seen freed from dust, as of late, and with let up from that unprecedented August heat, I say I find atmospheric results of marked individuality and perfection here, beyond Northern, Western, and farther Southern cities. (Our writers, writing, may pen as much as they please of Italian light, and of Rome and Athens. But this city, even in the crude state it is to-day, with its buildings of to-day, with its ample river and its streets, with the effects above noted, to say nothing of what it all represents, is of course greater, materially and morally, to-day than ever Rome or Athens.)

George H. Derby

(John Phoenix, Esq.)

Captain George H. Derby (1823–61), a West Pointer, was wounded in the Mexican War and subsequently billeted to gold-rush California. There for seven years he was both topographical engineer and contributor to frontier newspapers, first as Squibob, then as John Phoenix (see Introduction). In 1857, he was appointed a lighthouse engineer in Florida, where a sunstroke clouded his mind and shortened his life; he was thirty-eight when he died in New York City. Derby's bouncing style, recalling Sterne and Smollett, artfully mingled classical tags and Western slang. While in San Diego, Derby was entrusted with the Herald *when the paper's editor, J. J. Ames, was on a business trip. The results follow. His sketches were collected in* Phoenixiana *(New York, 1856) and* The Squibob Papers *(New York, 1865). George R. Stewart's* John Phoenix, Esq. *(New York, 1937) includes a bibliography. A fresh selection of* Phoenixiana, *edited by Francis P. Farquhar, was elegantly assembled in 1937 by San Francisco's Grabhorn Press.*

PHOENIX AS EDITOR*

The San Diego Herald *August 24, 1853*

"*Facilis decensus Averni,*" which may be liberally, not literally translated, It is easy to go to San Francisco. Big Ames has gone; departed in the *Goliah,* in the hope of obtaining new advertisements for this interesting journal, perchance, hoping also to be paid for the old ones—I hope he

*The San Diego *Herald* of August 20, 1853, contained the following notice:
"Next week, with the Divine assistance, a new hand will be applied to the bellows of this establishment, and an intensely interesting issue will possibly be the result. The paper will be published on Wednesday evening; and, to avoid confusion, the crowd will please

may succeed in both endeavors. During his absence, which I trust will not exceed two weeks, I am to remain in charge of the *Herald,* the literary part thereof—I would beg to be understood—the *responsible* portion of the editorial duties falling upon my friend Johnny, who has, in the kindest manner, undertaken "the fighting department," and to whom I hereby refer any pugnacious or bellicose individual who may take offence at the tone of any of my leaders. The public at large, therefore, will understand that I stand upon "Josh Haven's platform," which that gentleman defined some years since to be the liberty of saying any thing he pleased about anybody, without considering himself at all responsible. It is an exceedingly free and independent position, and rather agreeable than otherwise; but I have no disposition whatever to abuse it.

It will be perceived that I have not availed myself of the editorial privilege of using the plural pronoun in referring to myself. This is simply because I consider it a ridiculous affectation. I am a "lone, lorn man," unmarried (the Lord be praised for his infinite mercy), and though blessed with a consuming appetite "which causes the keepers of the house (Hooff & Tebbetts) to tremble," I do not think I have a tape worm, therefore I have no claim whatever to call myself "we," and I shall by no means fall into that editorial absurdity.

San Diego has been usually dull during the past week, and a summary of the news may be summarily disposed of.—There have been no births, no marriages, no arrivals, no departures, no earthquakes, nothing but the usual number of drinks taken, and an occasional "small chunk of a fight" (in which no lives have been lost), to vary the monotony of our existence. Placidly sat our village worthies in the arm-chairs in front of the "Exchange," puffing their short clay pipes, and enjoying their *"otium cum dignitate"* a week ago, and placidly they sit there still.

* * *

The only topic of interest now discussed among us is the approaching election, and on this subject I desire to say a few words:

Let the voters of this country remember, that on election of their legislative officers depends their peace, prosperity and well-being, and it may be, the life and liberty of some among them during the ensuing year. With this serious reflection, let them consider well before depositing a vote for a man to fill a responsible station, whether he is qualified or not for that station, and then act upon their convictions uninfluenced by gra-

form in the plaza, passing four abreast by the City Hall and *Herald* office, from the gallery of which Johnny will hand them their papers. *'E pluribus unum,'* or *A* word to the wise is *bastante.''*

In an approaching election the *Herald* was supporting the Democratic ticket headed by John Bigler, candidate for re-election as governor. Upon the departure of Judge Ames, however, Phoenix promptly switched to "Phoenix Independent Ticket," headed by William Waldo for Governor.

tuitous drinks of wretched whiskey, or the fact that he is called by one party name or another.

Party lines, my brother voters of San Diego, have become obliterated. Since the last general election, the only distinction between the great Democratic party of the United States and their opponents, is in the name. Difference of principle no longer exists. The National Bank has long since been levelled low; the great Tariff question has been finally put at rest, and the important Whig policy of Internal Improvements has become a plank of the Democratic platform. . . .

* * *

Frank, our accomplished compositor, who belongs to the fighting wing of the unterrified Democracy "groans in spirit, and is troubled," as he sets up our here local doctrines and opinions. He says "the Whigs will be delighted with the paper this week."

We hope so. We know several respectable gentlemen who are Whigs, and feel anxious to delight them, as well as our Democratic friends, (of whose approval we are confident,) and all other sorts and conditions of men always excepting Biglerites and Abolitionists. "Ah!" sighs the unfortunate Frank, "but what *will* Mr. Ames say when he gets back?" Haven't the slightest idea; we shall probably ascertain by reading the first *Herald* published after his return. Meanwhile, we devoutly hope that event will not take place before we've had a chance to give Mr. Bigler one *blizzard*, on the subjects of "Water-front extension," and "State printing." We understand these schemes fully, and are inclined to enlighten the public of San Diego with regard to them. Ah! Bigler, my boy, "old is B. but cunning, sir, and *devilish sly.*" Phœnix is after you, and you'd better pray for the return of the Editor *de facto*, to San Diego, while yet there is time, or you're a *goner*, as far as this county is concerned.

ILLUSTRATED NEWSPAPERS

San Diego Herald *October 1, 1853*

A year or two since, a weekly paper was started in London, called the "*Illustrated News.*" It was filled with tolerably executed wood cuts, representing scenes of popular interest, and though perhaps better calculated for the nursery than the reading room, it took very well in England, where few can read, but all can understand pictures, and soon attained an immense circulation. As when the inimitable London *Punch* attained its worldwide celebrity, supported by such writers as Thackeray, Jerrold and Hood, would-be funny men on this side of the Atlantic attempted absurd imitations—the *Yankee Doodle*, the *John Donkee*, &c., which as a matter of course proved miserable failures—so did the success of this Illustrated affair inspire our money-loving publishers with hopes of dollars, and soon appeared from Boston, New York and other places, Pictorial and Illustrated Newspapers, teeming with execrable and silly effusions, and filled with the most fearful wood engravings, "got up regardless of expense" or any thing else; the contemplation of which was enough to make an artist tear his hair and rend his garments. . . .

It must not be supposed from the tenor of these remarks that we are opposed to the publication of a properly conducted and creditably executed Illustrated paper. "On the contrary, quite the reverse." We are passionately fond of art ourselves, and we believe that nothing can have a stronger tendency to refinement in society, than presenting to the public, chaste and elaborate engravings, copies of works of high artistic merit, accompanied by graphic and well written essays. It was for the purpose of introducing a paper containing these features to our appreciative community, that we have made these introductory remarks, and for the purpose of challenging comparison and defying competition, that we have criticised so severely the imbecile and ephemeral productions mentioned above. At a vast expenditure of money, time and labor, and after the most incredible and unheard of exertion, on our part, individually, we are at length able to present to the public an Illustrated publication of unprecedented merit, containing engravings of exceeding costliness and rare beauty of design, got up on an expensive scale, which never has been attempted before, in this or any other country.

We furnish our readers this week with the first number, merely premising that the immense expense attending its issue will require a corresponding liberality of patronage on the part of the Public, to cause it to be continued.

PHŒNIX'S PICTORIAL,

And Second Story Front Room Companion.

Vol. I.] **San Diego, October 1, 1853.** **[No. I.**

Portrait of His Royal Highness Prince Albert—Prince Albert, the son of a gentleman named Coburg, is the husband of Queen Victoria of England, and the father of many of her children. He is the inventor of the celebrated "Albert hat," which has been lately introduced with great effect in the U.S. Army. The Prince is of German extraction, his father being a Dutchman and his mother a Duchess.

Mansion of John Phoenix, Esq., San Diego, California.

House in which Shakespeare was born, in Stratford-on-Avon.

Abbotsford, the residence of Sir Walter Scott, author of Byron's Pilgrim's Progress, etc.

The Capitol at Washington.

Residence of Governor Bigler, at Benicia, California.

Battle of Lake Erie (*see remarks*, p. 96).

[p. 96.]

The Battle of Lake Erie, of which our Artist presents a spirited engraving, copied from the original painting, by Hannibal Carracci, in the possession of J. P. Haven, Esq., was fought in 1836, on Chesapeake Bay, between the U. S. frigates Constitution and Guerriere and the British troops under General Putnam. Our glorious flag, there as everywhere, was victorious, and "Long may it wave, o'er the land of the free and the home of *the slave*."

Fearful accident on the Camden & Amboy Railroad!! Terrible loss of life!!!

View of the City of San Diego, by Sir Benjamin West.

Interview between Mrs. Harriet Beecher Stowe and the Duchess of Sutherland, from a group of Statuary, by Clarke Mills.

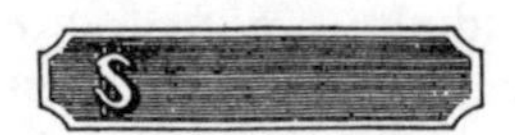

Bank Account of J. Phoenix, Esq., at Adams & Co., Bankers, Sasn Francisco, California.

Gas Works, *San Diego Herald* Office.

Steamer Goliah.

View of a California Ranch—Landseer.

RETURN OF THE EDITOR

San Diego Herald *October 1, 1853*

"Te Deum Laudamus" Judge Ames has returned! With the completion of this article my labors are ended; and wiping my pen on my coat-tail and placing it behind my sinister ear, with a graceful bow and a bland smile for my honored admirers, and a wink of intense meaning for my enemies, I shall abdicate, with dignity, the "Arm-Chair," in favor of its legitimate proprietor. . . .

During the period in which I have had control over the *Herald* I have endeavored to the best of my ability to amuse and interest its readers, and I cannot but hope that my good humored efforts have proved successful. If I have given offence to any by the tone of my remarks, I assure them that it has been quite unintentional, and to prove that I bear no malice, I hereby accept their apologies. Certainly no one can complain of a lack of versatility in the last six numbers. Commencing as an Independent journal, I have gradually passed through all the stages of incipient Whiggery, decided Conservatism, dignified Recantation, budding Democracy, and rampant Radicalism, and I now close the series with an entirely literary number, in which I have carefully abstained from the mention of Baldo and Wigler, I mean Wagler and Bildo, no—never mind—as Toodles says, I haven't mentioned *any of' em,* but been careful to preserve a perfect armed neutrality.

The paper this week will be found particularly stupid. This is the result of deep design on my part; had I attempted any thing remarkably brilliant, you would all have detected it and said, probably with truth: Ah, this is Phœnix's last appearance, he has tried to be very funny and has made a miserable failure of it. Hee! hee! hee! Oh! no, my Public, an ancient weasel may not be detected in the act of slumber, in that manner. I was well aware of all this, and have been as dull and prosy as possible to avoid it. Very little news will be found in the *Herald* this week: the fact is, there never is much news in it, and it is very well that it is so; the climate here is so delightful that residents, in the enjoyment of their *dolce far niente,* care very little about what is going on elsewhere, and residents in other places care very little about what is going on in San Diego, so all parties are likely to be gratified with the little paper, "and long may it wave."

So, farewell Public, I hope you will do well; I do, upon my soul. This leader is ended, and if there be any man among you who thinks he could write a better one, let him try it, and if he succeeds I shall merely remark that I could have done it myself if I had tried. Adios!

Respectably Yours,
John Phoenix

Samuel Clemens

(Mark Twain)

When Samuel Clemens (1835–1910) moved from Nevada to San Francisco in 1864, Derby was a local legend and ''Mark Twain'' a year old. Clemens had chosen the river term signifying two fathoms, or twelve feet, to sign his distinctive sketches in the Territorial Enterprise *in Virginia City. ''It has a richness about it,'' he told his editor, Joseph T. Goodman, ''it was always a pleasant sound for a pilot to hear on a dark night; it meant safe water.'' By then Clemens could make anything readable. In Carson City, he concocted these rules for ''Mark Twain's Hotel'' (later found in an old stable):*

This house will be considered strictly intemperate.
None but the brave deserve the fare.
Sheets will be nightly changed, once in six months, or more if necessary.
All moneys or other valuables are to be left in the care of the proprietor. This is insisted upon, as he will be responsible for no other losses.
Inside matter will not be furnished editors under any consideration.

In 1866, the Sacramento Union *sent Clemens to Hawaii for five months. So popular were his discursive letters that a more ambitious sequel was proposed. The* Alta California *of San Francisco paid his fare ($1,250) for a five-month cruise to Europe and the Holy Land. Clemens sailed to New York (via Nicaragua) for the June 8, 1867, departure of the* Quaker City; *his letters were recast as* Innocents Abroad. *What follows is from Franklin Walker and G. Ezra Dane,* Mark Twain's Travels with Mr. Brown *(New York, 1940); and Daniel M. McKeithan,* Traveling with the Innocents Abroad *(Norman, Oklahoma, 1958).*

TWAIN IN NEW YORK

Alta California, San Francisco *June 1867*

New York, June 5th, 1867.

Editors Alta:

I have at last, after several months' experience, made up my mind that it is a splendid desert—a domed and steepled solitude, where the stranger is lonely in the midst of a million of his race. A man walks his tedious miles through the same interminable street every day, elbowing his way through a buzzing multitude of men, yet never seeing a familiar face, and never seeing a strange one the second time. He visits a friend once—it is a day's journey—and then stays away from that time forward till that friend cools to a mere acquaintance, and finally to a stranger. So there is little sociability, and, consequently, there is little cordiality. Every man seems to feel that he has got the duties of two lifetimes to accomplish in one, and so he rushes, rushes, rushes, and never has time to be companionable—never has any time at his disposal to fool away on matters which do not involve dollars and duty and business.

All this has a tendency to make the city-bred man impatient of interruption, suspicious of strangers, and fearful of being bored, and his business interfered with. The natural result is, that the striking want of heartiness observable here, sometimes even among old friends, degenerates into something which is hardly even chilly politeness towards strangers. A large party of Californians were discussing this matter yesterday evening, and one said he didn't believe there was any genuine fellow-feeling in the camp. Another said: "Come, now, don't judge without a full hearing—try all classes; try everybody; go to the Young Men's Christian Association." But the first speaker said: "My son, I have been to the Young Men's Christian Association, and it isn't any use; it was the same old thing—thermometer at 32°, which is the freezing notch, if I understand it. They were polite there, exasperatingly polite, just as they are outside. One of them prayed for the stranger within his gates—meaning me—but it was plain enough that he didn't mean his petition to be taken in earnest. It simply amounted to this, that he didn't know me, but would recommend me to mercy, anyhow, since it was customary, but didn't wish to be misunderstood as taking any personal interest in the matter."

Of course that was rather a strong exaggeration, but I thought it was a pretty fair satire upon the serene indifference of the New Yorker to everybody and everything without the pale of his private and individual circle.

There is something about this ceaseless buzz, and hurry, and bustle, that keeps a stranger in a state of unwholesome excitement all the time, and makes him restless and uneasy, and saps from him all capacity to enjoy anything or take a strong interest in any matter whatever—a something which impels him to try to do everything, and yet permits him to do nothing. He is a boy in a candy-shop—could choose quickly if there were but one kind of candy, but is hopelessly undetermined in the midst of a hundred kinds. A stranger feels unsatisfied, here, a good part of the time. He starts to a library; changes, and moves toward a theatre; changes again and thinks he will visit a friend; goes within a biscuit-toss of a picture-gallery, a billiard-room, a beer-cellar and a circus, in succession, and finally drifts home and to bed, without having really done anything or gone anywhere. He don't go anywhere because he can't go everywhere, I suppose. This fidgetty, feverish restlessness will drive a man crazy, after a while, or kill him. It kills a good many dozens now—by suicide. I have got to get out of it.

There is one thing very sure—I can't keep my temper in New York. The cars and carriages always come along and get in the way just as I want to cross a street, and if there is any thing that can make a man soar into flights of sublimity in the matter of profanity, it is that. You know that, yourself. However, I must be accurate—I must speak truth, and say there is one thing that is more annoying. That is to go down West Tenth street hunting for the Art building, No. 51. You are tired, and your feet are hot and swollen, and you wouldn't start, only you calculate that it cannot be more than two blocks away, and you almost feel a genuine desire to go and see the picture on exhibition without once changing your mind. Very well. You come to No. 7; and directly you come to 142! You stare a minute, and then step back and start over again—but it isn't any use—when you are least expecting it, comes that unaccountable jump. You cross over, and find Nos. 18, 20, 22, and then perhaps you jump to 376! Your gall begins to rise. You go on. You get on a trail, at last, the figures leading by regular approaches up toward 51—but when you have walked four blocks they start at 49 and begin to run the other way! You are perspiring and furious by this time, but you keep desperately on, and speculate on new and complicated forms of profanity. And behold, in time the numbers become bewilderingly complicated: on one door is a 3 on a little tin scrap, on the next a 17 in gold characters a foot square, on the next a 19, a 5 and a 137, one above the other and in three different styles of figuring! You do not swear any more now, of course, because you can't find any words that are long enough or strong enough to fit the case. You feel degraded and ignominious and subjugated. And there and then you say that you will go away from New York and start over again; and that you will never come back to settle permanently till you have learned to swear with the utmost fluency in seventeen different languages. You become more tranquil, now, because you see your way clearly before you, how that, when you are properly accomplished, you can live in this great city and still be happy; you feel that in that day, when a subject shall defy English, you can try the Arabic, the Hungarian, the Japanese, the Kulu-Kaffir, and when the worst comes to the worst, you can come the Hindostanee on it and conquer. After this, you go tranquilly on for a matter of seventeen blocks and find 51 sandwiched in between Nos. 13 and 32,986. Then you wish you had never been born, to come to a strange land and suffer in this way.

Well, I intended, when I started out, to give my views of the pleasant side of New York, but I perceive that I have wandered into the wrong vein, and so I will stop short and give it up until I find myself in a more fortunate humor. I do not think that I could twist myself around now any easier than I could turn myself inside out.

TWAIN MEETS CZAR ALEXANDER II

Alta California, San Francisco *November 6, 1867*

Yalta, August 27th.

We anchored here day before yesterday. To me the place was a vision of California. The tall, gray mountains that back it, their sides bristling with pines—cloven with ravines—here and there a hoary rock towering into view—long, straight streaks sweeping down from the summit to the sea, marking the passage of some avalanche of former times—all these were as like what one sees in the Sierras as if the one were a portrait of the other. The little village of Yalta nestles at the foot of an amphitheatre which slopes backward and upward to the wall of hills, and looks as if it might have sunk quietly down to its present position from a higher elevation. This depression is covered with the great parks and gardens of noblemen, and through the mass of green foliage the bright colors of their palaces bud out here and there like flowers. It is a beautiful spot.

The first thing we did was to send a Committee on shore to confer with the Governor-General concerning our reception, and to present to him a brief address for the Emperor which our Consul had advised us to prepare, and which a Committee of the cheekiest of us had been ordered to draft. Why they should have made me Chairman of a Committee whose main talent was to consist of cheek, was an injustice which to me was as strange as it was painful. I accepted the office, but I did it under protest. I did it partly because I was more familiar with Emperors than the other passengers, and therefore able to write to such people with an easier grace than they, and partly because I thought that if I could spread it around that I had been corresponding with the Emperor of Russia, may be it would make my photograph sell.

Well, the Governor-General said that his Majesty would receive our whole party the next day at noon, at his summer palace—that etiquette would be waived and the address read and presented to him in person, at that time—that the Grand Duke Michael (his brother) had extended an invitation to the party to visit him at his palace on the same day, and that on the following day they and their families desired to visit the ship if the sea were smooth. He said we must disembark at half past ten or eleven and ride to the palace (three miles) in carriages which would be provided for us.

So the whole ship's company turned out at about 7 o'clock yesterday morning, and dressed from that time until 11. We got ashore, then, and drove to the Czar's mansion. It stood in the midst of a mixture of lawn, flower-garden and park, and was as snugly located as possible, almost. Its architecture is simple, but handsome and attractive, and its porches, stairways and windows are so clothed with vines and flowers, that the place looked like a cosy home, not a chilly palace.

The Reception. We formed a circle under the trees before the door, for there was no one room in the house able to accommodate our seventy-five persons comfortably, and in a few minutes the imperial family came out bowing and smiling, and stood in our midst. A number of great dignitaries of the Empire, in undress uniforms, came with them. With every bow, his Majesty said a word of welcome. I copy these speeches. There is character in them—Russian character—which is politeness itself, and the genuine article. The French are polite, but it is often mere ceremonious politeness. A Russian always imbues his polite things with a heartiness, both of phrase and expression, that compels belief in their sincerity. As I was saying, the Czar punctuated his speeches with bows: "Good morning—I am glad to see you—I am gratified—I am delighted—I am happy to receive you!" If he had said he was proud to receive that gang, I would not have believed a word of it. But he might have been happy—he looked it. All hands took off their hats, and the Consul inflicted the Address on him. He bore it with unflinching fortitude; then took the rusty looking document and handed it to some great officer or other, to be filed away among the archives of Russia—in the stove, perhaps. He thanked us for the Address, and said he was very much pleased to see us, especially as such friendly relations existed between Russia and the United States.

The Empress said the Americans were favorites in Russia, and she hoped the Russians were similarly regarded in America. These were all the speeches that were made, and I recommend them to parties who present the San Francisco police with gold watches, as models of brevity and point. After this the Empress went and talked sociably (for an Empress) with various ladies around the circle, several gentlemen entered into a disjointed general conversation with the Emperor, the Dukes and

Princes, Admirals and Maids of Honor dropped into free-and-easy chat with first one and then another of our party, and whoever chose stepped forward and spoke with the modest little Grand Duchess Marie, the Czar's daughter, who is fourteen years old, light-haired, blue-eyed, unassuming and pretty. Everybody talks English. Being after information more than anything else, I captured a fine old gentleman who seemed perfectly willing to be bored with questions, and bored him good. I kept it up, and bored him at many times and in many places during the afternoon. But I did not know he was the Lord High Admiral of Russia. I took him for a lieutenant in the army. But he was very affable and polite, and liked to talk. He was posted on everything, too.

If these dignitaries had come out in their trotting harness, and blazing with orders and decorations, and had assumed a courtly grandeur of bearing and speech, they would have put our light out in the twinkling of an eye—they would have extinguished us like setting a church down on a tallow candle. But they knew better. They guessed our style—they gauged us, and came at us accordingly. They dressed for the occasion. When, in the course of that half-hour's chat, the Countesses and Baronesses and Duchesses of the household got mixed up with our ladies, you could not tell them apart, except that our ladies were the finest dressed and looked the most showy. All the Admirals, and Dukes, and Princes, and Generals were Lieutenants to me; just as in Italy, I couldn't tell the policemen from the Marshals of the Kingdom. You couldn't tell which was the Empress of Russia without having her pointed out; the only way to find the Emperor was to hunt for the man that had the plainest clothes. I think any question asked with a kingly air would have stricken any of our party speechless; but the homespun simplicity of voice and manner of all the imperial party broke the ice at once and set every tongue going with a cheerful vivacity that had no suspicion of embarrassment about it. They were not five minutes in forgetting that they were in a helpless and desperate situation.

The Imperial Wardrobe. The Emperor wore a cap, frock coat and pantaloons, all of some kind of plain white drilling—cotton or linen—and sported no jewelry or any insignia whatever of rank. No costume could be less ostentatious. He is very tall and spare, and a determined looking man, though a very pleasant looking one, nevertheless. It is easy to see that he is kind and affectionate. There is something very noble in his expression when his cap is off. There is none of that cunning in his eye that all of us noticed in Louis Napoleon's.

. . . A strange, new sensation is a rare thing in this hum-drum life, and verily I had it here. There was nothing stale of worn out about the thoughts and feelings the situation and circumstances created. It seemed strange—stranger than I can tell—to think that the central figure in the cluster of men and women, chatting here under the trees like the most ordinary people in the land, was a man who could open his lips and ships would fly through the waves, locomotives would speed over the plains, couriers would hurry from village to village, a hundred telegraphs would flash the word to the four corners of an Empire that stretches its vast proportions over a seventh part of the world, and a countless multitude of men would spring to do his bidding. I had a sort of vague desire to examine his hands and see if they were of flesh and blood, like other men's. Here was a man who could do this wonderful thing, and yet if I chose to do it I could knock him down. The case was plain, but it seemed preposterous, nevertheless—as preposterous as trying to knock down a mountain or wipe out a continent. If this man sprained his ankle, a million miles of telegraph would carry the news over mountains—valleys—uninhabited deserts—under the trackless sea—and ten thousand newspapers would prate of it; if he were grievously ill, all the nations would know it before the sun rose again; if he dropped lifeless where he stood, his fall might shake the thrones of half a world! If I could have stolen a button off his coat, I would have done it. When I meet a man like that, I want something to remember him by.

The Address. As it has been sent to various Russian papers for publication —so I am told—you may as well print it yourself. Inasmuch as the Emperor approved the document, I hope you will allow me as much for it as if you had ordered me to write it yourself. The passengers approved it also—all except one. He objected to "your Majesty"—said it might be right enough, but still it looked like we were fishing for an invitation to dinner:

Traveling with the Innocents Abroad
To His Imperial Majesty, Alexander II, Emperor of Russia:

We are a handful of private citizens of the United States, travelling simply for recreation—and unostentatiously, as becomes our unofficial state—and, therefore, we have no excuse to tender for presenting ourselves before your Majesty save the desire of offering our grateful acknowledgments to the lord of a realm which, through good and through evil report, has been the steadfast friend of the land we love so well.

We could not presume to take a step like this, did we not know well that the words we speak here, and the sentiments wherewith they are freighted, are but the reflex of the thoughts and the feelings of all our countrymen, from the green hills of New England to the shores of the far Pacific. We are few in number, but we utter the voice of a nation!

One of the brightest pages that has graced the world's history, since written history had its birth, was recorded by your Majesty's hand when it loosed the bonds of twenty millions of men; and Americans can but esteem it a privilege to do honor to a ruler who has wrought so mighty a deed. The lesson taught us then we have profitted by, and are free in truth, to-day, even as we were before in name. America owes much to Russia—is indebted to her in many ways—and chiefly for her unwavering friendship in seasons of our greatest need. That that friendship may still be hers, we confidently pray; that she is and will be grateful to Russia and to her sovereign for it, we know full well; that she will ever forfeit it by any premeditated unjust act, or unfair course, it were treason to believe.

Sam. L. Clemens,
Wm. Gibson,
A. N. Sanford,
Timothy D. Crocker,
Col. P. Kinney, U.S.A.,
Committee on behalf of the passengers of the American steam yacht Quaker City.

This address will be copied into the various newspapers of Europe, and so I am perfectly satisfied, now, that my photographs will sell.

Our little unpretending visit of a few untitled American scrubs, instead of being [of] no consequence, except as a fifteen minutes' bore to the Czar, which was all we expected, begins to assume a national importance. I will observe, in this connection, that the price of photographs in Constantinople is twenty francs a dozen—say four dollars.

Frank G. Carpenter

(Carp)

This perceptive writer escaped limbo because his Washington sketches in the 1880s were preserved in a scrapbook by his bride, Joanna. Decades later, his daughter, Frances Carpenter, was able to gather a selection, published as Carp's Washington *(New York, 1960), with an introduction by Cleveland Amory. Frank G. Carpenter (1855–1924) was assigned to the Capital in 1882, when he was twenty-seven, as correspondent of the* Cleveland Leader. *Signing himself ''Carp,'' he coaxed a daily sketch from a clanking Calligraph typewriter. He later formed his own syndicate for travel articles that were recycled for classroom use into* Carpenter's Geographical Readers.

CONGRESSIONAL GREATNESS - FAUGH!

Cleveland Leader *December 1892*

I am sitting in the press gallery of the Senate. The Senators, in their various degrees of disorder, are carrying on their usual antics in the great pit, known as the Senate Chamber, just below me. I jot down, as I look on, such items of gossip as strike my fancy.

As I write, Coke, of Texas, has come to the chair of the President of the Senate, and his great bald dome shines directly beneath me as I lean over the gallery rail. I reach out my pen; the ink quivers on its point; a moment more and a black drop would spatter that rounded whiteness. But such a disaster would set the Senate in an uproar, so I draw back my pen over my paper.

Nearly one half of today's Senate is bald. The parchment pates of the great men shine up at the press gallery, and I see enough bare skin on their crowns to cover half a dozen bass drums.

Senator Joe Brown, of Georgia, is bald from forehead to crown, and the top of his skull is as level as that of a Flathead Indian. Circling his bald dome, the long iron-gray hair falls straight, then wraps itself into a thin roll which caresses his collar and looks like the haircut of a schoolgirl from the country. The Senator shaves his upper lip and his chin, leaving his luxuriant beard fastened to his lower jaw, like that of a billy goat. This beard contains enough hair to stuff a small pillow, and when Mr. Brown grows excited his only sign of emotion appears in his continual clutching of it.

Jones, of Nevada, has ten thousand dollars for every hair on the crown of his head. He would give hundreds if he could thicken the fuzz at the top, but hair is one of things not bought with money, and like Senator Brown, millionaire Jones also consoles himself with a long gray beard. Senator Hoar, of Massachusetts, has very fine hair, but the spot at his crown is as bare as a billiard ball, and it is as white as Boston beans are before baking.

Luke P. Poland is the only relic of the old statesman of the days of Clay, Calhoun, and Webster. He still clings to the swallowtail coat, with the bright gold buttons which were used by all of the fashionable men of that time. It is cleaner and prettier than the Prince Albert or the cutaway of the present. In fact the dress of men has been growing simpler and simpler during the past century. Washington, who died in 1799, had ruffles on his sleeves, and half of the men who signed the Declaration of Independence wore powdered wigs. Jefferson sported knee breeches, and Madison was proud of his originality in having worn an inauguration suit of clothes of American make.

Daniel Webster usually chose a suit of snuff brown, with a large soft necktie. Martin Van Buren was fastidious, appearing during the summer in the whitest of white linen. His suits were cut in the very latest style and he wore high stock neckties, which almost engulfed his standing collar. Andrew Jackson also dressed well, though he did not make his clothes a great matter. Henry Clay wore a swallowtail, and a standing collar, extravagantly high, and James Buchanan was precise, always appearing in full dress.

Today half of our public men go about in suits as businesslike as those of a bank cashier. William Walter Phelps, with all his millions, wears clothes which cost about twenty-five dollars a suit, and he likes a red necktie. Lyman, of Massachusetts, wears a green scarf; Orlando Potter's gray clothes would not sell for five dollars to a second-hand-clothes man; and the black ones of Colonel Woolford, of Kentucky, are shiny with age.

Carlisle and Randall wear little inch-wide neckties, and both are generally dressed in black. Both stoop a little and neither is very careful as to his appearance. Judge Reagan, of Texas, appears in a black diagonal Prince Albert coat and a turnover collar, while Perry Belmont's small frame is clothed in closely buttoned black broadcloth. And "Calamity" Weller's clothes are as rough and crazy as his brain. His wild-eyed face usually looks as though it needed a razor, and his hair stands on end.

George Hearst, the millionaire Senator from California, has been represented as an illiterate man, but Senator Frye tells a story that illustrates both his education and his humor. He has not played cards with the bluffers of California for nothing, and like most of his brother Senators, he is by no means averse to a bet. Not long ago he entered a restaurant in San Francisco and on the blackboard at the back of the bar he saw the word *bird* among the items of the day's bill of fare. Hearst called out to the keeper of the restaurant, who was a noted California character,

"See here, Sam, that's a devil of a way to spell *bird!* Don't you know any better than that? You ought to spell it *b-u-r-d.*"

"I would have you understand, George Hearst," replied the restaurant keeper, "that I am just as good a speller as you. I'm willing to leave it to the best scholar in the room that you don't know any more than I do. I bet you a basket of champagne that you can't spell *bird* the right way."

"Done," said Hearst.

"All right," said the man, "and there is a piece of paper for you to put it down in black and white."

He handed over a sheet of brown paper and with a stubby pencil Hearst wrote out the letters *b-i-r-d.*

"But," said the restaurant keeper in dismay, "you spelled it with a *u* before."

The Senator threw himself back and looked at the restaurant man with twinkling eyes. "Now, Sam," said he, "did you really think that I was damn fool enough to spell *bird* with a *u* when there was any money in it?" . . .

Who has not heard of "Pig-Iron" Kelley? He is sitting in his seat there in the center of the House. It is the best seat in the Chamber; at least Mr. Kelley thinks so, for he has chosen it of all the others. Mr. Kelley is the oldest member of the House, and as such he is the only one who has the right to a choice of seats. The other members obtain theirs by drawing lots.

Mr. Kelley is very tall and very lean, being the nearest approach to a skeleton in all the House. He has a red face, grisly hair and beard, and blue eyes surrounded by rose-red eyelids. His shoulders are stooping, and he bends over as he sits at his desk. Though sixty-nine years of age, he is still strong and active, and he is one of the most indefatigable workers at Washington.

The most noted thing about Mr. Kelley is his monomania on the subject of the tariff. He has studied tariff all his life. He comes here as representative of the tariff interests at Philadelphia, which he has represented for the last twenty years. His constituents have implicit faith in him, and he will continue to represent them to the day of his death, I suppose. If America should ever go for free trade, Mr. Kelley would die instanter. His soul is so wrapped up and saturated with "protection" that its withdrawal would cause an immediate collapse. Mr. Kelley thinks tariff, talks tariff, and writes tariff every hour of the day; a roommate of his tells me that he mumbles it over in his dreams during the night. When he goes into society he backs women into corners and asks them their opinion of the duty on steel rails. Such a case actually occurred here a few nights ago, and though the lady informed him she knew nothing about the subject, he persisted in giving her his views at length.

Mr. Kelley started out as a jeweler, working five years as an apprentice in a watch shop in Boston. He after studied and practiced law in Philadelphia, where he served for ten years as Judge of the Court of Common Pleas. He was elected first to the Thirty-seventh Congress, where he showed such strong tariff predilections that Philadelphia has kept him here ever since.

"Richelieu" Robinson, of Brooklyn, is also a hobby-rider of national note. His nickname, "The Twister of the Tail of the British Lion," has made his hobby-riding famous. Mr. Robinson is a striking character. Dressed in black, with a bronzed face and bushy iron-gray locks like those of the Negro statesman Fred Douglass, he sits hunched in his seat, watching the actions of the House with his keen eyes. The Congressman is a little deaf, but he can hear the words "England" or "Ireland," even when uttered in a whisper in the lobby, and such sounds call him at once to their place of utterance. He denounces England on every occasion in the House, for his whole soul is wrapped up in the cause of Ireland.

Ireland is Robinson's native country. He was born fifty-nine years ago in County Tyrone and was educated at Belfast. He left school to come to America, where he graduated at Yale. He served under Horace Greeley in the first days of the New York *Tribune,* and forty years ago he acted here as its Washington correspondent, signing his articles "Richelieu." He has since been continually connected with the chief newspapers of the United States, to which he is still a frequent contributor.

I watched Richelieu Robinson the other day during the reading of his buncombe ironical resolution proposing that the United States buy Ireland and annex it. His gray locks quivered with ecstasy, and his ironclad features lit up with smiles of satisfaction, as the clerk rolled his tongue around the denunciatory words regarding England's tyranny and Ireland's oppression. He rejoiced all the way from the beginning of the reading to the close, then left the chamber feeling he had done his duty for the day.

Down in Brooklyn, the Irish really think Richelieu is going to buy Ireland himself some day. They say they will keep him in Congress until he accomplishes its annexation, and they look on this resolution as one step in that direction.

What a homely old fellow is Senator William M. Evarts, our former Secretary of State. He is by all odds the ugliest man known to fame. Thin and bony, with slightly stooping shoulders, he has a face far worse than Frank Blair's, who was once given a knife because he was the ugliest man in the United States. If Blair were living now, he could rightly hand that knife over to Evarts.

Yet, Evarts' face, with its big rough nose, its sharply protruding chin and wrinkled cheeks, is full of character, and his keen eyes are extraordinarily bright. Though now sixty-five years of age, Mr. Evarts is strong and healthy; he could, I believe, run again the foot race he ran down Thirteenth Street in the days of President Hayes, and I am convinced he would win today.

There is no man in Washington who enjoys a good dinner more than Senator Evarts. He is one of the highest livers in the capital. Notwithstanding the fact that he is six feet tall and that he does not weigh 125 pounds, he can eat all around Piletus Sawyer, who weighs 300 and has a stomach so large you could roll Evarts up like a watch spring and coil him inside it with room to spare. Once when he was speaking at an assemblage in New York, a late comer asked the name of the man on the platform. "What," he exclaimed, "that lean little thing Evarts? Why, he looks as if he boarded!"

Major William McKinley takes a prominent place in the Forty-eighth Congress. He is one of the finest-looking men in the House, of medium height, very straight, and well proportioned. His complexion is dark, his face closely shaven. His head is not unlike that of Napoleon Bonaparte, with high, broad brow, short black hair, and a heavy jaw. Major McKinley, now thirty-nine years of age, was born in Ohio, enlisted in the Army as a private and came out a major. At the close of the War he practiced law in Canton, where he married the daughter of one of the richest men in the northern part of the state. He has now been six years in Congress, having been re-elected last fall. His Democratic opponent, Major Wallace, however, is now contesting the vote count, and it may be that McKinley will be deprived of his seat.

In addition to his salary, each member of Congress gets forty cents a mile for the distance he travels from his home town to the capital, which pay ranges all the way from $3.80 for those whose homes are in Maryland or Virginia, to $1600 for the Territorial Delegate from Arizona. When it is remembered that the latter sum is enough to pay steamship passage around the world, it seems a good deal.

Our Congressmen are each paid salaries of $5000 a year. There are 325 of them, and about 25 of that number are worth their salaries. Most could not make more than half as much by the sale of their talents in any other capacity. If some of them were forced to live by the sweat of their brows, outside of politics, I imagine that many of them would go on a low diet, and not from choice either.

Look over the men of your acquaintance. How many of them earn $5000 a year? Pick out 325 men from any part of the Union—men whose brain and muscle alone are worth $5000 a year on the market—and I will show you that they are of a far higher grade than those making up the body of Congress.

How did these men get here? In various ways. Some bought their seats, it is charged, and some hold them through their friendship with great corporations. Some got them by drinking in barrooms to cultivate the slum voters. Some hypocritically slid into them by praying in the churches at the same time. Others hold their places by the favor of certain district rings, and the mainspring which runs the successful machinery of still others is the sending out of seeds and government documents to their farming constituents.

A few members of Congress are really great men, but these I can count on my fingers. A few more are noble and upright, and now and then you will find one who casts his vote for his country's good, and not just because it will benefit himself. Most of the others swell about and pose as great men. I suppose they feel great, except at election time when they drink, truckle, and bootlick to maintain their greatness. Congressional greatness—faugh!

Lafcadio Hearn

The polymath Lafcadio Hearn (1850–1904) came to America almost as an extraterrestrial: born on an Ionian island of Maltese and Anglo-Irish parents, adopted by an aunt, and privately schooled in Ireland, England, and France. At nineteen, Hearn writes, ''after my people had been reduced from riches to poverty by an adventurer, and before I had seen anything of the world except a year of London among the common folk, I was dropped moneyless on the pavement of an American city to begin life.''

In 1871, he left New York for Cincinnati, finding a job there as a typesetter and proofreader, so meticulous that he was nicknamed ''Old Semicolon.'' Before long he was writing for local newspapers, using a distinctive vocabulary for sketches favoring such Hearnian words as spectral, arabesque, piteous, ghoulish, eldritch, darksome, elfin, dismal, monstrous, *and* fantastic. *Moving on to New Orleans in 1877, Hearn spent twelve years with* The Item *and* Times-Democrat, *contributing to the former five columns, ''Wayside Notes,'' ''Our Book Table,'' ''Odds and Ends,'' ''Varieties,'' and ''The Item Miscellany.'' He won a national audience with his accounts of Creole life and letters, his translations, short stories, and Caribbean travel pieces. In 1890, Hearn resettled in Japan, whose civilization he evoked in a dozen enraptured books. The sketches that follow are from* An American Miscellany *(New York, 1924), edited by Albert Mordell;* Lafcadio Hearn's American Days *(New York, 1924), edited by Edward Larocque Tinker; and* Creole Sketches *(Boston, 1924), edited by Charles W. Hutson.*

THE VOICES OF DAWN

New Orleans Item *July 22, 1881*

> A dreadful sound is in his ears.
> JOB XV, 21

There have never been so many fruit-peddlers and viand-peddlers of all sorts as at the present time—an encouraging sign of prosperity and the active circulation of money.

With the first glow of sunlight the street resounds with their cries; and, really, the famous "Book of London Cries" contains nothing more curious than some of these vocal advertisements—these musical announcements, sung by Italians, negroes, Frenchmen, and Spaniards. The vendor of fowls pokes in his head at every open window with cries of "Chick-EN, Madamma, Chick-EN!" and the seller of "Lem-ONS—fine Lem-ONS!" follows in his footsteps. The peddlers of "Ap-PULLS!" of "Straw-BARE-eries!" and "Black-Brees!"—all own sonorous voices. There is a handsome Italian with a somewhat ferocious pair of black eyes, who sells various oddities, and has adopted the word "lagniappe" for his war-cry—pronouncing it Italianwise.

He advances noiselessly to open windows and doors, plunges his blazing black glance into the interior, and suddenly queries in a deep bass, like a clap of thunder, "LAGNIAPPA, Madam-a!—lagniap-PA!" Then there is the Cantelope Man, whose cry is being imitated by all the children:

> "Cantel-lope-ah!
> Fresh and fine,
> Jus from the vine,
> Only a dime!"

There are also two peddlers, the precise meaning of whose cries we have never been able to determine. One shouts, or seems to shout, "A-a-a-a-ah! SHE got." Just what "SHE got" we have not yet been able to determine; but we fancy it must be disagreeable, as the crier's rival always shouts—"I-I-I!—I want nothing!" with a tremendous emphasis on the I. There is another fellow who seems to shout something which is not exactly proper for modest ears to hear; but he is really only announcing that he has fine potatoes for sale. Then there is the Clothespole Man, whose musical, quavering cry is heard at the distance of miles on a clear day, "Clo-ho-ho-ho-ho-ho-ho-ho-se-poles!" As a trilling tenor he is simply marvelous. The "Coaly-coaly" Man, a merry little Gascon, is too well known as a singer to need any criticism; but he is almost ubiquitous. There is also the fig-seller, who crieth in such a manner that his "Fresh figs!" seems to be "Ice crags!" And the fan-sellers, who intend to call, "Cheap fans!" but who really seem to yell "Jap-ans!" and "Chapped hands!" Then there is the seller of "Towwels" and the sellers of "Ochre-A" who appear to deal in but one first-class quality of paint, if we dare believe the mendacious sounds which reach our ears; neither must we forget the vendors of "Tom-ate-toes!" Whose toes? we should like to know.

These are new cries, with perhaps three exceptions;—with the old cries added to the list—the "calas" and the "plaisir" and other Creole calls, we might "spread out" over another column. If any one has a little leisure and a little turn for amusement, he can certainly have plenty of fun while listening to the voices of the peddlers entering his room together with the first liquid gold of sunrise.

COMPLAINT OF A CREOLE BOARDING-HOUSE-KEEPER

New Orleans Item *September 27, 1879*

O la canaille! la canaille! All time after dis I will make dem to pay in advance.

De first dat I have, say he vas a capitaine. I know not if he vas a capitaine; but he vas a misérable. After he have eat and sleep here six week and not pay me, I tell him, "Monsieur, I must money have."

He say: "Madame, you take me for tief?"

I say: "Monsieur, it is right dat you pay; I have wait long time assez."

He den say: "I learn you how to speak me in a manner so much insolent. *Now,* I not pay you till when I be ready, and I not hurry myself."

"Go out from my house!" I say.

"I go out, madame, from your dirty house when it me please"—dat how he speak me. And I could not force him to part till when I had take all de furniture out from his room. He owe me not more as seventy dollaire!

. . . After, I have one Frenchman, I tink him well elevated—le coco. He nail his valise on de floor for make me tink heavy; and he dispar one night—owing me forty-nine dollaire! I find noting in his valise only one *syringe.*

. . . After, I have two married. Dey pay me enough well, until when de woman run away wit some oder man. Her husban' stay till when he owe me eighty dollaire. After, he go too; and write me letter as dis:

"Madame, I cheat you of eighty dollaire; and I not wish only I could cheat you of eighty thousand dollaire. It was for cause of you dat my wife have run away."

After, I find out she was not his wife.

. . . Den I have a sick man. He fall on de banquette in face of my house, and I take him in to nurse. When dat he get well he tell me he vas one professor of langedge. He eat and sleep here four mont; and first he pay a little. He complain much from noise. He vas what you call nerveux—so like I was oblige for to make my daughter walk witout shoes in naked foots; and we to speak in dumb and deaf langedge by fear of make him trouble. He smoke in de bed and burn de cover; also he break de pot and de cradle-chair, and after, de window, an' de armoire an' de —vat you call de pendule;—he let fall ink on de carpet, and he spit tobacc' on de wall, and he vomit in de bed. But I noting say, as he not'ave baggage;—ainsi, wen he owe me forty dollaire I not want turn him out for dat I get my money more late. When at de end I tell to go out, he tell me he have receive a checque and pay me on Monday. But I nevaire see him after. He owe me one hundred and sixty-seven dollaire —and seventy cent vat I lend him for medicine to buy.

. . . After, I have one woman, species of camel (espèce de chameau) and one doctor, her husband (tout ce qu'il y avait d'abominable). She pretend to be—and you call dat?—sage femme; and he is not so much doctor as my cat; but for all dey doctor me for two hundred and fifty dollaire, and I not ever obtain of it not one sou.

. . . After, I have tree familee—all vat vas of rough and ugly; for one mont I not receive of rent. So I serve to dem notice of quit. But dey tell me dey not me pay nevaire, and not quit until when I make law-suit. Eh bien, de rent of de house vas not more as fifty dollaire, and de law cost me perhaps one affair of more like one hundred dollaire. Ainsi, I quit de house, an' leave dem all dere to do like dey would please. But before dat I could leave, dey steal me two buckets, and one stove, and one broom, and one clock, and one iron, and one coffee-mill, and one hen, and one leetle cat vat I much vas fond of, and one plate, and some linen of womans vat to me not belong.

THE STRANGER

New Orleans Item *April 17, 1880*

The Italian had kept us spellbound for hours, while a great yellow moon was climbing higher and higher above the leaves of the bananas that nodded weirdly at the windows. Within the great hall a circle of attentive listeners—composed of that motley mixture of the wanderers of all nations, such as can be found only in New Orleans, and perhaps Marseilles

—sat in silence about the lamplit table, riveted by the speaker's dark eyes and rich voice. There was a natural music in those tones; the stranger chanted as he spoke like a wizard weaving a spell. And speaking to each one in the tongue of his own land, he told them of the Orient. For he had been a wanderer in many lands; and afar off, touching the farther horn of the moonlight crescent, lay awaiting him a long, graceful vessel with a Greek name, which would unfurl her white wings for flight with the first ruddiness of morning.

"I see that you are a smoker," observed the stranger to his host as he rose to go. "May I have the pleasure of presenting you with a Turkish pipe? I brought it from Constantinople."

It was moulded of blood-red clay after a fashion of Moresque art, and fretted about its edges with gilded work like the ornamentation girdling the minarets of a mosque. And a faint perfume, as of the gardens of Damascus, clung to its gaudy bowl, whereon were deeply stamped mysterious words in the Arabian tongue.

* * *

The voice had long ceased to utter its musical syllables. The guests had departed; the lamps were extinguished within. A single ray of moonlight breaking through the shrubbery without fell upon a bouquet of flowers, breathing out their perfumed souls into the night. Only the host remained—dreaming of moons larger than ours, and fiercer summers; minarets white and keen, piercing a cloudless sky, and the many-fountained pleasure-places of the East. And the pipe exhaled its strange and mystical perfume, like the scented breath of a summer's night in the rose-gardens of a Sultan. Above, in deeps of amethyst, glimmered the everlasting lamps of heaven; and from afar, the voice of a muezzin seemed to cry, in tones liquidly sweet as the voice of the stranger—"All ye who are about to sleep, commend your souls to Him who never sleeps."

Joel Chandler Harris

Joel Chandler Harris (1848–1908) was a youngster in rural Georgia when he responded to this ad in a country weekly: "Wanted: An active, intelligent white boy 14 or 15 years of age to learn the printer's trade." The paper was published on a large plantation called Turnwold, where young Joe spent evenings in the slave quarters listening to yarns by Uncle George Terrell, Old Harbart, and Aunt Crissy, models for the main characters in the Uncle Remus tales.

After the Civil War, Harris took on a new job at the Monroe Advertiser: *"I set all the type, pulled the press, kept the books, swept the floor, and wrapped the papers for mailing; my mechanical, accounting and menial duties being concealed from the vulgar hilarity of the world by the honorable and impressive title of* Editor.*" Harris was a proficient country paragrapher when he was hired in 1876 by the* Atlanta Constitution. *Asked to try his hand at dialect sketches, he devised a character named "Remus," who appeared first as an urban sage and was finally transplanted to middle Georgia in Harris's first plantation fable, "The Story of Mr. Rabbit and Mr. Fox." In twenty-seven-years Harris wrote more than one hundred and eighty Remus stories. In 1880, the first selection,* Uncle Remus: His Songs and His Sayings, *sold 10,000 copies in four months. The stories attest to Harris's ear, memory, and lack of condescension; their success spurred serious interest in folklore. The first Uncle Remus tale was identical in its essentials with the version that appeared under the head, "Negro Folk Lore," on the* Constitution's *editorial page.*

THE STORY OF MR. RABBIT AND MR. FOX

Atlanta Constitution *July 20, 1879*

One evening recently, the lady whom Uncle Remus calls "Miss Sally" missed her little six-year-old. Making search for him through the house and through the yard, she heard the sound of voices in the back plaza, and, looking through the window, saw the child sitting by Uncle Remus. His head rested against the old man's arm, and he was gazing with an expression of the most intense interest into the rough, weather-beaten face, that beamed so kindly upon him. This is what "Miss Sally" heard:

"Bimeby, one day, arter Brer Fox bin doin' all dat he could fer ter ketch Brer Rabbit, en Brer Rabbit bin doin' all he could fer ter keep 'im fum it, Brer Fox say to hisse'f dat he'd put up a game on Brer Rabbit, en he ain't mo'n got do wuds out'n his mouf twel Brer Rabbit come a lopin' up de big road, lookin' des ez plump, en ez fat, en ez sassy ez a Moggin hoss in a barley-patch.

" 'Hol' on dar, Brer Rabbit,' sez Brer Fox, sezee.

" 'I ain't got time, Brer Fox,' sez Brer Rabbit, sezee, sorter mendin' his licks.

" 'I wanter have some confab wid you, Brer Rabbit,' sez Brer Fox, sezee.

" 'All right, Brer Fox, but you better holler fum whar you stan.' I'm monstus full er fleas dis mawnin',' sez Brer Rabbit, sezee.

" 'I seed Brer B'ar yistiddy,' sez Brer Fox, sezee, 'en he sorter rake me over de coals kaze you en me ain't make frens en live naberly, en I tole 'im dat I'd see you.'

"Den Brer Rabbit scratch one year wid his off hine-foot sorter jub'usly, en den he ups en sez, sezee:

" 'All a settin', Brer Fox. Spose'n you drap roun' termorrer en take dinner wid me. We ain't got no great doin's at our house, but I speck de ole 'oman en de chilluns kin sorter scramble roun'en git up sump'n fer ter stay yo' stummuck.'

"I'm 'gree'ble, Brer Rabbit,' sez Brer Fox, sezee.

" 'Den I'll pen' on you,' sez Brer Rabbit, sezee.

"Nex' day, Mr. Rabbit an' Miss Rabbit got up soon, fo' day, en raided on a gyarden like Miss Sally's out dar, en got some cabbiges, en some roas'n years, en some sparrer-grass, en dey fix up a smashin' dinner. Bimeby one er de little Rabbits, playin' out in de backyard, come runnin' in hollerin', 'Oh, ma! oh, ma! I seed Mr. Fox a comin'!' En den Brer Rabbit he tuck de chilluns by der years en make um set down, en den him en Miss Rabbit sorter dallo roun' waitin' for Brer Fox. En dey keep on waitin', but no Brer Fox ain't come. Atter 'while Brer Rabbit goes to de do', easy like, en peep out, en der, stickin' out fum behime de cornder, wuz de tip-een' er Brer Fox tail. Den Brer Rabbit shot de do' en sot down, en put his paws behime his years en begin fer ter sing:

" 'De place wharbouts you spill de grease,
Right dar youer boun' ter slide,
An' whar you fine a bunch er ha'r,
You'll sholy fine de hide.'

"Nex' day, Brer Fox sont word by Mr. Mink, en skuse hisse'f kaze he wuz too sick fer ter come, en he ax Brer Rabbit fer ter come en take dinner wid him, en Brer Rabbit say he wuz 'gree'ble.

"Bimeby, w'en de shadders wuz at der shorte's, Brer Rabbit he sorter brush up en santer down ter Brer Fox's house, en w'en he got dar, he yer somebody groanin', en he look in de do' en dar he see Brer Fox settin' up in a rockin' cheer all wrop up wid flannil, en he look mighty weak. Brer Rabbit look all 'roun', he did, but he ain't see no dinner. De dish-pan wuz settin' on de table, en close by wuz a kyarvin' knife.

" 'Look like you gwineter have chicken for dinner, Brer Fox,' sez Brer Rabbit, sezee.

" 'Yes, Brer Rabbit, deyer nice, en fresh, en tender,' sez Brer Fox, sezee.

"Den Brer Rabbit sorter pull his mustarsh, en say: 'You ain't got no calamus root, is you, Brer Fox? I done got so now dat I can't eat no chicken 'ceppin she's seasoned up wid calamus root.' En wid dat Brer Rabbit lipt out er de do' and dodge 'mong de bushes en sot dar watchin' fer Brer Fox; en he ain't watch long, nudder, kaze Brer Fox flung off de flannil en crope out er de house en got whar he could cloze in on Brer Rabbit, en bimeby Brer Rabbit holler out: 'Oh, Brer Fox! I'll des put yo' calamus root out yer on dish yer stump. Better come get it while hit's fresh,' and wid dat Brer Rabbit gallop off home. En Brer Fox ain't never kotch 'im yit, en w'at's mo', honey, he ain't gwineter."

Ernest L. Thayer

(Phinn)

Ernest L. Thayer (1863–1940) did not strike out in his only remembered poem, the folkloric ''Casey at the Bat.'' As a student at Harvard, where he edited The Lampoon, *Thayer befriended William Randolph Hearst, soon to be expelled and then to acquire the* San Francisco Examiner. *Hearst recruited Thayer, whose duties included filling a weekly column with doggerel. Casey appeared on June 3, 1888, over the signature ''Phinn,'' Thayer's school nickname.*

A visitor from New York, the writer A. C. Gunter, so liked the poem that he folded it in his wallet. Back home, Gunter heard that a local troupe was planning a ''baseball night.'' The creased clipping found its way to the actor DeWolf Hopper, who recited Casey to an ecstatic audience. ''I thought at the time,'' he recalled years later, ''that I was merely repeating a poem, a fatherless waif clipped from a San Francisco newspaper. As it turned out, I was launching a career, a career of declaiming those verses up and down this favored land the balance of my life.'' When Hopper declaimed in Worcester, Mass., a few years later, Thayer (who was employed there in the family textile business) identified himself as the begetter.

Martin Gardner, in The Annotated Casey at the Bat *(New York, 1967, rev. ed., 1984), gathers twenty-five sequels and parodies. Eugene Murdock adds more in* Mighty Casey: All-American *(Westport, Conn., 1984). The Library of Congress has reissued DeWolf Hopper's recorded recitation (Washington, n.d.), and Boston's Godine Press published a centennial edition illustrated by Barry Moser. Here is Thayer's original version, as disinterred by Gardner, with a sequel by the venerated sports columnist, Grantland Rice (see Red Smith's appreciation, pp. 286-88).*

CASEY AT THE BAT

A Ballad of the Republic, Sung in the Year 1888

San Francisco Examiner *June 3, 1888*

The outlook wasn't brilliant for the Mudville nine that day;
The score stood four to two with but one inning more to play.
And then when Cooney died at first, and Barrows did the same,
A sickly silence fell upon the patrons of the game.

A straggling few got up to go in deep despair. The rest
Clung to that hope which springs eternal in the human breast;
They thought if only Casey could but get a whack at that—
We'd put up even money now with Casey at the bat.

But Flynn preceded Casey, as did also Jimmy Blake,
And the former was a lulu and the latter was a cake;
So upon that stricken multitude grim melancholy sat,
For there seemed but little chance of Casey's getting to the bat.

But Flynn let drive a single, to the wonderment of all,
And Blake, the much despis-ed, tore the cover off the ball;
And when the dust had lifted, and the men saw what had occurred,
There was Johnnie safe at second and Flynn a-hugging third.

Then from 5,000 throats and more there rose a lusty yell;
It rumbled through the valley, it rattled in the dell;
It knocked upon the mountain and recoiled upon the flat,
For Casey, mighty Casey, was advancing to the bat.

There was ease in Casey's manner as he stepped into his place;
There was pride in Casey's bearing and a smile on Casey's face.
And when, responding to the cheers, he lightly doffed his hat,
No stranger in the crowd could doubt 'twas Casey at the bat.

Ten thousand eyes were on him as he rubbed his hands with dirt;
Five thousand tongues applauded when he wiped them on his shirt.
Then while the writhing pitcher ground the ball into his hip,
Defiance gleamed in Casey's eye, a sneer curled Casey's lip.

And now the leather-covered sphere came hurtling through the air,
And Casey stood a-watching it in haughty grandeur there.
Close by the sturdy batsman the ball unheeded sped—
"That ain't my style," said Casey. "Strike one," the umpire said.

From the benches, black with people, there went up a muffled roar,
Like the beating of the storm-waves on a stern and distant shore.
"Kill him! Kill the umpire!" shouted some one on the stand;
And it's likely they'd have killed him had not Casey raised his hand.

With a smile of Christian charity great Casey's visage shone;
He stilled the rising tumult; he bade the game go on;
He signaled to the pitcher, and once more the spheroid flew;
But Casey still ignored it, and the umpire said, "Strike two."

"Fraud!" cried the maddened thousands, and echo answered fraud;
But one scornful look from Casey and the audience was awed.
They saw his face grow stern and cold, they saw his muscles strain,
And they knew that Casey wouldn't let that ball go by again.

The sneer is gone from Casey's lip, his teeth are clenched in hate;
He pounds with cruel violence his bat upon the plate.
And now the pitcher holds the ball, and now he lets it go,
And now the air is shattered by the force of Casey's blow.

Oh, somewhere in this favored land the sun is shining bright;
The band is playing somewhere, and somewhere hearts are light,
And somewhere men are laughing, and somewhere children shout;
But there is no joy in Mudville—mighty Casey has struck out.

MUDVILLE'S FATE

By Grantland Rice

Base–Ball Ballads, Nashville, *Tenn., 1910*

I wandered back to Mudville, Tom, where you and I were boys,
And where we drew in days gone by our fill of childish joys;
Alas! the town's deserted now, and only rank weeds grow
Where mighty Casey fanned the air just twenty years ago.

Remember Billy Woodson's place, where, in the evening's shade,
The bunch would gather and discuss the home runs Casey made?
Dog fennel now grows thick around that "joint" we used to know,
Before old Casey whiffed the breeze some twenty years ago.

The grandstand, too, has been torn down; no bleachers met my gaze
Where you and I were wont to sit in happy bygone days;
The peanuts which we fumbled there have sprouted in a row
Where mighty Casey swung in vain just twenty years ago.

O how we used to cheer him, Tom, each time he came to bat!
And how we held our breath in awe when on the plate he spat;
And when he landed on the ball, how loud we yelped! But O
How loud we cursed when he struck out some twenty years ago!

The diamond is a corn patch now; the outfield's overgrown
With pumpkin vines and weedy plots; the rooters all have flown—

They couldn't bear to live on there, for nothing was the same
Where they had been so happy once before that fatal game.

The village band disbanded soon; the mayor, too resigned.
The council even jumped its graft, and in seclusion pined;
The marshal caught the next train out, and those we used to know
Began to leave in flocks and droves some twenty years ago.

For after Casey fanned that day the citizens all left,
And one by one they sought new lands, heartbroken and bereft;
The joyous shout no more rang out of children at their play;
The village blacksmith closed his shop; the druggist moved away.

Alas for Mudville's vanished pomp when mighty Casey reigned!
Her grandeur has departed now; her glory's long since waned.
Her place upon the map is lost, and no one seems to care
A whit about the old town now since Casey biffed the air.

Ambrose Bierce

*Ambrose Gwinnett Bierce (1842–1914?) was born in Horse Cave Creek, Meigs County, Ohio (''*Birth, *the first and direst of all disasters''). He came to journalism as a printer's devil on the* Northern Indianan. *Enlisting in the Union Army in 1861, he fought at Shiloh and Chickamauga and earned a lieutenant's commission; near Chattanooga in June 1864, he was grazed at the temple by a musket ball. In San Francisco in 1867, Bierce found a calling as his sardonic squibs began appearing in the weekly ''Town Crier'' column of the* News-Letter. *He was already known as ''Almighty God Bierce'' when in 1872 he launched his ''Prattle'' column, the venue for the first entries in his* Devil's Dictionary. *The column moved from* Argonaut *and* Wasp *to Hearst's* Examiner *in 1887. There it appeared regularly until 1901, irregularly until 1909 when Bierce ceased writing and vanished into Mexico's civil wars. Bierce edited a chaotic edition of his* Works, *and no really good text now exists.* Skepticism and Dissent *(Ann Arbor, Mich., 1986) edited by Lawrence I. Berkove, offers a selection of his journalism, 1898–1901. I have drawn from this volume, and from the earliest* Devil's Dictionary *entries unearthed by Richard O'Connor in* Ambrose Bierce *(Boston, 1967).*

FIRST ENTRIES IN THE *DEVIL'S DICTIONARY*

The Wasp *Circa 1880s*

ALONE in bad company.

AMBITION an overmastering desire to be vilified by the living and made ridiculous by friends when dead.

BRIDE a woman with a fine prospect of happiness behind her.

BRUTE see husband.

CONSUL in American politics, a person who having failed to secure an office from the people is given one by the Administration on condition that he leave the country.

EGOIST a person of low taste, more interested in himself than in me.

FRIENDSHIP a ship big enough to carry two in fair weather, but only one in foul.

GALLOWS a stage for the performance of miracle plays.

HANDKERCHIEF a small square of silk or linen used at funerals to conceal a lack of tears.

HUSBAND one who, having dined, is charged with the care of the plate.

LITIGANT a person ready to give up his skin in the hope of retaining his bones.

LOVE a temporary insanity curable by marriage.

MARRIAGE a master, a mistress and two slaves, making in all, two.

MERCY an attribute beloved of offenders.

MISFORTUNE the kind of fortune that never misses.

MOUTH in man, the gateway to the soul; in woman, the outlet of the heart.

NEPOTISM appointing your grandmother to office for the good of the party.

NOVEL a short story padded.

OPPOSITION in politics, the party that prevents the government from running amuck by hamstringing it.

PICTURE a representation in two dimensions of something wearisome in three.

PLATITUDE a moral without a fable.

PLATONIC a fool's name for the affection between a disability and a frost.

POLITENESS acceptable hypocrisy.

POSITIVE mistaken at the top of one's voice.

PRAY to ask that the laws of the universe be annulled in behalf of an unworthy petitioner.

PREJUDICE a vagrant opinion without visible means of support.

OPTIMIST a proponent of the doctrine that black is white.

QUILL an implement of torture yielded by a goose and wielded by an ass.

RIOT a popular entertainment given to the police by innocent bystanders.

SAINT a dead sinner revised and edited.

SIREN any lady of splendid promise and disappointing performance.

SUCCESS the one unpardonable sin.
TRICHINOSIS the pig's reply to pork chops.
VIRTUES certain abstentions.
YEAR a period of 365 disappointments.

OUR REAL WAR AIMS IN CUBA

San Francisco Examiner *July 31, 1898*

In a paper read before the Chit-Chat Club and afterward printed in *The Evening Post* W. H. Mills shows by the utterances of several generations of American statesmen, and by the general trend of public opinion through a century of discussion, that the present war is a consequence of causes long antedating the Cuban insurrection and more irresistible than the anger that followed the destruction of the "Maine." The disposition to regard Cuba as rightfully belonging to the United States and its actual possession as necessary to their security and welfare is one of the permanent factors of American politics, ever underrunning American political thought and feeling. To nearly all the dominating intelligences of our country acquisition of Cuba has presented itself as a question of when and how, never as a question of why. We are at war with Spain today merely in obedience to a suasion that has been gathering head from the beginning of our national existence. The stress of the current became at last so strong as to be uncontrollable by political expedient: the river came out of its banks and swept away the deftly drawn distinction between pretext and opportunity. While the President was officially prating of the privations of the reconcentrados whom his military policy has now exterminated, and the press was reciting the virtues and wrongs of the insurgents whom our armies treat with disdain—while we were all calling on high Heaven and the European Powers to witness the unselfishness, the purity, the holiness of our motive and purpose, there was that in the national consciousness "which said as plain as whisper in the ear": "Annexation." And annexation it will indubitably be—annexation peacefully and righteously accomplished, as in Hawaii, by revenue laws making that the only salvation of the main commercial interests, and by judicious corruption of the mongrel Government which we stand pledged to set up in that distracted isle. In fixing the price of peace to Spain it will be no more than fair to deduct the value of Cuba from the gross amount. She is a deferred payment.

Mr. Mills has gone into the matter of the causes underlying this war a trifle more deeply than most writers, but below his lowest deep there is a lower deep which he did not sound. To the shallow understanding of that god of the press and the politician, "the average man," it seems that we are fighting Spain, partly to give independence to Cuba, partly to avenge the sailors of the "Maine." Below these pretexts Mr. Mills discerns the unconscious national purposes, conquest and acquisition. But why do we want Cuba? The true answer to that question is not to be found where he and the illustrious statesmen whom he quotes profess to find it, namely, among the military and commercial considerations of which they make so much. That Cuba would have a military and commercial value to us is true; but that is neither great enough nor obvious enough to have excited so general and strong a cupidity in a people notoriously indifferent to its military security and commercial interests.

Nearly all modern wars are caused by internal stress and pressure: they have for ultimate cause the restlessness that comes of overpopulation. Howsoever those who make the wars may be pleased to explain them; with whatsoever assent others may signify acceptance of the explanation; how little soever of conscious deception there may be in either performance, yet behind it all is the great hand that moves the figures of the show—the hand of Need. Civilized nations feel—most frequently without knowing it—the necessity of additional territory upon which to settle their surplus people under their own flag and their familiar institutions.

It is needless to define the word "overpopulation." A nation is overpeopled when under existing conditions, not hypothetically possible ones, this centrifugal tendency is felt counteracting the patriotic instinct to remain at home. Of two countries equal in habitable area and natural resources, one may be overpeopled with half the population of the other. Much depends on the character of the people and the institutions under which they carry on their struggle for existence. During the present century the population of the chief European countries has nearly doubled despite the wars resulting from the increase. Of late years the blind instinct of territorial expansion has driven them with an incessant compulsion, not only to encroach upon one another's boundaries in Europe, but to effect the conquest of nearly the whole of Africa. And at this moment they are openly attempting the partition of Asia—a performance at which, with a newborn ambition, we are not content to assist as audience.

. . . For augmented population means poverty, which means discontent, which means territorial aggression, which means war; and war checks population. But during this century the interplay and counterplay of these forces have left an enormous balance for Death to draw against —which he seems about to do with a liberality fitly foreshadowed by the gigantic armaments of our time.

Rapid and fateful as has been the increase of population in Europe it is here in the United States that the portentous phenomenon is observed in most conspicuous manifestation. Our census totals grow with startling rapidity, and already the pressure is felt everywhere east of the Missouri river—the section whose public opinion and public feeling not only infect the rest of the country but find ready expression in Congressional action. The demand for this war was formulated among the huddled inhabitants of the Atlantic seaboard. It was formed out of the mere blind instinct (half-prophetic of a time of congestion) which for years had looked upon Cuba with a desire too mild to be translated into action. The passion for territory once roused rages like a lion; successive conquests only strengthen it. That is the fever that is now burning in the American blood. We came by the disorder honestly, and being natural it can hardly be called malign. But that it will run its course, like a seven years itch, or that we shall recover of our volition, is a little too much to expect. Either, like ancient Rome or modern Great Britain, we have entered upon a permanent policy of conquest, colonization and general expansion, or, like Japan and Turkey, we are to hear and heed a firmer and larger utterance of the still, small voice that has already persuaded us to hold in leash the thunder-dogs of Commodore Watson.

THE HERO BUSINESS

San Francisco Examiner *October 30, 1898*

I respectfully submit that we are making something too much of this "hero" business. It looks as if it would end, as it logically might, in declaring ourselves a nation of heroes, every individual an astonishing example of courage and devotion. Why should we not perpetuate the memory of this war by revising our titles of courtesy, and, instead of the mean and meaningless Mister, Master, Missis (or Missuz) and Miss call ourselves Hero Smith, Heroling Jones, Heroine Brown and Heroinette Robinson? Conveniently abbreviated these titles would be respectively, Hr., Hlg., Hne. and Hnt. This system would distinguish us from the other branches of the English-bragging race, the visiting card of an American proclaiming at once his nationality and his own opinion of it. It would tend to uniformity and simplicity, for even our "Judges" and "Colonels" ought gladly to surrender their present titles for the more glorious ones proposed. The "Generals" might perhaps hold out, especially those who earned their high distinction as attorneys-general, general agents and dealers in general merchandise; but eventually all these being gathered to their reputed fathers the title of "General" as a civilian honor would be "heard to cease."

The wake of this great social reform would of course be strewn with wrecks of "ancient and honorable orders," "fraternities" and the like. A man who could be a Hero without payment of dues would hardly care to give money and time (to say nothing of tossing in a blanket and sitting in a hot chair) to be a Knight of Paradise, a Noble of the Mystic Hoodoo, a Saint of the Expurgated Calendar, or a Janissary of Janissation. With the passing of the fraternities the income of many a zealous "organizer" would evanish, but perhaps they would deign to accept some light employment in the field of crime. Whatever the advantages or disadvantages of the suggested system, it would at least serve to mark the memory of a famous war in which we had the astonishing mischance to lose some of our soldiers, yet did not sue for peace.

There never was a war in which both or all the tribes engaged did not believe that their fighters showed themselves wonderfully courageous—never one in which the non-combatants did not beslubber the soldiery with silly adulation for qualities which are the common heritage of mankind and particularly conspicuous in the wild cock-sparrow and the domestic sow. Every nation has the conceit to believe, and most of its individuals have the indecency to affirm, an indubitable primacy in valor and devotion. There is nothing in it; one people is about as brave as another, their different degrees of military efficiency depending chiefly on organization and discipline—in which, by the way, our volunteers are horribly deficient. If the soldiers of two contending armies were as brave as they say they are many of the regiments would be destroyed to the last man; for in actual collision the command to retreat is seldom given and never heard. What is to prevent a man from fighting till he dies? Not courage.

Eugene Field

Amply published in his own time, Eugene Field (1850–1895) is generally forgotten in ours (even as author of "Wynken, Blynken and Nod"). Yet he was the forerunner in establishing Chicago as the innovative center of personal journalism until the 1920s (see pp. xxvii–xxx). A Westerner of New England antecedents, Field came to Chicago from Denver in 1883. His daily "Sharps and Flats" column appeared in the News *until his death. Field's friend and co-worker, Slason Thompson, published an affectionate* Life *(New York, 1928) and collated a two-volume selection of his journalism,* Sharps and Flats *(New York, 1900), from which the following is taken.*

A Play on Words (To Be Read Aloud Rapidly)

Chicago Daily News   *September 12, 1883*

ASSERT ten Barren love day made
  Dan woo'd her hart buy nigh tan day;
Butt wen knee begged she'd marry hymn,
  The crewel bell may dancer neigh.
Lo atter fee tin vein he side
  Ant holder office offal pane—
A lasses mown touched knot terse sole—
  His grown was sever awl Lynn vane,

"Owe, beam my bride, my deer, rye prey,
  And here mice size beef ore rye dye;
Oak caste mean knot tin scorn neigh way—
  Yew are the apple love me nigh!"

She herd Dan new we truly spoke.
  Key was of noble berth, and bread
Tool lofty mean and hie renown,
  The air too grate testates, 't was head.

"Ewe wood due bettor, sir," she bald,
  "Took court sum mother girl, lie wean—
Ewer knot mice stile, lisle never share
  The thrown domestic azure quean!"
"'T is dun, no farebutt Scilly won—
  Aisle waiste know father size on the!"
Oft tooth the nay bring porte tea flue
  And through himself into the see.

How Flaherty Kept the Bridge

Chicago Daily News *September 18, 1883*

Out spake Horatius Flaherty,—a Fenian bold was he,—
"Lo, I will stand at thy right hand and turn the bridge with thee!
So ring the bell, O'Grady, and clear the railway track—
Muldoon will heed the summons well and keep the street-cars back."

Forthwith O'Grady rang the bell, and straightway from afar
There came a rush of humankind and overloaded car.
"Back, back! a schooner cometh," the brave O'Grady cried;
"She cometh from Muskegon, packed down with horn and hide."
And "Back!" Muldoon demanded and Flaherty declaimed,
While many a man stopped short his course and muttered, "I'll be
    blamed!"
And many a horse-car jolted, and many a driver swore,
As the tother gangway of the bridge swung off from either shore.
And bold Horatius Flaherty a storm of curses heard,
But pushing bravely at his key, he answered not a word;
And round and round he turned the bridge to let the schooner through,
And round and round and round again O'Grady turned it too;
Till now at last the way is clear, and with a sullen toot
'Twixt bridge and short, ten rods or more, the tug and schooner shoot.

"Now swing her round the tother way," the brave O'Grady cried.
"'T is well!" Horatius Flaherty in thunder tones replied.
Muldoon waved high his club in air, his handkerchief waved high,
To see the stanch Muskegon ship go sailing calmly by;
And as the rafters of the bridge swung round to either shore,
Vast was the noise of men and boys and street-cars passing o'er.

And Flaherty quoth proudly, as he mopped his sweaty brow,
"Well done for you, and here's a chew, O'Grady, for you now."

They Call Things Differently in London

Chicago Daily News *March 10, 1890*

Our old friend P. T. Barnum has brought his London season to an end amid a blaze of glory. The crowds at the concluding performances of the Greatest Show on Earth were simply enormous, and I suspect that the old gentleman comes pretty near the truth when he says that one hundred thousand people were turned away from the ticket-office during the last week. At the final performance the wealth and fashion were present in full force. Barnum's private box was occupied by the Lord Mayor and his wife, Lord Chief Justice and Lady Coleridge, Consul-General New, Vice-Consul Johnstone, Dr. Playfair, and Mr. Fullerton of New York. There being a great clamor for Barnum, the crafty old showman made his appearance and spoke honeyed words. Later at night the Lord Mayor gave a swell dinner in honor of Mr. Barnum, and the interchange of compliments would have made a barrel of molasses turn sour with envy. You must pardon me for using that word "molasses." Having lived six months in Britain, I should have said "treacle." I study to be correct even in little matters of this kind, but I find it very hard to conform to English as it is spoken this side of the saline pool. Quite at random I make up a list of articles to which the English assign names differing from those we use.

That which we call a "bowl" is here known as a "basin." In England you ask for a "basin of bread and milk."

That which is known to us as a "pitcher" is here called a "jug." A donkey is here called a "moke"; in America a "moke" is a negro. Local slang for a cab-horse is "cat's-meat," because the meat of horses is peddled around the streets for feeding to cats. By the way, British cats average much larger than our American cats, and they are notorious chicken-killers. The brindle cat seems to be the commonest.

What we call "crackers" are here called "biscuit," and I suspect that this is strictly correct. What we call "shoes" are here known as "boots," and what we call "boots" are here known as "bluchers." There is one shoe called the "hilo," because it runs high from the heel up back of the ankle and is cut low in front.

Our "druggist" is here a "chemist," many of the old practitioners retaining the old spelling "chymist."

What we call "ale" is here known as "bitter beer."

What is here known as a "hash" we should call a "stew," and what we call a "hash" is here known as a "mince."

In England our "overcoat" becomes a "greatcoat," our "undershirt" becomes a "vest," and our "drawers" become "pantaloons." It is said that when Mr. George W. Childs of Philadelphia was in London a number of years ago he walked into a haberdashery and, seeking to appear to be a native, asked to be shown the styles in silk waistcoats. "Jeems," cried the proprietor to his assistant, "step this way and show this Hamerican gentleman our flowery weskits."

What we call "sick" the Englishman calls "ill"; "sickness" here implies nausea and vomiting. The British usage is wrong, but the late Richard Grant White settled that point pretty definitely. How came the British to fall into this perversion? It was, I think, because the British can go nowhere except by water; that travel by water induces unpleasant symptoms of nausea and retching, which condition, called "sickness," gradually came to be regarded as the correct definition of "sickness." I can't imagine how the British justify their use of the words "homesick," "heartsick," and "lovesick."

Here they call a "street-car" a "tram"; correct. Here, too, an "elevator" is a "lift," and that is right.

What we call a "telegram" is here called a "telegraph"; it will probably never be determined which of these usages is the better. Our "postal card" is here a "post-card"; "cuffs" become "wrists."

That material known to us as "Canton flannel" is here called "swan's-down," and our "muslin" is known hereabouts as "calico."

Our "locomotive" becomes "engine," and our "conductor" is here a "guard."

What we call "stewing" (culinary term) the British call "simmering." Our "lunch" is here a "luncheon," and our "baggage" becomes "luggage."

Our "wheat" is called "corn," and our "corn" is called "maize" or, sometimes, "Indian corn." "Pigs' feet" are called "trotters." By the way, a theatrical name for a bad actor is "rotter."

A "chill" is here called a "rigor," and the eruption commonly known among us as "hives" is here known as "nettle-rash." Candy is known variously as "sweets," "sweetmeats," and "lolly."

Writing to John Smith, your social equal, you are expected to address him as "John Smith, Esq."; if he be your tailor, grocer, etc., you address him as "Mr. John Smith."

The word "apt" is exceedingly popular here. It is "apt to rain," "apt to be muddy," a man is "apt to go down-town," a bank is "apt to suspend," etc. Even the best prints use this word as a synonym for "likely" and "like." Another barbarism everywhere prevalent in the United Kingdom is the use of "directly" for the conjunction "as soon as," e.g., "directly he went out I shut the door." Charles Dickens, who was quite slovenly at times, seems to have been addicted to this indefensible vice.

What does this British word "left-tenant" mean, I should like to know.

"Quite" is another hackneyed word here; it is edged in upon every occasion.

The first criticism I would pass upon the press of London would be for the indirectness of its speech. When a newspaper writer wishes to convey the idea that yesterday was a pleasant day, he says: "Yesterday was not an unpleasant day." A good play is "not half bad." A humorous speech is "not unrelieved by wit." A riotously applauded address is "not wholly unaccented by demonstrations of approval," and so on, *ad infin. et ad naus.* Now, all this sort of thing may be subtle and it may be conservative, but it is not in the spirit of the Anglo-Saxon, and it vexes me to find so little of the Anglo-Saxon in the literature, the speech, and the practice of the very people where I had thought to find so much.

George Ade

The fame of George Ade (1866–1944) has dwindled and his Fables in Slang *today seems relatively tame. It did not seem so in 1897, when Ade wrote the first, about Sister Mae who did as well as she could, for the* Chicago Record. *Ade used cant phrases adroitly capitalized and spoofed the Keen Young Men and Darned Swell Girls in small towns like his own Kentland, Indiana. His newspaper fables, parodies, and short stories led the way to James Thurber and S. J. Perelman. After his fables became bestsellers in books illustrated by his Purdue classmate, John McCutcheon, Ade devoted himself to musical comedy. With royalties from the successful* The Sultan of Sulu *(1902) and* The College Widow *(1904), he returned to the seclusion of Kentland. A friend and former Cleveland columnist, Fred C. Kelly, has written a biography,* George Ade, *and has edited* The Permanent Ade *(both Indianapolis, 1947). Jean Shepherd's anthology,* The America of George Ade *(New York, 1960), contains an informing introduction.*

FABLES IN SLANG

The Fable *of* Sister Mae, *Who* Did *as* Well *as* Could Be Expected

Chicago Record *September 17, 1897*

Two Sisters lived in Chicago, the Home of Opportunity.

Luella was a Good Girl, who had taken Prizes at the Mission Sunday School, but she was Plain, much. Her Features did not seem to know the value of Team Work. Her Clothes fit her Intermittenly, as it were. She was what would be called a Lumpy Dresser. But she had a good Heart.

Luella found Employment at a Hat Factory. All she had to do was to put Red Linings in Hats for the Country Trade; and every Saturday Evening, when Work was called on account of Darkness, the Boss met her as she went out and crowded three Dollars on her.

The other Sister was Different.

She began as Mary, then changed to Marie, and her Finish was Mae.

From earliest Youth she had lacked Industry and Application.

She was short on Intellect but long on Shape.

The Vain Pleasures of the World attracted her. By skipping the Long Words she could read how Rupert Bansiford led Sibyl Gray into the Conservatory and made Love that scorched the Begonias. Sometimes she just Ached to light out with an Opera Company.

When she couldn't stand up Luella for any more Car Fare she went out looking for Work, and hoping she wouldn't find it. The sagacious Proprietor of a Lunch Room employed her as Cashier. In a little While she learned to count Money, and could hold down the Job.

Marie was a Strong Card. The Male Patrons of the Establishment hovered around the Desk long after paying their Checks. Within a Month the Receipts of the Place had doubled.

It was often remarked that Marie was a Pippin. Her Date Book had to be kept on the Double Entry System.

Although her Grammar was Sad, it made no Odds. Her Picture was on many a Button.

A Credit Man from the Wholesale House across the Street told her that any time she wanted to see the Telegraph Poles rush past, she could tear Transportation out of his Book. But Marie turned him down for a Bucket Shop Man, who was not Handsome, but was awful Generous.

They were Married, and went to live in a Flat with a Quarter-Sawed Oak Chiffonier and Pink Rugs. She was Mae at this Stage of the Game.

Shortly after this, Wheat jumped twenty-two points, and the Husband didn't do a Thing.

Mae bought a Thumb Ring and a Pug Dog, and began to speak of the Swede Help as "The Maid."

Then she decided that she wanted to live in a House, because, in a Flat, One could never be sure of One's Neighbors. So they moved into a Sarcophagus on the Boulevard, right in between two Old Families, who had made their Money soon after the Fire, and Ice began to form on the hottest Days.

Mae bought an Automobile, and blew her Allowance against Beauty Doctors. The Smell of Cooking made her Faint, and she couldn't see where the Working Classes came in at all.

When she attended the theater a Box was none too good. Husband went along, in evening clothes and a Yachting Cap, and he had two large Diamonds in his Shirt Front.

Sometimes she went to a Vogner Concert, and sat through it, and she wouldn't Admit any more that the Russel Brothers, as the Irish Chambermaids, hit her just about Right.

She was determined to break into Society if she had to use an Ax.

At last she Got There; but it cost her many a Reed Bird and several Gross of Cold Quarts.

In the Hey-Day of Prosperity did Mae forget Luella? No, indeed.

She took Luella away from the Hat Factory, where the Pay was three Dollars a Week, and gave her a Position as Assistant Cook at five Dollars.

Moral: Industry and Perseverance bring a sure Reward.

The Fable *of the* Professor *who* Wanted *to be* Alone

Chicago Record *circa 1897–98*

Now it happens that in America a man who goes up hanging to a Balloon is a Professor.

One day a Professor, preparing to make a Grand Ascension, was sorely pestered by Spectators of the Yellow-Hammer Variety, who fell over the Say-Ropes or crowded up close to the Balloon to ask Fool Questions. They wanted to know how fur up he Calkilated to go and was he Afeerd and how often had he did it. The Professor answered them in the Surly Manner peculiar to Showmen accustomed to meet a Web-Foot Population. On the Q. T. the Prof. had Troubles of his own. He was expected to drop in at a Bank on the following Day and take up a Note for 100 Plunks. The Ascension meant 50 to him, but how to Corral the other 50? That was the Hard One.

This question was in his Mind as he took hold of the Trapeze Bar and signaled the Farm Hands to let go. As he trailed Skyward beneath the buoyant silken Bag he hung by his Knees and waved a glad Adieu to the Mob of Inquisitive Yeomen. A Sense of Relief came to him as he saw the Crowd sink away in the Distance.

Hanging by one Toe, and with his right Palm pressed to his Eyes, he said: "Now that I am Alone, let me Think, let me Think."

There in the Vast Silence He Thought.

Presently he gave a sigh of Relief.

"I will go to my Wife's Brother and make a Quick Touch," he said. "If he refuses to Unbelt I will threaten to tell his Wife of the bracelet he bought in Louisville."

Having reached this Happy Conclusion, he loosened the Parachute and quickly descended to the Earth.

Moral: Avoid Crowds.

The Fable *of the* Caddy *who* Hurt His Head While Thinking

Chicago Record *circa 1897–98*

One Day a Caddy sat in the Long Grass near the Ninth Hole and wondered if he had a Soul. His Number was 27, and he almost had forgotten his Real Name.

As he sat and Meditated, two Players passed him. They were going the Long Round, and the Frenzy was upon them.

They followed the Gutta Percha Balls with the intent swiftness of trained Bird Dogs, and each talked feverishly of Brassy Lies, and getting past the Bunker, and Lofting to the Green, and Slicing into the Bramble —each telling his own Game to the Ambient Air, and ignoring what the other Fellow had to say.

As they did the St. Andrews Full Swing for eighty Yards apiece and then Followed Through with the usual Explanations of how it Happened, the caddy looked at them and Reflected that they were much inferior to his Father.

His Father was too Serious a Man to get out in Mardi Gras Clothes and hammer a Ball from one Red Flag to another.

His Father worked in a Lumber Yard.

He was an Earnest Citizen, who seldom Smiled, and he knew all about the Silver Question and how J. Pierpont Morgan done up a Free People on the Bond Issue.

The Caddy wondered why it was that his Father, a really Great Man, had to shove Lumber all day and could seldom get one Dollar to rub against another, while these superficial Johnnies who played Golf all the Time had Money to Throw at the Birds. The more he Thought the more his Head ached.

Moral: Don't try to Account for Anything.

Finley Peter Dunne

Finley Peter Dunne (1867–1937) and his Dooley dialogues became an overnight success from their first appearance in 1894. In voicing doubts about an arrogant plutocracy and imperial cruelties, Dunne anticipated the mood that led to Progressivism, the income tax, and to first steps to social and racial justice. Because the wrongs he addressed continue to live, so does Dooley. The first Dooley dialogues can be found in Barbara C. Schaaf's collection, Mr. Dooley's Chicago *(New York, 1977). In an interesting experiment, Dooley was recast in standard English in* Mr. Dooley Remembers *(Boston, 1963), the informal memoirs of Finley Peter Dunne, edited with a commentary by his son, Philip Dunne. Some little known dialogues appeared in pirated editions, and I've taken one on the Monroe Doctrine from Louis Filler, ed.,* Mr. Dooley: Now and Forever *(Stanford, Calif., 1954). An annotated chronology has been compiled by Charles Fanning,* Finley Peter Dunne & Mr. Dooley *(Louisville, 1978).*

ON RUDYARD KIPLING

Chicago Evening Post *November 20, 1898*

"I think," said Mr. Dooley, "th' finest pothry in th' wurruld is wrote be that frind iv young Hogan's, a man be th' name iv Roodyard Kipling. I see his pomes in th' pa-aper, Hinnissy; an' they're all right. They're all right, thim pomes. They was wan about scraggin' Danny Deever that done me a wurruld iv good. They was a la-ad I wanst knew be th' name iv Deever, an' like as not he was th' same man. He owed me money. Thin there was wan that I see mintioned in th' war news wanst in a while—

th' less we f'rget, th' more we raymimber. That was a hot pome an' a good wan. What I like about Kipling is that his pomes is right off th' bat, like me con-versations with you, me boy. He's a minyit-man, a r-ready pote that sleeps like th' dhriver iv thruck 9, with his poetic pants in his boots beside his bed, an' him r-ready to jump out an' slide down th' pole th' minyit th' alarm sounds.

"He's not such a pote as Tim Scanlan, that hasn't done annything since th' siege iv Lim'rick; an' that was two hundherd year befure he was bor-rn. He's prisident iv th' Pome Supply Company—fr-resh pothry delivered ivry day at ye'er dure. Is there an accident in a grain illyvator? Ye pick up ye'er mornin' pa-aper, an' they'se a pome about it be Roodyard Kipling. Do ye hear iv a manhole cover bein' blown up? Roodyard is there with his r-ready pen. "Tis written iv Cashum-Cadi an' th' book iv th' gr-reat Gazelle that a manhole cover in anger is tin degrees worse thin hell." He writes in all dialects an' anny language, plain an' fancy pothry, pothry f'r young an' old, pothry be weight or linyar measuremint, pothry f'r young an' old, pothry be weight or linyar measuremint, pothry f'r small parties iv eight or tin a specialty. What's the raysult, Hinnissy? Most potes I despise. But Roodyard Kipling's pothry is aisy. Ye can skip through it while ye're atin' breakfuss an' get a c'rrect idee iv th' current news iv th' day—who won th' futball game, how Sharkey is thrainin' f'r th' fight, an' how manny votes th' pro-hybitionist got f'r gov'nor iv th' State iv Texas. No col' storage pothry f'r Kipling. Ivrything fr-resh an' up to date. All lays laid this mornin'.

"Hogan was in to-day readin' Kipling's Fridah afthernoon pome, an' 'tis a good pome.* He calls it 'Th' Thruce iv th' Bear.' This is th' way it happened: Roodyard Kipling had just finished his mornin' batch iv pothry f'r th' home-thrade, an' had et his dinner, an' was thinkin' iv r-runnin' out in th' counthry f'r a breath iv fr-resh air, whin in come a tillygram sayin' that th' Czar iv Rooshia had sint out a circluar letther sayin' ivrybody in th' wurrld ought to get together an' stop makin' war an' live a quite an' dull life. Now Kipling don't like the czar. Him an' th' czar fell out about something, an' they don't speak. So says Roodyard Kipling to himsilf, he says: 'I'll take a crack at that fellow,' he says. 'I'll do him up,' he says. An' so he writes a pome to show that th' czar's letter's not on

*Dooley' refers to Kipling's "Truce of the Bear" (1898) deriding a Czarist peace overture, this being the message:

> Yearly, with tent and rifle, our careless white men go
> By the Pass called Muttianee, to shoot in the vale below.
> Yearly by Muttianee he follows our white men in—
> Matun, the old blind beggar, bandaged from brow to chin.
>
> Eyeless, noseless, and lipless—toothless, broken of speech,
> Seeking a dole at the doorway he mumbles his tale to each;
> Over and over the story, ending as he began:
> "Make ye no truce with Adam-zad—the Bear that walks like a Man!"

th' square. Kipling's like me, Hinnissy. When I want to say annything lib-lous, I stick it on to me Uncle Mike. So be Roodyard Kipling. He doesn't come r-right out, an' say, 'Nick, ye're a liar!' but he tells about what th' czar done to a man he knowed be th' name iv Muttons. Muttons, it seems, Hinnissy, was wanst a hunter; an' he wint out to take a shot at th' czar, who was dhressed up as a bear. Well, Muttons r-run him down, an' was about to plug him, whin th' czar says, 'Hol'on,' he says—'hol' on there,' he says. 'Don't shoot,' he says. 'Let's talk this over,' he says. An' Muttons, bein' a foolish man, waited till th' czar come near him; an' thin th' czar feinted with his left, an' put in a right hook an' pulled off Muttons's face. I tell ye 'tis so. He jus' hauled it off th' way ye'd haul off a porous plasther—raked off th' whole iv Mutton's fr-ront ilivation. 'I like ye'er face,' he says, an' took it. An' all this time, an' 'twas fifty years ago, Muttons hasn't had a face to shave. Ne'er a one. So he goes ar-round exhibitin' th' recent site, an' warnin' people that, whin they ar-re shootin' bears, they must see that their gun is kept loaded an' their face is nailed on securely. If ye iver see a bear that looks like a man, shoot him on th' spot, or, betther still, r-run up an alley. Ye must niver lose that face, Hinnissy.

"I showed th' pome to Father Kelly," continued Mr. Dooley.

"What did he say?" asked Mr. Hennessy.

"He said," Mr. Dooley replied, "that I cud write as good a wan me-silf; an' he took th' stub iv a pencil, an' wrote this. Lemme see—Ah! here it is:—

> Whin he shows as seekin' frindship with paws that're thrust in thine,
> That is th' time iv pearl, that is th' thruce iv th' line.
> Collarless, coatless, hatless, askin' a dhrink at th' bar,
> Me Uncle Mike, the Fenyan, he tells it near and far,
>
> Over an' over th' story: 'Beware iv th' gran' flimflam,
> There is no thruce with Gazabo, th' line that looks like a lamb."

"That's a good pome, too," said Mr. Dooley; "an' I'm goin' to sind it to th' nex' meetin' iv th' Anglo-Saxon 'liance."

THE MONROE DOCTRINE*

Chicago Evening Post *October 19, 1895*

"Jawn," said Mr. Dooley, "Where's Venezwala?"*

"It's down beyond," said Mr. McKenna, pointing over his shoulder.

"In th' twinty-ninth?" asked Mr. Dooley with concern.

"Oh, no; in South America."

*During a boundary dispute between Britain and Venuezeula, Grover Cleveland's Secretary of State invoked the Monroe Doctrine, saying in a note to the British that the "United States is practically sovereign on this continent."

"Ye'er right," said Mr. Dooley. "I thought I'd thry ye. 'Tis in South America, and th' River Oorynoco r-runs through it. Well, sir, Hinnissy is all wur-rked up over it. 'Twas Cuba las' week. 'Oh,' says he, 'if I cud get ninety-nine min f'r to shoulder muskets,' he says, 'an' ma-arch acrost to Cuba,' he says, 'to free it fr'm th' roonous rule in Spain,' he says, 'I'd make th' hunderth.' 'Yis,' says I, 'an' if ye cud get nine hunderd an' ninety-nine thousand nine hundred an' ninety-nine, ye'd make th' millionth,' I says. 'Ye'd be the last, annyhow.'

"But las' night he'd forgot all about Cuba. Whin he come in, says I be way iv makin' spoort iv him, 'Well, Senior Hinnissy,' says I, 'how goes th' rivalution?' 'Which wan?' says he. 'Th' wan in Cuba,' I says. 'Oh, th' 'ell with that,' he says. 'They've begun their incroachmints, an' 'tis givin' me good time thinkin' iv thim. All week long they've been doin' nawthin' but gettin' beat,' he says.

" ' 'Tis no good stayin' on a dead card,' he says. 'I've been thinkin' iv this here Vin-ezwalan business,' he says. 'What's th' matter?' says I.

" 'Well, he says, 'th' British have been thryin' to plant thimsilves on Vin-ezwalan sile,' he says. 'They've begun their incroachmints, an' 'tis their intintion to get up a tyranny on this continent.' 'Who?' says I. 'Th' British,' says he. 'Now,' he says, 'th' quistion is whether we, as a nation, are prepared f'r to allow this aggression in vilation iv th' Mon-roe doctorin.' 'What did Doc Mon-roe prescribe?' says I. He got as mad as a hin at that. 'Ye have as much knowlidge iv history as me ol' shoe,' he says. 'What d'ye think Mon-roe was—a vithrinary surgeon?' he says. 'Monroe was a statesman, an' he laid down this principle that if annywan, anny iv th' Powers—' 'a mane fam'ly,' says I. 'Anny iv th' Eur-oyean powers attempted to throw th' boots into anny counthry on this continent, we'd throw th' boots into him.' 'I knowed a Eur-opyean Powers fam'ly,' says I, 'fr'm Watherford. A mane people. I'll give wan boot, if it's thim.'

" 'Yer crazy,' says he. 'These are th' powers like England, France an' Roosha. If they intherfere on this side we'll go at thim. An' England is puttin' in on Vin-ezwala," he says. 'Is Vin-ezwala wan iv th' new states?' says I. 'No,' says he,' it is not.' 'Is it pa-art iv th' Union?' says I. 'It's in South America,' says he. 'Have ye anny relitives there?' says I. 'No,' says he. 'Or anny frinds?' I says. 'No,' says he. 'Thin what's ailin' ye?' I says. 'It's th' Mon-roe doctorin',' says he. 'Well,' says I, 'I don't know Monroe,' I says, 'or whether he's a homypathy or on th' square or a farryer,' I says, 'but I know a better doctorin'. 'What's that?' says Hinnissy. 'It's th' Hoolihan doctorin'. Jerry Hoolihan was a polisman an' he held his job f'r thirty years. He was thravelin' beat wan night an' a woman put her head out iv a window an' says she, 'Officer?'—like that. 'What can I do f'r ye, lady?' says Hoolihan. 'Me husband's dhrunk,' says she. 'I thried to subdue him with a flat iron an' he's gone f'r an ax,' she says. 'I can do nawthin' for ye,' says Hoolihan. ''Tis on ye'e beat,' says the lady. 'I know it,' says Hoolihan. 'But there's pig's feet on me beat, too. 'I'm goin f'r some at this minyit,' he says.

"I don't see what that's got to do with Vin-ezwala,' says Hinnissy. 'Well,' says I, ' 'tis th' Hoolihan doctorin'.' 'Niver stop to fight whin ye'er goin' to supper,' says I.

"Do you think there'll be a war?" Mr. McKenna asked.

"If there is," said Mr. Dooley, "Hinnissy 'll go in my place. I stayed here durin' wan war an' I can stand a few more th' same way."

THE PHILIPPINE PEACE

Boston Globe *March 9, 1902*

Dooley rendered in standard speech, by the author's son, Philip Dunne

"It's strange we don't hear much talk about the Philippines," said Mr. Hennessy.

"The reason is," said Mr. Dooley, "that everything is perfectly quiet there. We don't talk about Ohio or Iowa or any of our other possessions because there's nothing doing in those parts. The people are going ahead, garnering the products of the soil, sending their children to school, worshipping on Sunday in the churches and thanking Heaven for the blessings of free government and the protection of the flag above them.

"So it is in the Philippines. I know, for my friend Governor Taft* says so, and there' a man that understands contentment when he sees it. The Filipinos, he says, are satisfied with our rule. And I believe him. A man that isn't satisfied when he's had enough is a glutton. They're satisfied and happy and slowly but surely they're acquiring that love for the government that floats over them that will make them good citizens without a vote or a right to trial by jury. I know it. Governor Taft says so.

"Says he, 'The Philippines is one or more of the beautiful jewels in the diadem of our fair nation. Formerly our fair nation didn't care for jewels, but did up her hair with side combs, but she's been abroad some since and she came back with beautiful golden hair that is better for having a tiara. She is not as young as she was. The simple home-loving maiden that our fathers knew has disappeared and in her place we find a Columbia, gentlemen, with maturer charms, a knowledge of European customs and not averse to a cigarette.

" 'The Philippines raise unkown quantities of produce, none of which fortunately can come into this country. My business kept me in Manila or I would tell you what they are. Besides, some of our loyal subjects are

*William Howard Taft, the future President, was the U.S. Governor in the Philippines after the archipelago's annexation; his main task was to suppress an insurrection.

getting to be good shots.

" 'Passing to the political situation, I will say it is good. Not perhaps as good as yours or mine, but good. Every once in a while when I think of it, an election is held. Unfortunately it usually happens that those elected have not yet surrendered. In the Philippines, the office seeks the man, but as he is also pursued by the solidery, it is not always easy to catch him and fit it on him. The country may be divided into two parts, politically —where the insurrection continues and where it will soon be. The brave but I fear not altogether cheery army controls the insurrected parts by martial law, but the civil authorities are supreme in their own house. The difference between civil law and martial law in the Philippines is what kind of coat the judge wears. The result is much the same. The two branches work in perfect harmony. We bag them in the city and they round them up in the country.

" 'It is not always necessary to kill a Filipino American right away. My desire is to educate them slowly in the ways and customs of the country. We are giving hundreds of these poor benighted heathen the well-known, old-fashioned American water cure. Of course you know how it is done. A Filipino, we'll say, never heard of the history of this country. He is met by one of our sturdy lads in black and blue who asks him to cheer for Abraham Lincoln. He refuses. He is then placed on the grass and given a drink, a bayonet being fixed in his mouth so he cannot reject the hospitality. Under the influence of the hose that cheers but does not inebriate, he soon warms or perhaps I might say swells up to a realization of the grandeur of his adoptive country. One gallon makes him give three groans for the Constitution. At four gallons, he will ask to be wrapped in the flag. At the dew point he sings "Yankee Doodle." Occasionally we run across a stubborn and rebellious man who would strain at my idea of human rights and swallow the Pacific Ocean, but I must say most of these little fellows are less hollow in their pretentsions.

" 'Naturally we have had to take a good many customs from the Spaniards, but we have improved on them. Among the most useful Spanish customs is reconcentration. Our reconcentration camps are among the most thickly populated in the world. I was talking with a Spanish gentleman the other day who had been away for a long time and he said he wouldn't know the country. Even the faces of the people on the streets have changed. They seemed glad to see him.

" 'I have not considered it advisable to introduce any fads like trial by jury of your peers into my administration. Plain straight-forward dealing is my motto. A Filipino at his best has only learned half his duty to mankind. He can be tried but he can't try his fellow man. It takes him too long. But in time I hope to have them trained to a point where they can be good men and true at the inquest.

" 'I hope I have told you enough to show you that the stories of disorder are greatly exaggerated. The country is progressing splendidly, the ocean still laps the shore, the mountains are there and apparently quite happy; the flag floats free and well-guarded over the government offices, and the cheery people go and come on their errands—go out alone and come back with the troops. Everywhere happiness, content and love of the stepmother country, except in places where there are people. Gentlemen, I thank you.'

"And there you are, Hennessy. I hope this lucid story will quiet the wagging tongues of scandal and that people will let the Philippines stew in their own happiness."

"But sure they might do something for them," said Mr. Hennessy.

"They will," said Mr. Dooley. "They'll give them a measure of freedom."

"But when?"

"When they'll stand still long enough to be measured," said Mr. Dooley.

William Cowper Brann

Writing in Waco, Texas, in 1898, J. W. Shaw said of his fellow townsman, William Cowper Brann (1855–98): "Beginning his literary career as a reporter, he was soon made an editorial writer, in which capacity he became well known throughout Illinois, Missouri and Texas. As such he was versatile, forceful and direct. There was no needless circumlocution in his composition." This seems a rare instance of Texas understatement. As editor from 1891 of Brann's Iconoclast, *he so infuriated readers that one of them shot him fatally in the back. In a Baptist stronghold, he assailed anti-Catholic bigotry and ridiculed the backwoods faithful. He praised Jews and women's rights. Yet he also wrote maliciously about Negroes and on occasion resorted to slander, as acknowledged by Roy Bedichek, who knew him. Bedichek's balanced estimate appears as the introduction to Charles Carver's* Brann and the Iconoclast *(Austin, 1947). These selections are from* The Writings of Brann the Iconoclast *(New York, 1938), preface by J. W. Shaw.*

CORONATION OF THE CZAR*

The Iconoclast *circa May 1896*

With more barbaric mummery, flummery and vulgar waste of wealth than characterized even the late Marlborough-Vanderbilt wedding, Nicholas Two-Eyes was crowned Emperor of the rag-tag and bob-tail of creation, officially known as "all the Russias." Nick has a nice easy job at a salary considerably in excess of ye average country editor, and he gets it all in

*Czar Nicholas II was crowned in Moscow on May 26, 1896.

gold roubles instead of post-oak cordwood and green water-melons, albeit his felicity is slightly marred by an ever-present fear that he may inadvertently swallow a few ounces of arsenic or sit down on an infernal machine.

Nick is emphatically an emperor who emps. He isn't bothered with do-nothing congresses or Populist politicians who want him impeached. When he saith to a man "come," he cometh p. d. q.; to another "go" he getteth a hustle on him that would shame a pneumatic tire. Nick is the greatest monarch "what they is." He is the divinely ordained Chief Gyasticutus of that motley aggregation of tallow-munchers and unwashed ignorami whose very existence is a menace to modern civilization. The Goths and Visigoths were models of cleanliness and avatars of intelligence compared with a majority of the seventy different breeds of bipedal brutes who acknowledge the rule of the Romanoffs. . . . The average Slav is as stupidly ignorant as an agency Indian. He respects no law but that of blind force. His Magna Charta is the dynamite bomb. He is courageous with the bravery of the brute, which has no conception of life's sacredness. Doubtless the rule of the bayonet is the only government possible for such a barbarous people—and the Romanoffs have not allowed it to rust.

The Czar is the immediate ruler of nearly 130,000,000 semi-savages, his lightest word their supreme law, while the chiefs of the robber hordes of Central Asia acknowledge him their official head. Such tremendous power in the hands of a weak-minded, vacillating monarch like Nicholas II—descended from Catherine the Courtesan, and having in his veins the blood of cranks—may well cause western Europe to lie awake. Bonaparte declared that in a hundred years the continent would be all Russian or all Republican—by which he meant that unless this nation of savages *in esse* and Vandals *in posse* were stamped out it would imitate the example of Alaric and Attila and precipitate such another intellectual night as that known as the Dark Ages. In western Europe Republicanism is making but slight progress, while in the east the power of the Great White Khan is rapidly increasing. In a struggle between the semi-savagery of the East and the civilization of the West, China and Turkey would be the natural and inevitable allies of the Czar. Small wonder that the Great First Consul trudged home from Moscow with a heavy heart!

Some faint idea of the savage ignorance of Russia may be had from the history of the Siberian exiles and the fiendish persecutions of the Jewish people. Siberia is the Ice Hell of the old Norse mythologists, into which men, women and children have been indiscriminately cast on the bare suspicion of desiring to better the wretched condition of the Russian people. Its horrors, which have long been a hideous nightmare to civilized men, need no description here. The very name of Siberia causes humanity to shudder—it casts a shadow on the sun! The experience of the Jews in Russia was akin to that of the early settlers in America, who were exposed to the unbridled ferocity of the Aborigines; yet the so-called Christian nations dared do no more than petition the Czar that these savage atrocities should cease—-futile prayers to the hog-headed god of the Ammonites!

The young man who has just been crowned at Moscow at an expense of some millions, and whose emblem of authority is ornamented with rubies as large as eggs and ablaze with 2,564 costly diamonds—while half his people are feeding on fetid offal—is a weak-faced pigmy who would probably be peddling Russia's favorite drunk promoter over a pine bar had he not chanced to be born in the purple. Having been spawned in a royal bed—perchance the same in which his great gran-'dame Catherine was wont to receive her paramours—he becomes the most powerful of princes—haloed with "that divinity which doth behedge a king"—and all the earth rejoices to do him honor. . . .

We have nothing in common with Russia. One government is the antithesis of the other. They are "on friendly terms" because they have practically no intercourse. Russia has no American possessions upon which we can pull the foolish manifesto of the erstwhile Monroe. There's no trade between the two countries—hasn't been since Russia unloaded her Alaskan glaciers upon us at a fancy price. It would have been eminently proper had Minister Breckinridge presented himself—togged out in his best Arkansas jeans instead of being attired like a troubadour—to wish Nick exemption from the Nihilists and express the hope that the occasion wouldn't swell his head; but there was absolutely no excuse for sending warships on an expensive cruise, and special envoys 5,000 miles to make unmitigated asses of themselves.

The unpalatable fact is that we are a nation of toadeaters. President Cleveland is, in this respect at least, eminently representative of the American people. . . .

THE COURAGE OF WOMANKIND.

The Iconoclast *circa 1897*

A gentleman of wide experience as frontiersman, soldier and purist, recently remarked to me: "Women are more truthful than men; they exhibit more gratitude; they are the superiors of men in physical courage."

This testimony will doubtless appear not a little startling to many who have ever regarded women as "the weaker vessel"; but I believe it will be confirmed by every careful student of humankind. That women possess more moral courage than men is generally admitted; but that physical is the necessary correlative of moral courage does not appear to have occurred to the average individual. The two virtues—if they be indeed binary—are so interdependent that divorcement is practically impossible. Why does a child accused of a misdemeanor confess its fault and accept punishment, when, by a subterfuge, it could escape castigation? Because it fears the scourge of conscience more than the maternal slipper. It must choose between two evils, and it elects the least. Without moral courage it would lie, because having naught to fear from conscience; without physical courage it would lie, because unable to accept bodily suffering that could be avoided. When the soldier rushes upon shotted guns we call it physical courage; when a voluptuous woman denies the improper importunities of her lover, we call it moral courage; yet the efficient causes of these actions are identical. The soldier does not desire to be killed or crippled, but dreads the deserter's shame more than the guns of the enemy, and the woman considers self-denial preferable to dishonor; hence courage, call it by what name you will, is but the balancing of one ill against another, and the acceptance of what the world elects to call a lesser. It regards the instincts of the dog and the ferocity of the savage as great physical and little moral courage; but such objections are idle without some knowledge of the ethics of dogdom and the moral concept of the savage. You cannot measure the moral courage of man or beast until you have ascertained the moral code applicable to the civilization or intelligence of his kind. The different races of men and the various orders of animals diverge widely in their natures, some being mild, others murderous. We cannot measure the heart by the standard of the hound, nor the gentle-souled Bengalee by the ferocious Britisher who finds pleasure in useless effusion of blood. . . .

Junius Henri Browne—angels and ministers of grace defend us! where did he find it?—has been telling the few old maids and anile Mugwumps who still worry through *Harper's Weekly,* what he doesn't know about "Woman's Courage." He assures us that women affect timidity they do not feel, because their supposed helplessness is considered a charm by the opposite sex. I trust that Junius Henri Browne will not topple his name over on me if I dissent from his very pretty but undigested dogma. It is neither cowardice nor affectation that makes so many ladies scream at sight of a bug or mouse, but sheer nervousness. You seldom see a healthy countrywoman clamber upon a table to avoid a harmless creeper. Our city-bred American women are neurotic; but this ill is being rapidly alleviated by the sensible outdoor exercise now coming generally into favor. A rightly constituted man glories in a courageous woman—a woman all womanly, without a suggestion of aggressive masculinity; but who could, if need be, go to the block as bravely as Marie Antoinette, strike as deadly as Hannah Dustin, or emulate the deeds of Saragossa's beautiful lioness when Spain's enemies were foiled by woman's hand before a battered wall.

Chappies like Junius Henri, whose sidewheel whiskers and trigeminous titles are giving them spinal curvature, may fancy hysterical maids; but a sure-enough American, competent to raise a crop of world-compellers all bearing his trade-mark, prefers a Grace Darling or a Molly Pitcher. He knows that, if mated with her equal, such a woman will never suckle worthless sons. Junius Henri next informs us that while woman may possess certain kinds of courage, she has not that which enables her to face physical danger with fortitude, to look on ghastly wounds or sustain great bodily suffering. I wonder who told him that? . . .

A woman is capable of grander heroism, greater self-sacrifice, nobler morality than man, and for the simple reason that her nature is, so to speak, of finer texture—because she is less a brute than her companion. Her courage differs from that of man much as the courage of a highbred gentleman differs from that of a Bowery bouncer. The latter really enjoys a brutal fisticuff, but a thread of cold steel will stop him. To the gentleman a bruising-bout is an abomination; but he will take his position before a dead shot with apparent indifference. It is this higher courage which woman possesses. If she starts more readily at trifles she will go with a steadier step to the supreme sacrifice. A finely constituted mind instinctively shrinks from the brutal; but where honor, principle, or the well-being of loved ones is concerned, it is ready to lead the forlorn hope. It faces inevitable death with a placid brow, while the "nerve" of a coarser nature either breaks completely down or indulges in mock bravado. Man is incapable of that intensity of suffering, mental or physical, which a delicate woman often uncomplainingly endures. If men instead of women were required to suffer the pains of parturition there wouldn't be enough people left on earth in two hundred years to organize a base-ball nine—and I'd be the first laddie-buck to take the vow of celibacy under such a dispensation. Courage! The sting of a hornet while severe pain to a man is madness to a woman; yet the former will dance the Highland fling and howl like a defeated candidate, while the latter, after one startled scream of surprise sets placidly about doctoring herself. The fact of the matter is that man wouldn't have amounted to much had not the Creator sent woman to steer the miserable savage against the tree of knowledge and shoulder nine-tenths of his trouble, while he's snorting like a mad bull over the remainder and imagining himself a hero. I'm very fond of the ladies—but I wouldn't be one of them for the world. In a moment of mental aberration I might marry some J. Henri Browne.

Bert Leston Taylor

(B.L.T.)

Bert Leston Taylor (1866–1921) was not only witty himself but earned his livelihood by provoking wit in others. B.L.T.'s own verse set a standard his Chicago readers were challenged to match, first in "A Little About Everything" in the Journal *from 1899 to 1901, then in "A Line-O'-Type or Two" in the* Tribune *for most of the years until his death. His column was a compote of paragraphs on the day's tidings, excerpts from the rural press, his own verse, and letters and poetry provided gratis by readers. Though his verse was collected in a half-dozen slim volumes, B.L.T. lives chiefly in anthologies. The following can be found in a selection of his verse,* Motley Measures *(New York, 1927), with an introduction by Ring Lardner.*

B.L.T.' LINE OF VERSE

Chicago Tribune *circa 1901–21*

Meditations by a Mossy Stone

"Give me ten accomplished men for readers,
and I am content"

WALTER SAVAGE LANDOR

What? Ten accomplished readers? That, meseems,
Puts much too high a value on a pen.
I never in my most presumptuous dreams
Have thought of ten!

Content, indeed! I should be flattered pink;
To please a smaller clientele I strive.

I've never thought, nor ever dared to think,
Of six—or five.

Why, five accomplished readers are a host;
So large a number quite abashes me.
If I have thought at all, I've thought, at most
Of two—or three.

And when I view this Motley Monument
Of jape and jingle, paragraph and pun.
I sometimes feel that I should be content
With one—or none.

The Season Opens [1913]

The tariff battle now is on,
  Wide-mouthed Revision sounds tantivy!
The tax will be removed anon
  From dragon's-blood and divi-divi.
And east and west you hear men say,
"Going to the baseball game to-day?"

Our frank and fearless President
  Is smashing this and that tradition,
And stuffing with astonishment
  The oldest living politician.
And east and west you hear men cry,
"Wait for a good one! That's the eye!"

Embattled dames in London Town,
  Forgetting they are perfect ladies,
Are blowing up and burning down,
  And raising every sort of Hades.
And east and west you hear the shout,
"The pitcher's rotten! Take him out!"

The peace of Europe is at stake,
  The cannons roar, the sabres rattle;
A dozen kingdoms are a-quake,
  And listening for the call to battle.
And east and west men yell, *"Keep cool!*
Sit down there! Let the umpire rule!"

The Dinosaur

Behold the mighty Dinosaur,
Famous in prehistoric lore,

Nor only for his weight and strength
But for his intellectual length.
You will observe by these remains
The creature had two sets of brains—
One in his head (the usual place),
The other at his spinal base.
Thus he could reason *a priori*
As well as *a posteriori.*
No problem bothered him a bit:
He made both head and tail of it.
So wise he was, so wise and solemn,
Each thought filled just a spinal column.
If one brain found the pressure strong
It passed a few ideas along;
If something slipped his forward mind
'Twas rescued by the one behind;
And if in error he was caught
He had a saving afterthought.
As he thought twice before he spoke
He had no judgments to revoke;
For he could think, without congestion,
Upon both sides of every question.

Oh, gaze upon this model beast,
Defunct ten million years at least.

The Bards We Quote

Whene'er I quote I seldom take
From bards whom angel hosts environ;
But usually some damned rake
Like Byron.

Of Whittier I think a lot,
My fancy to him often turns;
But when I quote 'tis some such sot
As Burns.

I'm very fond of Byrant, too,
He brings to me the woodland smelly;
Why should I quote that "village roo,"
P. Shelley?

I think Felicia Hemans great,
I dote upon Jean Ingelow;

Yet quote from such a reprobate
As Poe.

To quote from drunkard or from rake
Is not a proper thing to do.
I find the habit hard to break,
Don't you?

My Lady New York

O siren of tresses peroxide,
And heart that is hard as a flint,
Blue orbs of complacency ox-eyed,
That light at the mark of the mint,
Ears only for jingle of joybells,
A conscience as light as a cork—
You are wedded to follies and foibles,
My Lady New York.

True, you have (not enough, tho', to hurt you)
Your moods and your manners austere;
You have visions and vapors of virtue,
And "reform" for a time has your ear;
But of chaste Puritanic embraces
You soon have enough and to spare,
And then you kick over the traces,
And virtue forswear.

So go it, milady! Foot fleetly
The paths that are primrose and gay;
Abandon your fancy completely
To follies and fads of the day.
"Reform" is a something that throttles
The joys of the pace that's intense—
Smash hearts, reputations, and bottles,
And ding the expense!

Kurt M. Stein

(K.M.S.)

Among B.L.T.'s prize contributors was K.M.S., or Kurt M. Stein, who in 1913 began appearing in "the Line" in "die schönste Lengevitch." *The poet was initially inspired by a chance question on a Chicago street: "Pardong, sir, holds se tramway here?" Soon Stein was rendering grand opera, Shakespeare, and Longfellow in the dialect of German immigrants. Stein writes: "To me, the most interesting thing is the new meaning given to words through a similarity of sound or association of ideas. For instance, the most common: like (adv.-similar) :* gleich; *hence, to like -* gleichen. *Then verbal translations of idiomatic phrases: I've made up my mind*—Ich habe meine Meinung aufgemacht *(for* sich entscheiden *or* entschliesen). Or, Ich wunder *(I wonder), for* Ich mochte wisse." *Stein published* Die Schönste Lengevitch *(1925),* Gemixte Pickles *(1928), and* Limberger Lyrics *(1930), all brought together in an omnibus ("mit additions") published in New York by Crown in 1953.*

HAMLET, IN CHICAGO DEUTSCH

Chicago Tribune *circa 1920.*

Part I

"Ich hab a Hunch die Welt geht an die Bum,"
Sagt Hamlet zu Horatio, sei Chum.
"Da is was rotten hier in diesem State
Und 's is kei use dass es so weiter geht."
"Well, business wird schon starteh aufzupickeh."

Sagt Raish. " 'S tut anyhow kei gut zu kickeh."
"Von Business," answert Ham, "tu ich net talkeh.
Mich boddert nur der Ghost wo hier tut walkeh.
Iss das mei Pa sei Ghost und an die Level,
Or shust a fake, for fun geschickt vom Devil?"
"Das," sagt Horatio, "lässt sich hart decideh.
In mei opinion war er Bona fide.
So proud herumzustalkeh wie das Ding
Kann nur a moo-vieh-actor or a King."
"Ich wett," mused Ham, "sie haben ihm geburied
Weil er a bunch Insurance hat gecarried.
Belief me, Raish, je mehr ich tu reflecteh,
Je mehr tu ich mei royal Ma suspecteh
Und meinen Ohm. Die haben Pa gemurdered
Durch faule Means ins Jenseits ihn befördert!
Net accidentlich starb der King der Däne!
'S war diesen Weg—der Ghost tat mir's explaineh:

Einen Tag—es war Pa's habit—
Setzt nach lunch er sei bequeme
Alte Crown auf, um im Garten
Schnell a kurze Nap zu nehme.
Denn nach Meals a kleines Restchen
Is a gut Ding for Digestion.

Wie er so tut sweetly nappeh,
In dem coolen shade der Bäume
Sei Gesicht in Smiles gerwrinkelt
Weil er von sei Queen tut träume,
Kommt mei Uncle, ungehört,
Da er Rubbersohle weart.

In sei Hand trägt er a Bottle
Voll mit Gift. Das tut er poureh—
Dieses Luder von a Bruder—
In mei royal Vater's Ohre.
Und wie Pa aufwakeh will
Is er mausetod und still.

Natcherly, da war excitement,
Und der Coroner wollt wisse
Wie's gehappent; but mei Uncle
Schwor a snake hätt Pa gebisse.
Und before a month war hin
Heiratet er schon die Queen.

So sprach der Ghost, Raish. Das is jetzt die Question:
Wie find ich Proofs? Ich hab es! Bei Suggestion!
Wir tun heut Nacht vor'm King den Murder stageh!
Tu schnell a couple Barry mores engageh
Und watch mei Uncle bei der Show! Gib Acht,
Wir geben ihm den Third Degree heut Nacht.

Part II

"Hello, Laertes! When kamst du zurück?"
Exclaimed King Claude. "But why der schwarze Blick?
Dass du so mad bist wie a nasse Hen
Kann ich bei dei Misfortunes gut verstehn.
Erst wird dei Vater totgestabbed, denn geht
Dei Schwester crazy. So a saddes Fate!
Doch wir sein for dei Troubles net zu blemeh.
Drum tu es for a minute easy nehme,
Und please tu gegen uns dei grouch net nurseh:
Wir werden dich schon plenty reimburseh."
"Dass is allright, King," antwortet, Laertes,
"Doch in mei Sohn- und Bruder-herz da gährt es.
Es yelled nach Blut; es hollert for Revenge!
Sei doch a Sport, und sag mir welcher Mensch
Der Cause von alle meine Troubles is."
"Du sollst es höre," sagt der King, "Gewiss!
Der Hamlet war's! Geh, even's auf mit ihm.
Jedoch hold an! Ich hab a kleine Scheme
Ihn gut zu fixeh, und es soll appeareh,
Als sei's a accident. Geh du und schmiere
Den Stuff hier an dei Sword. Er acted quicker
Wie heimgebrewter prohibition liquor.
Ich send dem Ham denn Wort du wolltest heute
A friendly Bout mit ihm zusamme fighteh.
Du brauchst ihn nur mit deinem Foil zu scratcheh,
Der Stuff hier tut den Rest. 'S is soft, wie Quetsche.
Und um ganz sure zu sein dass wir ihn fixeh
Will ich in diesem Cup a drink noch mixeh.
Der finisht ihn." "You're an," schreit der Laertes,
"Ich nehm den Job. Wie wir's geschemed, so werd es."
"So lang, denn," smiled der King. "Der iss gehooked
Und Hamlet seine Goose iss sure gecooked.

Part III

"Laertes, wenn bei deiner Schwester Grab
Maybe dei feelings ich gehurted hab,"

Sagt Ham, "denn tu ich jetzt apologizeh.
Ich tat dei friendship allweil highly prizeh
Und tu es still. So for Ophelia's sake
Forgib mei unpoliteh acts und shake."
"Ham," sagt Laertes, "ich will net refuseh
Dei roughneck'ges Behavior zu excuseh.
Und shake auch for old time sake mit dir Hände.
Jedoch die Code of Honor tut demandeh
Dass wir a duel fighteh. Also, bitte,
Prepare. Das Picture-money tun wir splitteh."
"Du hast mei Wort," sagt Ham. "Ich bin bereit.
Horatio, act als Referee beim Fight,
Hier sein die Foils. Ich care net for reward.
Die Bell hat schon gerungen. An dei Guard!"

So tun denn die Beide in Combat engageh
Und Stahl flashed auf Stahl. Der Battle tut rageh.
Und ladies in silk und gearmorte Knights
Die watchen intenslich den Auskom des Fights.
Und unter 'nem schwellen Brocade Baldachin
In State sitzt King Claudius und Gertrude, die Queen.
Und vor ihm, wie raspberry soda so pink
Im goldenen Cupf der gepoisonte Drink.
Shust wait, Claude, du Villain, so weit ging's dir gut,
But lang is die Road wo kei Turn habe tut.
Dei Hash is gesettled. Revenge eilt dir nach.
Ach!—

"A hit for Hamlet! Goody, noch a hit!"
Exclaimed Queen Gert. "Ham, rest dich doch a bit.
Mei Sohn is fett. Gut gracious, tust du schwitze!
Hier, nimm a kleine Drink. Der kühlt dei Hitze.
No? Well, denn trink ICH auf dei Wohlsein. Prost!
Help!—Das is poison! Help! Ich bin gelost!"
"Die Queen tut fainteh! Hier is faules play,"
Schreit Ham. "Ich bin gewounded. Ha, ich seh—
Das scharfe Foil, mit Poison an der Kling'!
Das sein dei Tricks, du niederträcht'ger King,
(Er macht a pass.) Verreck, du Teufelssohn."
"Ouch!" schreit der King, und tumbelt von sei Throne.
"Horatio, ich geh tot," sagt Hamlet. "Please
Sag den Papieren wie's gehappent is
Und wer zu blameh is for all die Vi'lence.
Jetzt sterb ich. Well, gut bye."—Der rest is silence.

Ring Lardner

Virginia Woolf once reviewed You Know Me, Al, *"a story about baseball, a game which is not played in England, often in a language which is not English." Woolf found in Ring Lardner (1885–1933) talents of a "remarkable order," a writer whose quick strokes and sure touch brought to life a boastful, innocent athlete. She went on shrewdly to guess that baseball helped him solve a difficult problem for American writers: "It has given him a clue, a centre, a meeting place for divers activities of people whom a vast continent isolates, whom no tradition controls. Games give him what society gives his English brother."*

Lardner doted on baseball as a boy in Niles, Michigan; he covered the Central League as a novice in South Bend, Indiana; he wrote about the major leagues for the Chicago Tribune *until he became a sports columnist in 1913. Baseball was the staple of "In the Wake of the News," his busher letters made his name, and carried him east in 1919, just before the shattering Black Sox scandal. Nobody has yet troubled to make a book of Lardner's "Wake" or his later weekly columns, described in Donald Elder's* Ring Lardner *(New York, 1956) and Jonathan Yardley's* Ring *(New York, 1977). Ring Lardner, Jr., recalls the* Tribune *years in his memoir,* The Lardners: My Family Remembered *(New York, 1976). Lardner's newspaper writing is itemized in a bibliography by Matthew J. Bruccoli and Richard Layman (Pittsburgh, 1976). Here are three "Wakes," including a trial run for the story "Haircut," and a report on "A World's Serious."*

ONLY 20 M.P.H.

Chicago Trubune *September 2, 1915*

Frend Harvey:

Im sorry Harvey, but I will half to call off our golf game Fri P. M. on acct. of me haveing an other engagement and when I made the engagement with you I dident know nothing a bout this here other engagement because it come after words and I wouldent of made it my self but it was made for me by a couple motor sickle police mans and I would rather play golf but dident want to hurt there feelings. The engagement I got is to meet them up to the city Hall and I all ready been all threw the city Hall so I dont care nothing a bout going there again but these fellows insisted that thats where I got to meet them. I will tell you how it come off Harvey so your feelings wont be hurt a bout me not playing golf with you and probly you can get up an other match because most of the other good golf players from Chi will of come back from Det. by that unlest they got relitives to vissit up there.

I was driveing a long palmer Square over north west and I couldent of been driveing more than 6 miles per hr. when all of a sudden I seen these heer 2 motor sickles rideing a long besides me and at 1st. I dident think much a bout it but thot they probly seen my pitcher some wheres and wanted to look at me close up. But then 1 of them says whats your hurry and the other 1 says pull up to the curb on the right so I dont like to turn no body down so I pulled up to the curb and stoped and says what can I do for you. And then 1 of them says why was you in such a hurry and I says because I wanted to get to where I was going to. I dident know exactly what was he geting at because I couldent of been going more than 11 miles per hr. and that aint what youd call being in a hurry. But I all ways try to be pollite and humor evry body.

So then 1 of them says whats your name and I told him and I had to spell it out for him on acct. of he never hearing it beffore and I dont know where has he been all his life. So then he ast me where did I live and I give him the name of the place where I live at and he must of herd of it because this time I dident half to do no spelling so he says You shouldent ought to go a long like that and I says like what and he says like you was going so I says whats the matter wasent I going strate and he says yes but you was going to fast. What do you know a bout that Harvey and I couldent of been going more than 15 miles per hr.

So then 1 of them says we should ought to go to the station but I says I dident have no train to catch, only I dident say that out loud but to my self. So he says where do you work and I told him and he says are you a Ingraver and I says no and he says what do you do and I had to think a while but finely I came right out and told him. Then he says are you teling the truth and I says dident I just tell you where I worked at and then the other 1 says He looks like a pretty square guy so you see Harvey he had good eye site even if he did think I was going to fast and me not going more than may be 19 miles per hr.

So then the other 1 says well if your a square fella and telling the truth we will leave you go a long now, but we will see you Fri. P. M. and I says all right, because they was pretty good Co. and I dont mind talking to them some more and giveing them a treat. So then they ast me some more questions like how long had I drove and was I down to the office evry day but they dident ask me nothing a bout what I ett for breakfast and who would win the penant in the Nat. league. So then the 1 thats says I looked like a square guy says it over again and all so says he dont look like no joy rider and I guess he says that because I kept my face so sollum wile they was talking to me. So they give me a card and it says on it that I got to answer to the charge of V 8 10 S & L and I don't know what crime that is but I will higher the best legal tallent in the country if they try to R. R. me to the penitentiery because if I had to send my stuff from Jolliet the males might get delayed some times and 1st. thing you know the paper would half to go to the press with out nothing from me.

Well Harvey I guess you see why I got to call off our engagement and I wisht they had of picked out some other place to meet me like the art inst. or the Municple Peer or some place where I havent been all ready but for some reason they choose the city Hall and I don't want to criticise them. And I couldent of been going more than 22 miles per hr.

Respy.

R.

FIFTEEN CENTS' WORTH

Chicago Tribune *January 15, 1918*

VICTIM Just a shave.

BARBER Yes, sir. Funny weather for this time o' year. Unhealthy weather. 'Least everybody seems to be ailin'. Grip or influenza or whatever you want to call it. Reg'lar contagion. You had it yet?

V. (through a towel, darkly)—Burrh.

B. You're lucky if you ain't had it. But you better knock wood. I had it bad Christmas week. Great time to be sick! Felt rotten. Just my luck to have it come Christmas week. Did you have a good Christmas?

V. (chewing lather)—Tuh.

B. Christmas is all right in a way, but they's too much useless givin'. Man gets ten things he can't use for every one he wants. Was you in town for Christmas?

V. (through another towel) Caw.

B. Well, we're better off than them poor fellas over in the trenches. They must of had a fine Christmas! And most o' them don't even know what they're fightin' about. Nobody knows except the kaiser and the king and the fellas that started it, and they don't know themselves. All as I hope is that we don't get mixed up in it. But I guess they's no danger o' that with Wilson at the hellum. We'd been mixed up long ago if Rusefelt was president. We're pretty lucky to have a man like Wilson at the hellum just at this time. Razor hurt? But still and all Wilson lets them make a monkey out of him. Look at the Lusitaynia! He made Germany promise they wouldn't sink no more boats without givin' warnin' and they go right ahead in a week or two and sink a couple more. A nice massage would fresh you up. And every time they sink a boat and kill a few hundred people he writes them another note and they say it was a mistake and they was aimin' at a fish or somethin' and promise they won't do it no more and next thing you know you pick up a paper and they's been another boat sank and Wilson stands for it. You wouldn't never see them monkeyin' with ol' Teddy that way. First time they done somethin' he didn't like he'd say: "Here!" he'd say. "You fellas cut that out or you'll have the stars and stripes on your neck," he'd tell them. And they'd cut it out, too, because they'd know he was talkin' business. That's the kind o' guy we should ought to have at the hellum. They make a monkey o' Wilson. You must shave yourelf most o' the time; you got it all uneven up here. What do you think o' Henry Ford. That was a fine flivver! I wonder what he thought he was goin' to pull. Guess he thought they'd all jump out o' the trenches when they seen him comin'. Wanted some free advertisin', I guess. Well, he got it, but I'd hate to make a monkey o' myself for advertisin'. Still, I guess his intentions was all right. You know they laughed at Columbus. But he showed them. Ford's all right and a whole lot better than the people that's been laughin' at him. A nice massage'd clear that skin all up. Well, the Feds is all through. They'd ought to know better than buck organized ball. They was too much money against them. And they didn't have no real stars neither. Maybe one or two, but take them all through and they was no class to them. I guess Ward and them's sore they ever went into it. Weeghman'll come out all right, but it'll cost him some money. They certainly made organized ball quit at that. I'd hate to of lose all the dough that's been dropped in baseball these last two years. But next year'd ought to be good, and they'll be better ball played. Last year they wasn't enough good ball players to go round. The Cubs 'd ought to be good this year if Weeghman combines the two teams. I think he's a sucker, though, to play on the north side. It'll get a lot o' people sore that's been goin' to games on the west side for twenty years. But, o' course, the park's new out north and I guess they'll draw all right with a good club. They won't hurt the Sox none though. Old Commy's got the crowd with him; Commy and the hitless wonders. What do you think o' Thompson closin' up Sundays? They must of been some reason or he wouldn't done it. It made a lot o' people sore. I think a man'd ought to be able to get a drink Sunday if he wants it. Still I guess they's enough time durin' the week to get all that's good for you. I don't touch it myself. You got a couple o' blackheads there. Nice massage'd clean that all up. All right, you know best. Wet or dry?

(Victim leaves chair, puts on collar, coat, and hat, takes check to cashier, pays it, and walks out.)

BARBER (to himself) The cheap stiff!

THE WAKE OF THE WAKE

Chicago Tribune *June 20, 1919*

It is with mingled feelings of pain and astonishment that I stand before you tonight on the occasion of the sixth annual dinner of this organization. Arduous as the duties of this office have been at times, I have enjoyed the work as much as a person of my social position can enjoy work of any kind and have more than enjoyed the association with you gentlemen which my incumbency has afforded. In retiring from this office I wish to express my profound thanks to each and every one of you who have been of assistance to me in my carrying on this great work.

That reminds me of a story that I just read in a book called "Toasts and After-Dinner Stories": "A man who had just lost his wife said to a friend who happened to be fresh from a reminder at home that he was very much married: 'I tell you, old man, it is hard to lose a wife.' 'Hard,' said the henpecked husband. 'Why, man, it's impossible!' " The book does not state whether the two men were named Pat and Mike or what.

I have been asked by the committee to say a few words to you in regard to my successor to this high office.* This man, gentlemen, is a man who scarcely needs an introduction. In fact, I have seen him speak to

*This was Lardner's final column for the *Tribune*. He was succeeded by Jack Lait, later a columnist for Hearst's *New York Mirror*.

people whom he had never met. This man, gentlemen, is a man whose name is known from the short-ribbed shores of Maine to the grape juice highballs of California.

This man's name is usually pronounced as spelled, but in France, whence this man's ancestors came, the final t would be silent. The name, translated from the French, means Milk, which reminds me of another story which I will copy out of the book:

"A small boy who had been very naughty was first reprimanded, then told that he must take a whipping. He flew upstairs and hid in the far corner under a bed. Just then his father came home. The mother told him what had occurred. He went upstairs and proceeded to crawl under the bed toward the youngster, who whispered excitedly, 'Hello, pop; is she after you, too?' "

The person holding this high office from which I am retiring is supposed to know something of sport. I can say for my successor that he possesses a far greater knowledge of the manly art of poker than myself do and proved this several times as far back as 1908, when the both of us was single and employed on Market St. In those days, gentlemen, this man to whom I refer got a salary of $35 per week besides mine.

Gentlemen, I know there are a great many of you who live in the suburbs and who must catch your trains home so I will not take up any more of your time. Gentlemen, it is my pleasure to present to you Mr. Jacquin Lait, the new president of the Wake of the News.

I thank you.

A WORLD'S SERIOUS

San Francisco Examiner *October 1, 1920*

Advance Notice

All though they have been world serious practally every yr. for the last 20 yrs. this next world serious which is supposed to open up Wed. p.m. at the Polo grounds is the most important world serious in history as far as I and my family are conserned and even more important to us than the famous world serious of 1919 which was win by the Cincinnati Reds greatly to their surprise.

Maybe I would better exclaim myself before going any further. Well, a few days previous to the serious of 1919 I was approached by a young lady who I soon recognized as my wife, and any way this woman says would I buy her a fur coat as the winter was comeing on and we was going to spend it in Connecticut which is not genally considered one of the tropics.

"But don't do it," she says, "unless you have got the money to spare because of course I can get along without it. In fact," she added bursting into teers, "I am so used to getting along without this, that, and the other thing that maybe it would be best for you not to buy me that coat after all as the sight of a luxury of any kind might prove my undoing."

"Listen," was my reply, "as far as I am concerned you don't half to prove your undoing. But listen you are in a position to know that I can't spare the money to buy you one stoat leave alone enough of the little codgers skins to make a coat for a growed up girl like you. But if I can get a hold of any body that is sucker enough to bet on Cincinnati in this world serious, why I will borrow from some good pal and cover their bet and will try and make the bet big enough so as the winnings will buy you the handsomest muleskin coat in New England."

Well friends I found the sucker and got a hold of enough money to cover his bet and not only that but give him odds of 6 to 5 and that is why we did not go out much in Greenwich that winter and not for lack of invitations as certain smart Alex has let fall.

I might also mention at this junction that they was a similar agreement at that serious between Eddie Collins the capt. of the White Sox and his Mrs. only of course Eddie did not make no bet, but if his team win, why he should buy the madam a personal sedan whereas if his team lost, why she would half to walk all winter. Luckily the Collinses live in Lansdowne, Pa., where you can't walk far.

Well friends I do not know what is the automobile situation in the Collins family at the present writeing as have not saw them of late but the fur coat situation in my family is practically the same like it was in 1919 only as I hinted in the opening paragraph of this intimate article, it is a d-a-m sight worse.

Because this yr. they won't be no chance for the little woman to offset her paucity of outdoor raps by spending the winter in the house. She is going to need furs even there.

Therefore as I say this comeing serious is the most important of all as far as we are conserned for Mother ain't the same gal when she is cold and after all is said and done what is home with mother in her tantrums?

So I and my little ones is hopeing and praying that the boys on who I have staked my winters happiness this yr. will not have no meetings in no hotel rooms between now and Wednesday but will go into this serious determined to do their best which I once said was the best anybody could do and the man who heard me say it said "You are dead right Lardner" and if these boys do their best, why it looks to me like as if the serious should ought to be well over by Sunday night and the little woman's new fur coat delivered to our little home some time Monday and maybe we will get invited out somewheres that night and they will be a blizzard.

About That Fur Coat

San Francisco Examiner *October 6, 1920*

Well friends you can imagine my surprise and horror when I found out last night that the impression had got around some way another that as soon as this serious was over I was planning to buy a expensive fur coat for my Mrs. and put a lot of money into same and buy a coat that would probably run up into hundreds and hundreds of dollars.

Well I did not mean to give no such kind of a impression and I certainly hope that my little article was not read that way by everybody a specially around my little home because in the first place I am not a sucker enough to invest hundreds and hundreds of dollars in a garment which the chances are that the Mrs. will not wear it more than a couple times all winter as the way it looks now we are libel to have the most openest winter in history and if women folks should walk along the st. in expensive fur coats in the kind of weather which it looks like we are going to have why they would only be laughed at and any way I believe a couple can have a whole lot better time in winter staying home and reading a good book or maybe have a few friends to play bridge.

Further and more I met a man at supper last night that has been in the fur business all his life and ain't did nothing you might say only deal in furs and this man says that they are a great many furs in this world which is reasonable priced that has got as much warmth in them as high price furs and looks a great deal better. For inst. he says that a man is a sucker to invest thousands and thousands of dollars in expensive furs like Erminie, Muleskin, squirrel skin, and kerensky when for a hundred dollars or not even that much, why a man can buy a owl skin or horse skin or weasel skin garment that looks like big dough and practically prostrates people with the heat when they wear them.

So I hope my readers will put a quietus on the silly rumor that I am planning to plunge in the fur market. I will see that my Mrs. is dressed in as warm a style as she has been accustomed to but neither her or I is the kind that likes to make a big show and go up and down 5th ave sweltering in a $700 hogskin garment in order so as people will turn around and gap at us. Live and let live is my slocum.

So much for the fur coat episode and let us hear no more about it and will now go on with my article which I must apologize for it not being very good and the reason is on account of being very nervous after our little ride from the polo grounds to park row. It was my intentions to make this trip in the subway but while walking across the field after the game I run into Izzy Kaplan the photographer and he says would I like to ride down in a car which him and his friends had hired so I and Grantland Rice got in and we hadn't no sooner than started when one of our fellow passengers says that we ought to been with them coming up.

"We made the trip from park row in 24 minutes," he says, "and our driver said he was going to beat that record on the return trip."

So we asked what had held them back comeing up and one of them said that the driver had kept peeling and eating bananas all the way and that he did not drive so good when both his hands was off the wheel. Besides that, they had ran into a guy and had to wait til the ambulance come and picked him up.

Well friends I will not try and describe our flight only to say that we did not beat the record but tied it and the lack of bananas didn't prevent our hero from driving with his hands off the wheel as he used the last named to shake his fists at pedestrians and other riff raff that don't know enough to keep off the public highways during the rush hour.

Most of the things I was going to mention in this article was scared out of me during our little jaunt. One of them however was the man from Toronto that stood in line with his wife from 8 p.m. Tuesday night till the gates opened Wednesday morning so as to be sure of good seats. According to officials of the club, they could of got the same seats if they had not showed up till a couple hours before the game, but if they had of done that, why the lady would not of had no chance to brag when she got back home. The way it is, why she can say to her friend, "Charley may not be much for looks, but he certainly showed me the night life of New York."

Dividing interest with this couple was a couple of heel and toe pedestrians that done their base circling stunt just before the start of the game. One of them was the same guy that done it before the first game last fall, but this time he was accompanied by a lady hoofer and it is not too much to say that the lady was dressed practically as though for her bath. Casey Stengel expressed the general sentiment in the following words, "If that is just her walking costume I would hate to see her made up for tennis."

Don Marquis

How Don Marquis (1878–1937) of Walnut, Illinois, made his way from country printshop to the New York Sun *is related in "Confessions of a Reformed Columnist" (see pp. xxxv–xxxvii), written in 1928. By then his fears were realized: he was known chiefly as the creator of a cockroach. Contributing to the enduring appeal of Archy and Mehitabel were the inimitable illustrations later provided by George Herriman, creator of "Krazy Kat." The "Sun Dial" in which Archy first appeared follows. The whole corpus is in* The Lives and Times of Archy and Mehitabel *(New York, 1950), introduced by E. B. White. The columnist's other creations are represented in* The Best of Don Marquis *(New York, 1946). Edward Anthony has written a life,* O Rare Don Marquis *(New York, 1962).*

ARCHY'S FIRST BOW

New York Evening Sun *March 29, 1916*

Dobbs Ferry possesses a rat which slips out of his lair at night and runs a typewriting machine in a garage. Unfortunately, he has always been interrupted by the watchman before he could produce a complete story.

It was at first thought that the power which made the typewriter run was a ghost, instead of a rat. It seems likely to us that it was both a ghost and a rat. Mme. Blavatsky's ego went into a white horse after she passed over, and some one's personality has undoubtedly gone into this rat. It is an era of belief in communications from the spirit land—there is Patience Worth, and there is the author of the *Letters of a Living Dead Man,* and there are many other prominent and well-thought of ghosts in touch with the physical world today—and all the other ghosts are becoming encouraged by the current attitude of credulity and are trying to get into the game, too.

We recommend the Dobbs Ferry rat to the Psychical Research Society. We do not pretend to know anything about the Dobbs Ferry rat at first hand. But since this matter has been reported in the public prints and seriously received we are no longer afraid of being ridiculed, and we do not mind making a statement of something that happened to our own typewriter only a couple of weeks ago. We came into our room earlier than usual in the morning and discovered a gigantic cockroach jumping about upon the keys.

He did not see us, and we watched him. He would climb painfully upon the framework of the machine and cast himself with all his force upon a key, head downward, and his weight and the impact of the blow were just sufficient to operate the machine, one slow letter after another. He could not work the capital letters, and he had a great deal of difficulty operating the mechanism that shifts the paper so that a fresh line may be started. We never saw a cockroach work so hard or perspire so freely in all our lives before. After about an hour of this frightfully difficult literary labor he fell to the floor exhausted, and we saw him creep feebly into a nest of the poems which are always there in profusion.

Congratulating ourself that we had left a sheet of paper in the machine the night before so that all this work had not been in vain, we made an examination, and this is what we found:

expression is the need of my soul
i was once a vers libre bard
but i died and my soul went into the body of a cockroach
it has given me a new outlook upon life
i see things from the under side now
thank you for the apple peelings in the wastepaper basket
but your paste is getting so stale i cant eat it
there is a cat here called mehitabel i wish you would have
removed she nearly ate me the other night why dont she
catch rats that is what she is supposed to be for
there is a rat here she should get without delay

most of these rats here are just rats
but this rat is like me he has a human soul in him
he used to be a poet himself
night after night i have written poetry for you
on your typewriter
and this big brute of a rat who used to be a poet
comes out of his hole when it is done
and reads it and sniffs at it

he is jealous of my poetry
he used to make fun of it when we were both human
he was a punk poet himself
and after he has read it he sneers
and then he eats it

i wish you would have mehitabel kill that rat
or get a cat that is onto her job
and i will write you a series of poems showing how things
look
to a cockroach
that rats name is freddy
the next time freddy dies i hope he wont be a rat
but something smaller i hope i will be a rat
in the next transmigration and freddy a cockroach
i will teach him to sneer at my poetry then

dont you ever eat any sandwiches in your office
i havent had a crumb of bread for i dont know how long
or a piece of ham or anything but apple parings
and paste leave a piece of paper in your machine
every night you can call me archy

Franklin P. Adams

(F.P.A.)

In a tribute to the now dwindled fame of Franklin P. Adams (1881–1960), E. B. White recalled that as a young newcomer to New York, "the block seemed to tremble" when he walked by F.P.A.'s home on West Thirteenth Street "the way Park Avenue trembles when a train leaves Grand Central." This was in the 1920's, when Adams was perhaps the most avidly read columnist in The Evening World. *He came to New York from Chicago in 1904 as a protégé of B.L.T.'s, but outpaced his master. His "Conning Tower" contributors developed the competitive ferocity of a family as he favored one pet or another, the material reward being an invitation to an annual banquet which he declined to attend. Reputations flowered, but it was his fate to be overshadowed by his protégés, notably Dorothy Parker and George S. Kaufman. These selections are from Adams's own* Column Book of F.P.A. *(New York, 1928) and* The Melancholy Lute *(New York, 1936). Poetry by others is collected in* The Conning Tower Book *(New York, 1926) and* The Best of the World *(New York, 1973), edited by John K. Hutchens and George Oppenheimer. Sally Ashley has written a biography,* F.P.A. *(New York, 1986).*

TO HIS LYRE*

Ad Lyram

Horace: Book I, Ode 32
"Poscimur. Siquid vacui sub umbra"—

If ever, as I struck thy strings,
My song has sounded sempiternal,

*This verse appeared in FPA's first day on the *World,* January 2, 1922.

Help me, my Lyre, to glorious things
For this matutinary journal.

Thine erstwhile owner versified
War, Love, and Wine in panegyric;
And folks in Lesbos often cried,
"That kid can chuck a nasty lyric!"

Then aid me, Lute, beginning now!
Give me theme for colophon or leader;
And some day there may grace my brow
The laurel from some Grateful Reader.

THE MONUMENT OF Q. HORATIUS FLACCUS*

Ad Melpomenen

Horace: Book III, Ode 30
"Exegi monumentum aere perennius
Regalique situ pyramidum altius."

Reader, the monument that I've
Erected ever shall survive
As long as brass; and it shall stay
Despite the stormiest, wildest day.
Though winds assail, yet shall it stand
High as the pyramids, and grand.
Eternally my name will be
Triumphant in posterity.
Recurrent will my praises sound;
I shall be terribly renowned.
Born though I was of folk obscure,
Unknown, I spilled Some Lit'rature.
Now, O Melpomene, my Queen
Entwine the laurel on my bean!

*F.P.A.'s final column for the *Evening Mail,* December 31, 1913.

FROM THE CONNING TOWER

New York World *circa 1922–30*

The Pessimist's Forecast

Monday's child is sad of face;
Tuesday's child will lose the race;
Wednesday's child has a row to hoe;
Thursday's child is full of woe;
Friday's child has futile strife;
Saturday's child has a mournful life;
While the child that's born on the Sabbath day
Will find that life is dull and grey.

Lines on Reading Frank J. Wilstach's "A Dictionary of Similes"

As neat as wax, as good as new,
As true as steel, as truth is true,
Good as a sermon, keen as hate,
Full as a tick, and fixed as fate—

Brief as a dream, long as the day,
Sweet as the rosy morn in May,
Chaste as the moon, as snow is white,
Broad as barn doors, and new as sight—

Useful as daylight, firm as stone,
Wet as a fish, dry as a bone,
Heavy as lead, light as a breeze—
Frank Wilstach's book of similes.

Broadmindedness

How narrow his vision, how cribbed and confined!
How prejudiced all of his views!
How hard is the shell of his bigoted mind!
How difficult he to excuse!

His face should be slapped and his head should be banged;
A person like that ought to die!
I want to be fair, but a man should be hanged
Who's any less liberal than I.

Rarae Aves

Announce it here with triple leading
That once I heard a Noisy Wedding;

And accurately I recall
The day I saw a Sober Brawl.

I saw some burglars drive away
In a Low-Powered Car, the other day;
And yesterday, I'm pretty sure,
An Unknown Clubman died, obscure.

To a Thesaurus

O precious codex, volume, tome,
  Book, writing, compilation, work
Attend the while I pen a pome,
  A jest, a jape, a quip, a quirk.

For I would pen, engross, indite,
  Transcribe, set forth, compose, address,
Record, submit—yea, even write
  An ode, an elegy to bless—

To bless, set store by, celebrate,
  Approve, esteem, endow with soul,
Commend, acclaim, appreciate,
  Immortalize, laud, praise, extol.

Thy merit, goodness, value, worth,
  Expedience, utility—
O manna, honey, salt of earth,
  I sing, I chant, I worship thee!

How could I manage, live, exist,
  Obtain, produce, be real, prevail,
Be present in the flesh, subsist,
  Have place, become, breathe or inhale.

Without thy help, recruit, support,
  Opitulation, furtherance,
Assistance, rescue, aid, resort,
  Favour, sustention and advance?

Alas! Alack! and well-a-day!
  My case would then be dour and sad,
Likewise distressing, dismal, grey,
  Pathetic, mournful, dreary, bad.

* * *

Though I could keep this up all day,
  This lyric, elegiac song,
Meseems hath come the time to say
  Farewell! Adieu! Good-bye! So long!

ADDENDA BY DOROTHY PARKER

New York World *circa 1922–28*

Comment

Oh, life is a glorious cycle of song,
  A medley of extemporanea;
And love is a thing that can never go wrong;
  And I am Marie of Rumania.

Resume

Razors pain you;
  Rivers are damp;
Acids stain you;
  And drugs cause cramp.
Guns aren't lawful;
  Nooses all give;
Gas smells awful;
  You might as well live.

Social Note

Lady, lady, should you meet
One whose ways are all discreet,
One who murmurs that his wife
Is the lodestar of his life,
One who keeps assuring you
That he never was untrue,
Never loved another one . . .
Lady, lady, better run!

Coda

There's little in taking or giving,
  There's little in water or wine;
This living, this living, this living
  Was never a project of mine.
Oh, hard is the struggle, and spare is
  The gain of the one at the top,
For art is a form of catharsis,
  And love is a permanent flop,
And work is the province of cattle,
  And rest's for a clam in a shell,
So I'm thinking of throwing the battle—
  Would you kindly direct me to hell?

H. L. Mencken

H. L. Mencken (1880–1956) was both the historian and re-inventor of the American language. Here, inimitably, he describes a Presidential hopeful in the 1920s, the California Republican, Hiram Johnson: "Almost the ideal candidate—an accomplished boob-bumper, full of the sough and gush of the tin-horn messiah, and yet safely practical. He will give a good show if he is elected." Mencken drew on an immense vocabulary of 25,000 words, salting his paragraphs with slang, arch Germanisms (Kultur, Herr Professor-Doktor, Polizei) *and mock honorifics (the Archangel Woodrow, Calvin Coolidge, L.L.D., and Roosevelt Major and Minor). His Tory anarchism and joyous pugnacity made him, in Lippmann's words, "the most powerful personal influence on this whole generation of educated people."*

At twenty-five, Mencken was already editor of the Baltimore Herald; *when the paper failed in 1906, he moved to* The Sunpapers, *where he spent most of his working days. His early journalism is collected in* The Young Mencken *(New York, 1973), edited by Carl Bode, which includes samples of his "Free Lance" column, from 1911–15. In 1920 he launched his celebrated "Monday political articles," writing some seven hundred over several decades; a selection can be found in* A Carnival of Buncombe *(Baltimore, 1956), edited by Malcolm Moos. His columns for the* Chicago Tribune *from 1924 to 1927 are gathered in* The Bathtub Hoax *(New York, 1958), edited by Robert McHugh. His final pieces, on the 1948 election, are in* Mencken's Last Campaign *(Washington, 1976), edited by Joseph C. Goulden.*

THE TWO ENGLISHES

Baltimore Evening Sun *October 10, 1910*

Prof. Brander Matthews, that learned man, has been writing of late upon his favorite topic, the mutability of language. Words change and their meanings change. Idioms decay, dry up and blow away. Grammatical forms give birth to new grammatical forms. The same verb is conjugated differently in different ages, in different countries, on different sides of the street. The English people speak an English which differs enormously, in vocabulary and idiom, from the English spoken by Americans.

In proof of this last fact Professor Matthews cites the word university, which means one thing in England and quite a different thing in the United States—and yet other things in Germany and France. An English university would be called a group of colleges in the United States and a lunatic asylum in Germany. A German university, if set up among us, would be nominated a school of vice, a Chautauqua or an anti-Young Men's Christian Association. But enough of such subtleties. There are plenty of other words, more familiar to all of us, which show the great gaps which separate languages, and particularly the English and American languages.

Consider the common word shoe. In the United States it means—a shoe. In England it means a slipper. The thing we call a shoe is known in England as a boot, and the thing we call a boot is known there as a Blucher. Their shoe is our slipper. "Ladies' boots" is a familiar sign in London. They call shoe polish boot blacking.

To an Englishman all forms of grain are corn. The Corn laws were laws which placed a prohibitive duty upon foreign wheat. The thing we call corn is maize in England. In the English vocabulary a trolley car is a tram-car, a parlor is a drawing room (parlor, among the Britons, means the back room of a saloon), a saloon is a public house or "pub," a saloonkeeper is a licensed victualer, molasses is treacle, a locomotive engineer is an engine driver, a railway brakeman is a guard, a Methodist church is a chapel, the tin roof of a house is the leads, a trained nurse is a nursing sister, a newspaper reporter is a journalist, a chief of police is a chief constable and railway switches are points.

Even when the same word appears in both the English and American languages it is often spelled differently. The English word for jail, for example, is gaol. It is pronounced exactly like jail. So with the English word for tire, which is pronounced tire and spelled tyre. So with check, which is spelled cheque. They put an extra and useless "u" into labor, honor, harbor, color and ardor. They spell wagon with two "g's." They use an "x" instead of "ct" in such words as connection. They hang on, like grim death, to all superfluous "e's."

Their forms of address vary greatly from ours. They call a judge, not your Honor, but Your Worship or Your Lordship. A mayor is also Your Worship. A physician is always Doctor, as with us, but he is often without an M.D. degree. A surgeon, unless he has notoriously received an M.D., is always plain Mister—and it is seldom that he has an M.D. A nurse is addressed, not as Miss McGinnis, but as Nurse. A cook is called, not Maggie, but Cook. A ladies' maid is called, not Myrtle, but Brown or Simpson or Sweeny or whatever her surname happens to be. In England no preacher is called Doctor unless he is an actual doctor of divinity. The American custom of addressing dancing masters, magicians, oyster shuckers, school teachers and acrobats as professor is unknown.

Among the simple-minded barbarians over there a typewriter of either sex is known as a typist, a pass to the theatre is an order, the city editor of a newspaper is the chief reporter, the financial editor is the city editor, with a capital "C," a hardware dealer is an ironmonger, a delicatessen dealer is an Italian warehouseman, a dry goods merchant is a draper, a shoemaker is a bootmaker, a saleswoman is a lady clerk or shop assistant, the president of a corporation is its chairman, the corporation itself is a public company, a party platform is a constitution, a candidate stands instead of running for office, the orchestra seats in a theatre are stalls, the orchestra circle is the pit, a letter carrier is a postman, a letter box is a pillar box, keeping company is walking out, a poorhouse is a workhouse, instead of studying medicine a medical student walks the hospitals, a fish dealer is a fishmonger, a locomotive fireman is a stoker, a dishwasher is a slavey, coal is coals, the room clerk of a hotel is the secretary, a round-trip ticket is a return ticket, a policeman is a constable and an apple pie is an apple tart.

There are many English words in common use across the ocean which have never become naturalized in America. The word chambers is an example. It means a bachelor's lodgings. The word don is another. It means a college dignitary. A host of ecclesiastical terms familiar to all Englishmen are seldom heard in this country. Of such sort are vicar, curate, canon, verger and prebendary. An usher over there means an assistant teacher. The principal of a school is called the head master or head mistress. Such English terms as subaltern, civil servant and moor are practically unknown in the United States. But we have adopted weekend, jam, tub and music hall.

The Englishman is very careful to restrict the application of the word gentleman to men of some consideration, but he uses lady indiscriminately. Advertisements for lady typists, lady clerks and lady teachers fill the English papers. Barristers and solicitors differ vastly in England, both in function and in public estimation, but among us there is no difference. An American lawyer may call himself advocate, barrister, solicitor, attorney or counselor, as he pleases. Candy, to the Britisher, is sweets, the fire department is the fire brigade, a floorwalker is a shopwalker, a store is a shop (we have begun to borrow the English term) and a warehouse is a store; oatmeal is porridge, boards are deals, a railway car is a carriage, freight is goods, a sidewalk is a footway, a subway is a tube, vegetables are greens, a street cleaner is a crossing sweeper, a job holder is a public servant, a railroader is a railway servant, store fixtures are shop fittings, lumber is timber, flathouses are mansions, window panes are lights, a drummer is a bagman or traveler, baggage is luggage, a trunk is a box, a satchel is a bag, a druggist is a chemist or apothecary and a doctor's office is his consulting room. The English have no bartenders or mixologists. Barmaids do the work.

GAMALIELESE

Baltimore Sun *March 7, 1921*

On the question of the logical content of Dr. Harding's harangue of last Friday I do not presume to have views. The matter has been debated at great length by the editorial writers of the Republic, all of them experts in logic; moreover, I confess to being prejudiced. When a man arises publicly to argue that the United States entered the late war because of a "concern for preserved civilization," I can only snicker in a superior way and wonder why he isn't holding down the chair of history in some American university. When he says that the United States has "never sought territorial aggrandizement through force," the snicker arises to the virulence of a chuckle, and I turn to the first volume of General Grant's memoirs. And when, gaining momentum, he gravely informs the boobery that "ours is a constitutional freedom where the popular will is supreme, and minorities are sacredly protected," then I abandon myself to a mirth that transcends, perhaps, the seemly, and send picture postcards of A. Mitchell Palmer and the Atlanta Penitentiary to all of my enemies who happen to be Socialists.

But when it comes to the style of a great man's discourse, I can speak with a great deal less prejudices, and maybe with somewhat more competence, for I have earned most of my livelihood for twenty years past by translating the bad English of a multitude of authors into measurably better English. Thus qualified professionally, I rise to pay my small tribute to Dr. Harding. Setting aside a college professor or two and a half a dozen dipsomaniacal newspaper reporters, he takes the first place in my Valhalla of literati. That is to say, he writes the worst English that I have ever encountered. It reminds me of a string of wet sponges; it reminds me of tattered washing on the line; it reminds me of stale bean-soup, of college yells, of dogs barking idiotically through endless nights. It is so bad that a sort of grandeur creeps into it. It drags itself out of the dark abysm (I was about to write abscess!) of pish, and crawls insanely up the topmost pinnacle of posh. It is rumble and bumble. It is flap and doodle. It is balder and dash.

But I grow lyrical. More scientifically, what is the matter with it? Why does it seem so flabby, so banal, so confused and childish, so stupidly at war with sense? If you first read the inaugural address and then heard it intoned, as I did (at least in part), then you will perhaps arrive at an answer. That answer is very simple. When Dr. Harding prepares a speech he does not think it out in terms of an educated reader locked up in jail, but in terms of a great horde of stoneheads gathered around a stand. That is to say, the thing is always a stump speech; it is conceived as a stump speech and written as a stump speech. More, it is a stump speech addressed primarily to the sort of audience that the speaker has been used to all his life, to wit, an audience of small town yokels, of low political serfs, or morons scarcely able to understand a word of more than two syllables, and wholly unable to pursue a logical idea for more than two centimeters.

Such imbeciles do not want ideas—that is, new ideas, ideas that are unfamiliar, ideas that challenge their attention. What they want is simply a gaudy series of platitudes, of threadbare phrases terrifically repeated, of sonorous nonsense driven home with gestures. As I say, they can't understand many words of more than two syllables, but that is not saying that they do not esteem such words. On the contrary, they like them and demand them. The roll of incomprehensible polysyllables enchants them. They like phrases which thunder like salvos of artillery. Let that thunder sound, and they take all the rest on trust. If a sentence begins furiously and then peters out into fatuity, they are still satisfied. If a phrase has a punch in it, they do not ask that it also have a meaning. If a word slides off the tongue like a ship going down the ways, they are content and applaud it and wait for the next.

Brought up amid such hinds, trained by long practice to engage and delight them, Dr. Harding carries over his stump manner into everything he writes. He is, perhaps, too old to learn a better way. He is, more likely, too discreet to experiment. The stump speech, put into cold type, maketh the judicious to grieve. But roared from an actual stump, with arms flying and eyes flashing and the old flag overhead, it is certainly and brilliantly effective. Read the inaugural address, and it will gag you. But hear it recited through a sound-magnifier, with grand gestures to ram home its periods, and you will begin to understand it.

Let us turn to a specific example. I exhume a sentence from the latter half of the eminent orator's discourse:

> I would like government to do all it can to mitigate, then, in understanding, in mutuality of interest, in concern for the common good, our tasks will be solved.

I assume that you have read it. I also assume that you set it down as idiotic—a series of words without sense. You are quite right; it is. But now imagine it intoned as it was designed to be intoned. Imagine the slow tempo of a public speech. Imagine the stately unrolling of the first clause, the delicate pause upon the word "then"—and then the loud discharge of the phrase "in understanding," "in mutuality of interest," "in concern for the common good," each with its attendant glare and roll of the eyes, each with its sublime heave, each with its gesture of a blacksmith bringing down his sledge upon an egg—imagine all this, and then ask yourself where you have got. You have got, in brief, to a point where you don't know what it is all about. You hear and applaud the phrases, but their connection has already escaped you. And so, when in violation of all sequence and logic, the final phrase, "our tasks will be solved," assaults you, you do not notice its disharmony—all you notice is that, if this or that, already forgotten, is done, "our tasks will be solved." Whereupon, glad of the assurance and thrilled by the vast gestures that drive it home, you give a cheer.

That is, if you are the sort of man who goes to political meetings, which is to say, if you are the sort of man that Dr. Harding is used to talking to, which is to say, if you are a jackass.

The whole inaugural address reeked with just such nonsense. The thing started off with an error in English in its very first sentence—the confusion of pronouns in the *one-he* combination, so beloved of bad newspaper reporters. It bristled with words misused: *Civic* for *civil, luring* for *alluring, womanhood* for *women, referendum* for *reference,* even *task* for *problem.* "The *task* is to be *solved*"—what could be worse? Yet I find it twice. "The expressed views of world opinion—" what irritating tautology! "The expressed conscience of progress"—what on earth does it mean? "This is not selfishness, it is sanctity"—what intelligible idea do you get out of that? "I know that Congress and the administration will favor every wise government policy to aid the resumption and encourage continued progress"—the resumption of what? "Service is the supreme *commitment* of life"—*ach, du heiliger!*

But is such bosh out of place in a stump speech? Obviously not. It is precisely and thoroughly in place in a stump speech. A tight fabric of ideas would weary and exasperate the audience; what it wants is simply a loud burble of words, a procession of phrases that roar, a series of whoops. This is what it got in the inaugural address of the Hon. Warren Gamaliel Harding. And this is what it will get for four long years—unless God sends a miracle and the corruptible puts on incorruption. . . . Almost I long for the sweeter song, the rubber-stamps of more familiar design, the gentler and more seemly bosh of the late Woodrow.

WHY TRUMAN WON

Baltimore Sun *November 7, 1948*

The super- or ultra-explosion that staved in the firmament of heaven last Tuesday not only blew up all the Gallups of this great free Republic; it also shook the bones of all its other smarties. Indeed, I confess as much myself. Sitting among my colleagues of the *Sunpapers* as the returns began to come in, I felt and shared the tremors and tickles that ran up and down their vertebrae.

How could so many wizards be so thumpingly wrong? How could the enlightenment play so scurvy a trick upon its agents? Certainly there must have been very few Americans in the IQ brackets above 35 who actually expected Truman to win. I met a great many politicos during the campaign, some of whom whooped for the right hon. gentleman and some of whom whooped against him, and yet I can't recall a single one who showed any sign of believing that he could beat the rap.

Even his running-mate, the Hon. Alben W. Barkley, who hollered in Baltimore on October 22, was plainly full of doubts. A professional jobholder and rabble-rouser for 43 years, he did his stuff as in duty bound, but if there were any high hopes welling within him he certainly kept them concealed from the customers. His general air was that of a country pastor preaching the funeral services of a parishioner plainly bound for hell, and all the other jobholders present looked and acted a good deal more like pallbearers than wedding guests.

But, meanwhile, the Missouri Wonder was roving and ravaging the land, pouring out hope and promise in a wholesale manner. To the farmers he promised a continuance of the outrageous prices that are reducing the rest of us to eating only once a day. To the city proletariat he promised ever higher and higher wages, with the boss a beggar at his own gate. To the jobholders he promised more and more jobs, and juicier ones. To the colored brethren he promised the realization of their fondest hopes and hallucinations. And so on to the glorious end of the chapter.

What had Dewey to offer against all this pie in the sky? Virtually nothing. His plan in the campaign, as in that of 1944, was to chase what appeared to be the other fellow's ambulance. He seemed eager to convince everyone that he was for everything that Truman was in favor of, but with much less heat. Never once in his canvass, so far as I can recall, did he tackle Truman's buncombe and blather in a frank and forthright manner. His speeches were beautiful songs, but all of them were sung *pianissimo*.

His literati have been blamed for this tender gurgling, but it seems to me that the fault was all his own. The late Al Smith, in the campaign of

1928, was afflicted by literati even more literary, but he got rid of them by tearing up their speeches and striking out on his own. To be sure, he didn't win, but that was surely not because he did not make a good fight; it was simply because the Bible searchers everywhere had become convinced that if he got to the White House the Pope would move into the cellar.

Dewey had no such handicap and yet he came to grief in the grand manner. His defeat ran against all the probabilities and was complete, colossal and ignominious. Its springs, I believe, are to be sought in defects of his own personality. He is by nature cute but cautious. He is a good trial lawyer, but an incompetent rabble-rouser. He addresses great multitudes as if they were gangs of drowsing judges, all of them austere in their hangmen's gowns, but consumed inwardly by an expectant thirst.

Truman made no such mistake. He assumed as a matter of course that the American people were just folks like himself. He thus wasted no high-falutin rhetoric upon them, but appealed directly to their self-interest. Every one of them, he figured, was itching for something, and he made his campaign by the sempiternal device of engaging to give it to them. A politico trained in a harsh but realistic school, he naturally directed his most gaudy promises to the groups that seemed to be most numerous, and the event proved that he was a smart mathematician.

Neither candidate made a speech on the stump that will survive in the schoolbooks, but those of Truman at least had some human warmth in them. Like Al Smith, he frequently disregarded the efforts of his literati, and proceeded on his own, and it was precisely at such times that he was most effective. While Dewey was intoning essays sounding like the worst bombast of university professors, Truman was down on the ground, clowning with the circumambient morons. He made votes every time he gave a show, but Dewey lost them.

One of the most significant phenomena of the campaign was the collapse of Henry Wallace's effort to convert the plain people to the Russian whim-wham. They simply refused to be fooled, and if the battle had gone on for another month his support would have been reduced to the Communists. At the start Truman was plainly afraid of him, but only at the start. It soon appeared clearly that all the actual Progressives behind Wallace were made uneasy by the Communists' collaring of him, and in the end most of them were fetched by Truman.

He was out to get all the softheads, and he got them triumphantly. Unhampered by anything resembling a coherent body of ideas, he was ready to believe up to the extreme limits of human credulity. If he did not come out for spiritualism, chiropractic, psychotherapy and extra sensory perception it was only because no one demanded that he do so. If there had been any formidable body of cannibals in the country he would have promised to provide them with free missionaries fattened at the taxpayers' expense.

So we now have him for four years more—four years that will see the country confronted by the most difficult and dangerous problems presented to it since 1861. We can only hope that he will improve as he goes on. Unhappily, experience teaches that no man improves much after 60, and that after 65 most of them deteriorate in a really alarming manner. I could give an autobiographical example, but refrain on the advice of counsel. Thus we seem to be in for it. I can only say in conclusion that the country jolly well deserves it.*

*A week later, Menchen wrote his final article, an attack on the Ku Klux Klan, before being disabled by a stroke.

Will Rogers

"America is a great country, but you can't live in it for nothing." Will Rogers (1879–1935) heeded his adage. Lack of funds was rarely a problem from the year of his breakthrough, 1902, when he joined a Wild West Circus as "The Cherokee Kid," until his death in a plane crash with the aviator Wiley Post, in Point Barrow, Alaska. His aw-shucks slouch was for the stage; Rogers was a beaver of industry, and perhaps earned more than any columnist of his time.

He came to journalism out of vaudeville and the Follies, and succeeded by teasing his Oklahoma drawl into newsprint. In 1922, Rogers wrote the first of 667 weekly articles, syndicated to some 350 newspapers. Four years later, he began his daily "telegrams," a few paragraphs headed "Will Rogers Says." An industrious curator of Rogers's works, Bryan Sterling, calculates that he wrote 2,861 daily squibs and 511 columns headed "The Worst Story I've Heard Today," which, together with the weekly column, a half-dozen books, stacks of magazine pieces, and sixty-nine radio broadcasts, totals at least two million words. He also made twenty-one sound films, lectured in four hundred cities, and wrote and kept updating his part in a Broadway show, Three Cheers. *The column that follows appears in Richard M. Ketchum,* Will Rogers: His Life and Times *(New York, 1973). The Oklahoma State University Press has published* Will Rogers' Weekly Articles 1929–1931 *(1981), Steven K. Gragart, ed.; and* Will Rogers' Daily Telegrams 1926–1929 *(1978), preface by Will Rogers, Jr.*

BATTING FOR LLOYD GEORGE

New York Times *December 24, 1922*

I want to apologize and set the many readers of *The Times* straight as to why I am blossoming out as a weekly infliction on you all.

It seems *The Times* has Lloyd George signed up for a pack of his Memoirs. Well, after the late election Lloyd couldn't seem to remember anything, so they sent for me to fill in this space where he would have had his junk.

You see, they wanted me in the first place, but George came along and offered to work cheaper, and also to give to his charity. That benevolence on his part was of course before England gave him two weeks' notice.

Now I am also not to be outdone by an ex-Prime Minister donating my recipients from my Prolific Tongue to a needy charity. The total share of this goes to the civilization of three young heathens, Rogers by name, and part Cherokee Indian by breeding.

Now, by wasting seven minutes if you are a good reader—and ten to twelve if you read slow—on me every Sunday, you are really doing a charitable act yourself by preventing these three miniature bandits from growing up in ignorance. So please help a man with not only one little Megan, but three little Megans.

A great many people many think this is the first venture of such a conservative paper as *The Times* in using something of a semi-humorous nature, but that is by no means the case. I am following the Kaiser, who rewrote his life after it was too late. I realize what a tough job I have, succeeding a man who to be funny only has to relate the facts.

Please don't consider these my memoirs. I am not passing out of the picture, as men generally do who write those things. I want to warn you of a few pitfalls into which our poorly paid but highly costing politicians are driving us daily.

We pay an awful lot of dough in the course of a year to try to get our country run in such shape that a certain per cent of our citizens can keep out of the poorhouse. The shape we are in now, over and above all the taxes we pay, allows us to hang to about 8 per cent of our earnings.

Now, that's entirely too rich we are getting—too prosperous. So they are talking of lending Europe about a billion and a half more. I knew there would be something stirring when Morgan visited Washington last week.

He goes down once every year and lays out the following year's program.

Europe owes us now about eleven billions. Lending them another billion and a half would make it just even 12.50. You see it so much easier to figure the interest on 12.50 than 11. Of course the interest ain't going to be paid, but it's got to be figured.

The Government could charge it off on their income tax to publicity. I only hope one thing, and that is, if we make the loan, Europe will appreciate this one.

The Allied Debt Conference broke up last week in London.

It's getting harder every day for Nations to pay each other unless one of them has some money.

They called that an Economic Conference, and, as we didn't attend, it was. Why don't somebody lend Germany the money so they can pay France what France owes England so England can pay the money to lend Germany to pay France?

It only needs somebody to start it.

Senator Borah opened up and told the U.S. what he thought of this loan. For speaking right out in church he is the Clemenceau of America.

They are bringing over Ambassador Harvey. He don't know anything about it; over there he has been too busy learning speeches. If they don't have a Concert on the Ship coming over, his trip will be spoiled.

I see they have been holding another Peace Conference in some burg called Lausanne. They are having those things just like Chatauquas—you jump from one to the other.

This one must have been somewhere in Italy, as that is the stopping place of the Ambassador that we sent there. He didn't go officially as we don't belong to the League of Nations (we only finance it). Well, this fellow Child, as I say, he went as a kind of Uninstructed Delegate. He got into the game, but his efforts were more like a cheerleader at a football game. They heard him, but he had no direct effect on the game.

It seems that the Allies (that is, those of them that are speaking to each other) wanted Turkey to promise to protect the minor nations within her territory. Now this Turkey is a pretty foxy nation; she's got on her mind something besides wives and cigarettes.

Turkey says: "We'll agree to give minor nations the same protection that you all give yours."

Well, that was not exactly what the Allies wanted, but they took it as a compromise and hope at some future time to get *full* protection for them.

[Mr. Will Rogers promises further contributions to the *New York Times,* if he can tame his typewriter.]

Milt Gross

In the final years of the New York World, *readers turned reflexively to the dementedly comic work of Milt Gross (1895–1953), rendered in an American Yiddish dialect. Raised in the Bronx, Gross at twelve was a copy boy on Hearst's* New York American, *where he soon moved to the art department. In 1922, he joined the* World, *developed two comic strips, ''Banana Oil'' and ''Count Screwloose,'' then found his métier with Mrs. Feitlebaum, usually pictured feeding her baby ''utmeal,'' ''crembarry suss,'' or ''shradded whit'' while telling her tales. The feature died with the* World, *but is preserved in* Nize Baby *(New York, 1926),* Dunk Esk!! *(New York, 1927), and* Hiawatta, Witt No Odder Poems *(New York 1928).*

THE STURRY FROM POCAHANTAS

New York World *1925*

Oohoo, Nize baby, itt opp all de shradded whit, so momma'll gonna tell you a sturry from Pocahantas witt Keptain John Smeet. Wance oppon a time was a Hindian Chiff from a whole tripe from Hindians, wot he hed it a dudder wot her name was Pocahantas! Hm! sotch a byooty wot she was, witt a grazeful forum witt a feegure witt raving bleck hair, witt a holive skeen witt fleshing heyes—mmmmm—a ragular pitch!! (Nize baby, take annodder spoon shradded whit.) So in de minntime, it came over from Hingland a band from oily tsettlers wot dey called dem columnists. So di lidder from de columnists was a Ganeral from a name from Keptain John Smeet.

Pot II. So one hefternoon he was taking a leedle strull in de woots, so it was lurching dere in a hambush dem doidy Hindians so, so soon wot he pessed by dey gave queeck a yell, "Hends opp!—odder we'll cot you off de scallop wid a tommyhuck!" So he saw wot it was against him de hodds so he compiled gredually witt de requast!!!

Pot III. So dey brut him in de front from de tripe so de chiff witt de Sockems from de tripe held a mitting, wot should be from him. So de chiff sad, "Wot kind beezness you in??" So he replite, "I'm a columnist wot we tsettling de colonizz!!!" So de chiff sad, "Hm! Trapezing on priwate property, ha?? So for diss, we'll gonna chop you off de had!!" So jost dey leefted opp de hexes it should chop him off de had—it jomped opp gredually Pocahantas, wot she sad, "Cis!!!"

So de chiff sad, "Wot's de rizzon we should ciss?"

So Pocahantas sad, "I dun't want wot you should do diss didd!!"

So de chiff sad, "Somebody esked you for a hopinion?? Why you don't want??"

So Pocahantas sad, "Jost bickuss!!!"

So de chiff sad, "Aha!! You took already maybe a lightning to diss goot-for-notting, ha?? Hm! Sotch hideas! Is here a whole tripe from brave witt nubble worriers wot itch one it would make a idill hosband so you got to fall yat for a dope wot he billives wot Sitting Bull is a cow's hosband!! I tink wot I'll make him run yet a gimlet foist!"

So Pocahantas said, "I luff him."

So de chiff sad, "Hm! A son-in-law I nidd yat wot he wears a monologue in de heye!! Go beck queeck in de weeckwam und don't meex opp in mind beezness odder you'll gat from me wid a strep."

So Pocahantas sad, "If you'll wouldn't lat him go I'll tell momma wot it was going on lest night in de Hell Fay Clob witt you witt dot wemp from a Minnie-Ha-Ha!! Hm, you gatting pale, ha?? Noo, so wot'll gonna be? Queeck!"

So de chiff said, "Blast you, mine cheeldren!!!"

So dey gredually got merried so in a shut time de chiff was a grenpapa from a leedle caboose! (Hm!! Sotch a dollink baby ate opp all de shradded whit!!)

Ben Hecht

Ben Hecht (1894–1964) was a journalist, novelist, Hollywood writer-producer, and militant Zionist. Yet his memory is kept alive by the endlessly revived and recast comedy he wrote with Charles MacArthur, The Front Page, *deriving from their experience in Chicago journalism. The play opened on Broadway August 14, 1928, as the rowdy, "gimme rewrite!" tradition was already fading. Hecht joined the* Chicago Daily News *in 1910, the noontime of the city's literary renaissance. Harriet Monroe was launching* Poetry, *and among local presences were Carl Sandburg (also of the* Daily News*), Sherwood Anderson, Vachel Lindsay, and Edgar Lee Masters. In 1921, having finished his then-daring novel,* Erik Dorn, *Hecht proposed writing a daily short story in the* News, *to which the editor, Henry Justin Smith, agreed. The series, "One Thousand and One Afternoons," was published in a book of the same title (New York, 1922); a second selection followed,* Broken Necks *(Chicago, 1926). Years later, Hecht revived the idea for the experimental New York afternoon tabloid,* PM, *the results collected in* 1001 Afternoons in New York *(New York, 1941). Two tales from the first series follow. Hecht's autobiography,* A Child of the Century *(New York, 1954), is notable for its liveliness, length, and its one-sentence reference to* The Front Page.

DON QUIXOTE'S LAST WINDMILL

Chicago Daily News   *Circa 1921*

Sherwood Anderson, the writer, and I were eating lunch in the back room of a saloon. Against the opposite wall sat a red-faced little man with an elaborate mustache and a bald head and happy grin. He sat alone at a tilted round table and played with a plate of soup.

"Say, that old boy over there is trying to wigwag me," said Anderson. "He keeps winking and making signs. Do you know him?"

I looked and said no. The waiter appeared with a box of cigars.

"Mr. Sklarz presents his compliments," said the waiter, smiling.

"Who's Sklarz?" Anderson asked, helping himself to a cigar. The waiter indicated the red-faced little man. "Him," he whispered.

We continued our meal. Both of us watched Mr. Sklarz casually. He seemed to have lost interest in his soup. He sat beaming happily at the walls, a contagious elation about him. We smiled and nodded our thanks for the cigars. Whereupon after a short lapse, the waiter appeared again.

"What'll you have to drink, gentlemen?" the waiter inquired.

"Nothing," said Anderson, knowing I was broke. The waiter raised his continental eyebrows understandingly.

"Mr. Sklarz invites you, gentlemen, to drink his health—at his expense."

"Two glasses," Anderson ordered. They were brought. We raised them in silent toast to the little red-faced man. He arose and bowed as we drank.

"We'll probably have him on our hands now for an hour," Anderson frowned. I feared the same. But Mr. Sklarz reseated himself and, with many head-bowings in our direction, returned to his soup.

"What do you make of our magnanimous friend?" I asked. Anderson shrugged his shoulders.

"He's probably celebrating something," he said. "A queer old boy, isn't he?"

The waiter appeared a third time.

"What'll it be, gentlemen?" he inquired, smiling. "Mr. Sklarz is buying for the house."

For the house. There were some fifteen men eating in the place. Then our friend, despite his unassuming appearance, was evidently a creature of wealth! Well, this was growing interesting. We ordered wine again.

"Ask Mr. Sklarz if he will favor us by joining us at our table for this drink," I told the waiter. The message was delivered. Mr. Sklarz arose and bowed, but sat down again. Anderson and I beckoned in pantomime. Mr. Sklarz arose once more, bowed and hesitated. Then he came over.

As he approached a veritable carnival spirit seemed to deepen around us. The face of this little man with the elaborate black mustache was violent with suppressed good will and mirth. He beamed, bowed, shook hands and sat down. We drank one another's health and, as politely as we could, pressed him to tell us the cause for his celebration and good spirits. He began to talk.

He was a Russian Jew. His name was Sklarz. He had been in the Russian army years ago. In Persia. From a mountain in Persia you could see three great countries. In Turkey he had fought with baggy-trousered soldiers and at night joined them when they played their flutes outside the coffeehouses and sang songs about women and war. Then he had come to America and opened a box factory. He was very prosperous and the factory in which he made boxes grew too small.

So what did he do but take a walk one day to look for a larger factory. And he found a beautiful building just as he wanted. But the building was too beautiful to use for a factory. It should be used for something much nicer. So what did he do then but decide to open a dance hall, a magnificent dance hall, where young men and women of refined, fun-loving temperaments could come to dance and have fun.

"When does this dance hall open?" Anderson asked. Ah, in a little while. There were fittings to buy and put up first. But he would send us special invitations to the opening. In the meantime would we drink his health again? Mr. Sklarz chuckled. The amazing thing was that he wasn't drunk. He was sober.

"So you're celebrating," I said. Yes, he was celebrating. He laughed and leaned over the table toward us. His eyes danced and his elaborate mustache made a grotesque halo for his smile. He didn't want to intrude on us with his story, but in Persia and Turkey and the Urals he had found life very nice. And here in Chicago he had found life also very nice. Life was very nice wherever you went. And Anderson quoted, rather imperfectly, I thought:

> Oh, but life went gaily, gaily
> In the house of Idah Dally;
> There were always throats to sing
> Down the river bank with spring.

Mr. Sklarz beamed.

"Yes, yes," he said, "down the river benk mit spring." And he stood up and bowed and summoned the waiter. "See vat all the gentlemen vant," he ordered, "and give them vat they vant mit my compliments." He laughed, or, rather, chuckled. "I must be going. Excuse me," he exclaimed with a quick bow. "I have other places to call on. Good-by. Remember me—Sam Sklarz. Be good—and don't forget Sam Sklarz when there are throats to zing down the river benk mit spring."

We watched him walk out. His shoulders seemed to dance, his short legs moved with a sprightly lift.

"A queer old boy," said Anderson. We talked about him for a half-hour and then left the place.

Anderson called me up the next morning to ask if I had read about it in the paper. I told him I had. A clipping on the desk in front of me ran:

Sam Sklarz, forty-six years old and owner of a box factory on the West Side, committed suicide early this morning by jumping into the drainage canal. Financial reverses are believed to have caused him to end his life. According to friends he was on the verge of bankruptcy. His liabilities were $8,000. Yesterday morning Sklarz cashed a check for $700, which represented the remains of his bank account, and disappeared. It is believed that he used the money to pay a few personal debts and then wandered around in a daze until the end. He left no word of explanation behind.

THE GREAT TRAVELER

Chicago Daily News *Circa 1921*

Alexander Ginkel has been around the world. A week ago he came to Chicago and, after looking around for a few days, located in one of the less expensive hotels and started to work as a porter in a well-known department store downtown.

A friend said, "There's a man living in my hotel who should make a good story. He's been around the world. Worked in England, Bulgaria, Russia, Siberia, China and everywhere. Was cook on a tramp steamer in the South Seas. A remarkable fellow, really."

In this way I came to call on Ginkel. I found him after work in his room. He was a short man, over thirty, and looked uninteresting. I told him that we should be able to get some sort of story out of his travels and experiences. He nodded.

"Yes," he said, "I've been all around the world."

Then he became silent and looked at me hopefully.

I explained, "People like to read about travelers. They sit at home themselves and wonder what it would be like to travel. You probably had a lot of experiences that would give people a vicarious thrill. I understand you were a cook on a tramp steamer in the South Seas."

"Oh, yes," said Ginkel, "I've been all over. I've been around the world."

We lighted pipes and Ginkel removed a book from a drawer in the dresser. He opened it and I saw it was a book of photographs—mostly pictures taken with a small camera.

"Here are some things you could use," he said. "You wanna look at them."

We went through the pictures together.

"This one here," said Ginkel, "is me in Vladivostok. It was taken on the corner there."

The photograph showed Ginkel dressed just as he was in the hotel room, standing near a lamppost on a street corner. There was visible a part of a store window.

"This one is interesting," said Ginkel, warming up. "It was taken in the archipelago. You know where. I forget the name of the town. But it was in the South Seas."

We both studied it for a space. It showed Ginkel standing underneath something that looked like a palm tree. But the tree was slightly out of focus. So were Ginkel's feet.

"It is interesting," said Ginkel, "But it ain't such a good picture. The lower part is kind of blurred, you notice."

We looked through the album in silence for a while. Then Ginkel suddenly remembered something.

"Oh, I almost forgot," he said. "There's one I think you'll like. It was taken in Calcutta. You know where. Here it is."

He pointed proudly toward the end of the book. We studied it through the tobacco smoke. It was a photograph of Ginkel dressed in the same clothes as before and standing under a store awning.

"There was a good light on this," said Ginkel, "and you see how plain it comes out."

Then we continued without comment to study other photographs. There were at least several hundred. They were all of Ginkel. Most of them were blurred and showed odds and ends of backgrounds out of focus, such as trees, streetcars, buildings, telephone poles. There was one that finally aroused Ginkel to comment:

"This would have been a good one, but it got light struck," he said. "It was taken in Bagdad."

When we had exhausted the album Ginkel felt more at ease. He offered me some tobacco from his pouch. I resumed the original line of questioning.

"Did you have any unusual adventures during your travels or did you get any ideas that we could fix up for a story?" I asked.

"Well," said Ginkel, "I was always a camera bug, you know. I guess that's what gave me the bug for traveling. To take pictures, you know. I got a lot more than these, but I ain't mounted them yet."

"Are they like the ones in the book?"

"Not quite so good, most of them," Ginkel answered. "They were taken when I hadn't had much experience."

"You must have been in Russia while the revolution was going on, weren't you?"

"Oh, yes. I got one there." He opened the book again. "Here," he said, "This was in Moscow. I was in Moscow when this was taken."

It was another picture of Ginkel slightly out of focus and standing against a store front. I asked him suddenly who had taken all the pictures.

"Oh, that was easy," he said. "I can always find somebody to do that. I take a picture of them first and then they take one of me. I always give them the one I take of them and keep the one they take of me."

"Did you see any of the revolution, Ginkel?"

"A lot of monkey business," said Ginkel. "I seen some of it. Not much."

The last thing I said was, "You must have come in for a lot of sights. We might fix up a story about that if you could give me a line on them." And the last thing Ginkel said was:

"Oh, yes, I've been around the world."

Heywood Broun

Born well-to-do in a New York brownstone, schooled at Harvard, and bred to privilege, Heywood Broun (1888–1939) devoted much of his life to the underdog. In "It Seems to Me" he fought for Sacco and Vanzetti and women's rights; he was the founding president of the American Newspaper Guild and a Socialist candidate for Congress in 1930. Before joining the New York World *as a columnist in 1921, Broun spent a decade as baseball writer, war correspondent, theater critic, and book reviewer. His column adorned the* World's *Op-Ed page until a quarrel with the proprietor, Ralph Pulitzer, over Broun's denunciation of Harvard's complicity in the execution on murder charges of two immigrant anarchists. He shifted in 1928 to the* Telegram, *later the* World-Telegram, *where his column appeared until shortly before his death. These are from the* Collected Edition of Heywood Broun *(New York, 1941), compiled by Heywood Hale Broun. There are two biographies, by Dale Kramer (New York, 1949) and Richard O'Connor (New York, 1975), and a Newspaper Guild memorial tribute,* Heywood Broun: As He Seemed to Us *(New York, 1940), continuity by Morris Watson and Ernest L. Meyer.*

HOW I BECAME A RED

The New Republic *November 17, 1937*

"Tell me, Mr. Broun," said the young lady to my right, "how did you happen to become a Red?"

Assuming that her question was sincere, I proceeded to tell her in my own words which, by a fortunate coincidence, happened to be just sufficient to fill one page of the *New Republic* when printed in this type.

The first person to tempt me to the left was a Vermont Republican named John Spargo who wrote mildly Marxian tracts years ago when I was very young. Had there been more violence in his views or in his presentation I might have scurried home, but he was as soft-spoken as any curate and he succeeded in sneaking up on me. Later, in college, Lippmann asked me to join the Socialist Club and, as I had failed to make either Gas House or Fly, I yielded to his blandishments.

But commitment was not yet complete. The way of retreat still lay open. It was Professor Carver of the Harvard faculty and Tris Speaker, centerfielder of the Boston Red Sox, who closed the door and left me locked in the hall of heresy. Possibly an assist should be scored for a particularly salubrious Spring which came to New England in 1908. The good bald economist gave a course which might be roughly described as "radical panaceas and their underlying fallacies.' Professor Carver introduced these crack-brained notions in the Fall and Winter semesters and then proceeded to demolish them in the Spring and early Summer.

Faithful to the Harvard tradition of fair play, the Professor gave the revolutionaries an ample amount of rope. Indeed, he did not undertake to state the case himself for the various aberrant philosophies, but invited a leader of each school of thought to tell all from his particular point of view. The soap boxes were carried openly into the classroom and mounted by the orators. Men with burning causes are generally more eloquent then cloistered dons and there was not a single visitor who could not talk rings around Professor Carver.

We had an anarchist, a socialist, a syndicalist, a single-taxer and a few other theorists whose special lines escape me now with the passing of the years. The single-taxer was a little on the dull side, but all the others sparkled. It may be that I was more susceptible than my fellow students, for I must report that I hit the sawdust trail at each and every lecture in the creation of a united front against the capitalist system.

In other courses it was my custom to draw starfish in my notebook and jot down reminders such as : "Be sure and get Osmond Fraenkel for the poker game tonight." But under Carver I took notes as long as the cavalcade of visiting firemen was on. Later I would peruse the book and find entries such as "Born 1839—died 1897," "unearned increase in land values," "P. J. Proudhon," "Five dollars ($5) with Horseface McCarthy Dartmouth doesn't beat us by two touchdowns." And I used to wonder what some of the notations could possibly mean. But I paid closer attention than ever until the last galoot had reached the shore and Carver went up into the pilot house.

At this point Spring broke through. It had been a tough Winter. Dartmouth did beat us by two touchdowns and Osmond Fraenkel, who was a mathematical shark, had just become wise to the fact that it doesn't pay to draw to inside straights. I had failed to make the *Harvard Crimson* after my third try.

But gone was the Winter of our discontent and now it was Professor Carver's turn to haul us back from the pie counter of the sky and put our feet again upon the good earth and the law of supply and demand.

Carver was at bat but, as luck would have it, so was Tris Speaker. Spring can come up like thunder in Massachusetts. One day all the elms of the yard may be cased in ice and the next you read that the Boston Red Sox are opening the season against the Athletics. Tris Speaker, just up from Texas league, was the particular star of the greatest outfield trio that has ever been gathered together.

Speaker and Duffy Lewis and Harry Hooper—that was a united front. For a little while I tried to give Professor Carver an even break. Naturally I missed his first blast for the cause of conservatism, because that particular lecture conflicted with the opening game. Still I did ask a fellow scholar what went on and he lent me his notebook. He seemed to have boiled the discourse down to the bone because all he had written was, "Adam Smith was Scotch and died before the industrial revolution—look this up in the library. He approached economics from the side of ethics and believed in laissez-faire, which means 'anything goes.' Has anybody here seen Kelly?"

I decided that in all fairness to my parents, who were paying for my education, I should attend some of the lectures myself and so in the beginning I made it a rigorous rule to give only three afternoons a week to the Red Sox and the other three to Professor Carver. But on a certain Wednesday, which was a baseball afternoon, Tris Speaker made a home run, two triples and a double. In addition he went all the way back to the flagpole in center field and speared a drive with his gloved hand while still on the dead run. Two innings later he charged in for a low liner and made the catch by sliding the last twenty feet on his stomach. Professor Carver couldn't do that. I had seen him stand on his head in the classroom but never did he slide on his stomach. Even at such times as he got to base it was my impression that he had profited by an error or made a lucky-handle hit. Tris Speaker was batting .348 and Carver wasn't hitting the size of his hat. At times he did dig his toes in and take his full cut at the ball, but never when I was around did he succeed in knocking socialism over the fence.

Adam Smith had died in 1790 and Professor Carver wasn't getting any younger. I got hold of *The Wealth of Nations* and learned that to Smith laissez-faire meant "the obvious and simple system of natural liberty." And so in the books I read no more then, and not very much later.

I just went completely laissez-faire and never missed any of the remaining games. The arguments for the radical theories I had heard, but I never got around to hearing the answers. Tris Speaker had thrown Carver out at the plate because the Professor forgot to slide. And I went out into the world the fervent follower of all things red, including the Boston Red Sox.

EVEN TO JUDAS

New York World-Telegram *December 24, 1938*

> "Last night before I went to sleep, I chanced to read in an evening paper a story by a columnist which appealed to me so much as a Christmas sermon that I am going to read to you from it. Here is his parable."*

We were sitting in a high room above the chapel and although it was Christmas Eve my good friend the dominie seemed curiously troubled. And that was strange, for he was a man extremely sensitive to the festivities of his faith.

The joys and sorrows of Jesus were not to him events of a remote past but more current and living happenings than the headlines in the newspapers. At Christmas he seems actually to hear the voice of the herald angels.

My friend is an old man, and I have known him for many years, but this was the first time the Nativity failed to rouse him to an ecstasy. He admitted to me something was wrong. "Tomorrow," he said, "I must go down into that chapel and preach a Christmas sermon. And I must speak of peace and good-will toward men. I know you think of me as a man too cloistered to be of any use to my community. And I know that our world is one of war and hate and enmity. And you, my young friend, and others keep insisting that before there can be brotherhood there must be the bashing of heads. You are all for good-will to men, but you want to note very many exceptions. And I am still hoping and praying that in the great love of God the final seal of interdiction must not be put on even one. You may laugh at me, but right now I am wondering about how Christmas came to Judas Iscariot."

It is the habit of my friend, when he is troubled by doubts, to reach for the Book, and he did so now. He smiled and said, "Will you assist me in a little experiment? I will close my eyes and you hold out the Bible to me. I will open it at random and run my fingers down a page. You read me the text which I blindly select."

I did as he told me and he happened on the twenty-sixth chapter of St. Matthew and the twenty-fourth verse. I felt sorry for him, for this was no part of the story of the birth of Christ but instead an account of the great betrayal.

"Read what it says," commanded the dominie. And I read: "Then Judas, which betrayed Him, answered and said, 'Master, is it I?' He said unto him, 'Thou has said.' "

*With this preface, Franklin D. Roosevelt read Broun's column on the president's annual Christmas broadcast.

My friend frowned, but then he looked at me in triumph. "My hand is not as steady as it used to be. You should have taken the lower part of my finger and not the top. Read the twenty-seventh verse. It is not an eighth of an inch away. Read what it says." And I read, "And He took the cup and gave thanks and gave it to them, saying, 'Drink ye all of it.' "

"Mark that," cried the old man exultantly. "Not even to Judas, the betrayer, was the wine of life denied. I can preach my Christmas sermon now, and my text will be 'Drink ye all of it.' Good-will toward men means good-will to every last son of God. Peace on earth means peace to Pilate, peace to the thieves on the cross, and peace to poor Iscariot."

I was glad, for he had found Christmas and I saw by his face that once more he heard the voice of the herald angels.

SACCO, VANZETTI AND HARVARD

New York World *August 5, 1927*

When at last Judge Thayer in a tiny voice passed sentence upon Sacco and Vanzetti, a woman in the court room said with terror: "It is death condemning life!"

The men in Charlestown Prison are shining spirits, and Vanzetti has spoken with an eloquence not known elsewhere within our time. They are too bright, we shield our eyes and kill them. We are the dead, and in us there is not feeling nor imagination nor the terrible torment of lust for justice. And in the city where we sleep smug gardeners walk to keep the grass above our little houses sleek and cut whatever blade thrusts up a head above its fellows.

"The decision is unbelievably brutal," said the chairman of the Defense Committee, and he was wrong. The thing is worthy to be believed. It has happened. It will happen again, and the shame is wider than that which must rest upon Massachusetts. I have never believed that the trial of Sacco and Vanzetti was one set apart from many by reason of the passion and prejudice which encrusted all the benches. Scratch through the varnish of any judgment seat and what will you strike but hate thick-clotted from centuries of angry verdicts? Did any man ever find power within his hand except to use it as a whip?

Governor Alvan T. Fuller never had any intention in all his investigation but to put a new and higher polish upon the proceedings. The justice of the business was not his concern. He hoped to make it respectable. He called old men from high places to stand behind his chair so that he might seem to speak with all the authority of a high priest or a Pilate.

What more can these immigrants from Italy expect? It is not every prisoner who has a president of Harvard University throw on the switch for him. And Robert Grant is not only a former judge but one of the most popular dinner guests in Boston. If this is a lynching, at least the fish peddler and his friend the factory hand may take unction to their souls that they will die at the hands of men in dinner coats or academic gowns, according to the conventionalities required by the hour of execution.

Already too much has been made of the personality of Webster Thayer. To sympathizers of Sacco and Vanzetti he has seemed a man with a cloven hoof. But in no usual sense of the term is this man a villain. Although probably not a great jurist, he is without doubt as capable and conscientious as the average Massachusetts judge, and if that's enough to warm him in wet weather by all means let him stick the compliment against his ribs.

Webster Thayer has a thousand friends. He has courage, sincerity and convictions. Judge Thayer is a good man, and when he says that he made every effort to give a fair trial to the anarchists brought before him, undoubtedly he thinks it and he means it. Quite often I've heard the remark: "I wonder how that man sleeps at night?" On this point I have no first hand information, but I venture to guess that he is no more beset with uneasy dreams than most of us. He saw his duty and he thinks he did it.

And Governor Fuller, also, is not in any accepted sense of the word a miscreant. Before becoming Governor he manufactured bicycles. Nobody was cheated by his company. He loves his family and pays his debts. Very much he desires to be Governor again, and there is an excellent chance that this ambition will be gratified. Other governors of Massachusetts have gone far, and it is not fantastic to assume that some day he might be President. His is not a master mind, but he is a solid and substantial American, chiming in heartily with all our national ideals and aspirations.

To me the tragedy of the conviction of Sacco and Vanzetti lies in the fact that this was not a deed done by crooks and knaves. In that case we would have a campaign with the slogan "Turn the rascals out," and set up for a year or two a reform administration. Nor have I had much patience with any who would like to punish Thayer by impeachment or any other process. Unfrock him and his judicial robes would fall upon a pair of shoulders not different by the thickness of a fingernail. Men like Holmes and Brandeis do not grow on bushes. Popular government, as far as the eye can see, is always going to be administered by the Thayers and Fullers.

It has been said that the question at issue was not the guilt or innocence of Sacco and Vanzetti but whether or not they received a fair trial. I will admit that this commands my interest to some extent, but still I think it is a minor phase in the whole matter. From a Utopian point of view the trial was far from fair, but it was not more biased than a thousand which take place in this country every year. It has been pointed out that the public prosecutor neglected to call certain witnesses because their testimony would not have been favorable to his case. Are there five district attorneys, is there one, in the whole country who would do otherwise?

Again Professor Frankfurter has most clearly shown that the prosecution asked a trick question in regard to the pistol and made the expert seem to testify far more concretely than he was willing to commit himself. That was very wrong, but not unique. Our judicial processes are so arranged that it is to the interest of district attorneys to secure convictions rather than to ascertain justice, and if it would profit his case, there is not one who would not stoop to confuse the issue in the minds of the jurymen.

Eleven of the twelve who convicted Sacco and Vanzetti are still alive, and Governor Fuller talked to them. He reports somewhat naïvely that they all told him that they considered the trial fair. Did he expect them to report, "Why, no, Governor, we brought in a verdict of guilty just out of general depravity"?

By now there has been a long and careful sifting of the evidence in the case. It is ridiculous to say that Sacco and Vanzetti are being railroaded to the chair. The situation is much worse than that. This is a thing done cold-bloodedly and with deliberation. But care and deliberation do not guarantee justice. Even if every venerable college president in the country tottered forward to say "guilty" they could not alter facts. The tragedy of it all lies in the fact that though a Southern mountain man may move more quickly to a dirty deed of violence, his feet are set no more firmly in the path of prejudice than a Lowell ambling sedately to a hanging.

I said of Calvin Coolidge that I admired his use of "I do not choose," but he was dealing with a problem wholly personal, and had every right to withhold his reasons. For Governor Fuller I can't say the same. These are the lives of others with which he is dealing. In his fairly long statement he answers not a single point which has been made against the justice of the conviction. The deliberations of himself and his associates were secret, and seemingly it is his intention that they shall remain secret. A gentleman does not investigate and tell.

I've said these men have slept, but from now on it is our business to make them toss and turn a little, for a cry should go up from many million voices before the day set for Sacco and Vanzetti to die. We have a right to beat against tight minds with our fists and shout a word into the ears of the old men. We want to know, we will know-"Why?"

A CONTROVERSY

New York World *August 12, 1927*

Regarding Mr. Broun

The World has always believed in allowing the fullest possible expression of individual opinion to those of its special writers who write under their own names. Straining its interpretation of this privilege, *The World* allowed Mr. Heywood Broun to write two articles on the Sacco-Vanzetti case, in which he expressed his personal opinion with the utmost extravagance.

The World then instructed him, now that he had made his own position clear, to select other subjects for his next articles. Mr. Broun, however, continued to write on the Sacco-Vanzetti case. *The World,* thereupon, exercising its right of final decision as to what it will publish in its columns, has omitted all articles submitted by Mr. Broun.

Ralph Pulitzer, Editor, The World.

Heywood Broun Replies

Naturally I was interested in the column which Ralph Pulitzer wrote and which appeared in my old shop window. I was grateful to him for writing it. This seemed to me a fair and frank statement of the issue. But upon one or two points I would like the privilege of stating my own attitude. "*The World,*" wrote Mr. Pulitzer, "then instructed him, now that he made his own position clear, to select other subjects for his next articles."

My recollection is that no official notice was issued. An executive of the paper remarked rather casually that it might be better for me not to write any more about Sacco and Vanzetti. The point is not important. Even though my instructions had been definite I would still have been unable at that time to write on anything but this case. Mr. Pulitzer unintentionally does me an injustice when he suggests that I should have been satisfied to make my own position clear and then keep silent. I felt and I feel passionately about the issue. The men were not yet dead. I was not simply trying to keep my own record straight. That's not good enough.

"When Pilate saw that he could prevail nothing, but that rather a tumult was made, he took water, and washed his hands before the multitude, saying, I am innocent of the blood of this just person: see ye to it."

The judgment of the world has been that Pilate did not do enough. There is no vigor in expressing an opinion and then washing your hands.

And after all Pilate was only a sort of Governor and I'm a newspaperman. *The World* would not be satisfied to declare itself upon some monstrous injustice and then depart saying, "See ye to it." If Mr. Pulitzer believes that I have any respect for the traditions of his paper, how could he expect me to behave like that?

I do respect the traditions of *The World*. It has carried on fine fights and will continue to do so. That it is in every respect superbly a liberal newspaper I cannot say. Still, it more nearly approaches this ideal than any other New York daily in its field.

The curious part about the commotion lies in the fact that fundamentally *The World* and I were on the same side. The responsible heads of the paper were disturbed, I think, not so much at my opinion as at my manners and my methods of controversy.

Mr. Pulitzer has said that I expressed my personal opinion "with the utmost extravagance." I spoke only to the limit of my belief and passion. This may be extravagance, but I see no wisdom in saving up indignation for a rainy day. It was already raining. Besides, fighters who pull their punches lose their fights.

Once there was a pitcher on the Giants who was sued for breach of promise, and fortunately this suit resulted in his love-letters being made public. The memory of one of these, I have always treasured. He wrote, "Sweetheart, they knocked me out of the box in the third inning, but it wasn't my fault. The day was cold and I couldn't sweat. Unless I can sweat I can't pitch."

I realize that I have been a special writer who sometimes embarrassed his newspaper. However, I wish to correct the impression that in the Sacco-Vanzetti case I went completely roaring mad after twice being generously afforded the privilege of vehement expression. I am sorry to say that the two subsequent columns which *The World* refused contained no fiery phrases. Although I would like it very much, I have been around long enough to realize that no columnist can possibly be accorded the right to say whatever comes into his mind. There is libel, there is obscenity, there is blasphemy; and there are policies and philosophies which his paper happens to hold especially dear. I do not anticipate that *The World* or any other newspaper would give me license to scoff at every campaign to which it committed itself. And I have not. Of course, I have always contended that in "It Seems to Me" I expressed my own opinions and did not commit the paper. This, to be sure, would not be true in the case of libel, but that has already been referred to and I did not ever involve *The World* in any suit. Ralph Pulitzer said that *The World* has always believed in "allowing the fullest possible expression of individual opinion to those of its special writers who write under their own names." And yet he also says that I was ordered not to write about the Sacco-Vanzetti case at all after I had twice gone on record. In other words, in this case the paper was prepared to censor what I might have to say even before it had been written. How full is the fullest?

There is no use in my pretending that I do not believe myself right and *The World* wrong in the present controversy. As far as Sacco and Vanzetti went, both the paper and the individual wanted an amelioration of the sentence. Nothing less than a pardon or a new trial was satisfactory to me. Apparently, *The World* believed that if life imprisonment was all that could possibly be won from Governor Fuller, that would be better than nothing. Here an interesting point in tactics arises. The editorial strategy of *The World* seemingly rested upon the theory that in a desperate cause it is well to ask a little less than you hope to get. I think you should ask more.

Rigorously *The World* excluded from its editorial columns all invectives. To call names, *The World* felt, would merely stiffen the resistance of Fuller and his advisers. I did call names, and this might possibly have embarrassed *The World* in the precise sort of campaign which it deemed it wise to make. Again, *The World* undoubtedly felt anxious about the bomb outrages. With passions so high, sparks were undoubtedly to be avoided. But in spite of silly crimes of violence I felt and feel that the most tragic factor of the Sacco-Vanzetti case is the general apathy. In ten minutes' time I will guarantee to fetch from the streets of New York one hundred persons who have never heard of the case and thousands who have not the slightest idea what it is all about. This could have been a duet with the editorial page carrying the air in sweet and tenor tones while in my compartment bass rumblings were added.

By now, I am willing to admit that I am too violent, too ill-disciplined, too indiscreet to fit pleasantly into *The World*'s philosophy of daily journalism. And since I cannot hit it off with *The World* I would be wise to look for work more alluring. I am still a member of Actors' Equity, the top floor is well stocked with early Brouns and I know a card-trick. In farewell to the paper I can only say that in its relations to me it was fair, generous and gallant. But that doesn't go for the Sacco-Vanzetti case.

Ernest L. Meyer

Born in Denver, bred in Milwaukee, Ernest L. Meyer (1892–1952) broke into journalism as a writer-compositor for The Warden Herald *in backcountry Washington. He was a police reporter in Chicago, edited the literary magazine at the University of Wisconsin, and from 1920 to 1935 was managing editor, then telegraph editor and columnist for the* Capital Times, *advocate of the LaFollette Progressives. His ''Making Light of the Times'' was an editorial page miscellany of stories, sketches, verse, and commentary. He renamed the column ''As the Crow Flies'' for the* New York Post, *where it appeared until 1941. While telegraph editor at the* Daily News, *he continued writing a weekly column for* The *(Madison, Wis.)* Progressive. *Meyer's daily journalism formed the core of three books,* Making Light of the Times *(Madison, 1928);* Hey, Yellowbacks! *(New York, 1934), the wartime diary of a conscientious objector, prefaced by William Ellery Leonard; and* Bucket Boy: A Milwaukee Legend *(New York, 1947). I have drawn from the first of these, and from family scrapbooks in the selection that follows from my father's work.*

THE GOSPEL OF FORD

Things always go on all right.—Henry Ford, in an interview in the current issue of *Collier's Weekly.*

Capital Times *November 1, 1934*

Henry Ford perused the headlines:
Millions shivering in breadlines. . . .
Drought and famine . . . Balkan war-clouds. . . .
Experts vision more and more clouds. . . .

Hunger marchers clash with cops. . . .
Hoppers ruin Midwest crops. . . .
World in race for strongest arms. . . .
Banks foreclose on ninety farms. . . .
Thirty dead in week-end crashes. . . .
Mob gives Negro ninety lashes. . . .
Jobless war vet murders bride. . . .
Broker, broke, tries suicide. . . .
Meat-shops stormed by hungry horde. . . .
Things always go on all right, said Ford.

Henry Ford laid down the headlines
Blaring news of wars and breadlines,
Laid the paper on the polished top
Of the polished table of his polished shop,
And he flecked some dust from the polished tips
Of his polished shoes, with a smile on his lips;
And he mused of his millions, his many many millions,
His old-time fiddlers and old-time cotillions,
His museum, his mansion, his country estate,
His mile-long plant and his pet V-8;
And he looked from his window at the chimney tops
Of the mortgaged homes of the men in his shops;
And he looked without seeing at the job-hunting throng
Scowling at the pavement as they shuffled along;
And he felt the shiver of his giant plants
Where ant-men scuttled at the work of ants;
And he left the paper on the polished top
Of the polished table of the polished shop,
And he made a little prayer on the goodness of the Lord:
Things always go on all right, said Ford.

BEGINNINGS OF A CREDO

Capital Times *January 1,1934*

ITEM: There is nothing I profess to know with perfect assurance, for nothing is perfect, not even the conviction of one's own superiority, though there are some who will admit no flaw in that proposition. These are interesting people to observe, though difficult, to get along with.

ITEM: As a corollary, I believe in skepticism. For a statement universally acknowledged as true at once loses its fascination, and too many such acknowledgments make a dull world. Thus, the formula that two and two makes four is bereft of interest as soon as stated. But the skeptic, having seen kindred dogmatisms overthrown in his lifetime, is willing to postulate that two and two may make six, or 26, or an applecart. Besides, it's more fun.

ITEM: And that is the end of life: happiness. Personal, hedonistic happiness, here and now. If the world were more convinced of this simple notion, there would be fewer going moonstruck with ascetic, astral ideals.

ITEM: As a footnote, it is well to observe that happiness, like a bay-tree, cannot thrive in sterile soil. That is, no man unless he is abnormal can be happy if those about him are unhappy. This is hedonism with a social check, and leads the hedonist to wage war on whatever is bringing misery to his neighbors. Which resolves itself to the simple thought that even the sound man can't dance around the May-pole if his partners in the dance are crippled.

ITEM: I believe in a Utopia, which, defined, means a place where nobody is constrained by unsatisfied desires to talk about a Utopia. Since this may mean a frightfully dull and unimaginative place, I prefer, on second thought, not to believe in Utopia.

ITEM: I am in favor of war, providing that it is fought solely by those who delight in it or who profit from it, and so long as it does not injure bystanders or interfere with my pleasures. Since this is hardly possible, I am not in favor of war.

ITEM: As an afterthought on war, I believe in birth control as a more humane method of racial decimation. Nor has it ever been quite clear to me how some people can condone the one and condemn the other, since having one's life terminated as a germ cell is less annoying than growing up to have one's intricate and interesting bodily mechanism voided by a shrapnel.

ITEM: This leads to a pleasant speculation, namely: that much might be said for universal, complete birth control. Which would mean that after this generation the world would be depopulated and given over to the owl, the prairie dog, and the rattle-snake, all of whom, it is reported, get along quite amiably in the same little hole. Mankind, though sharing a roomy world, has not attained such felicity.

ITEM: I believe in freedom of speech, with the proviso that what one says should be said gracefully of things that matter or amuse. This would automatically rule out radio dialogists, creators of comic strips, and half the debates in congress.

ITEM: I believe in love, which requires no elaboration, except by puritans and psychoanalysts, who are excellent people, if avoided.

ITEM: As a footnote, I believe the poet who wrote "never seek to tell thy love" is wrong. It should be told again and again, for the world is a misty place in which people go about never showing the suns they carry with them. This is miserliness, with shame as an insufficient motive, and propriety as a sorry reward.

ITEM: There is nothing I profess to know with perfect assurance, not even the statement that there is nothing I profess to know with perfect assurance.

GREELEY STATUE, NEW YORK

[Horace Greeley: editor, publicist, reformer.]

Capital Times *November 13, 1927*

The man of stone on Greeley Square
Sits grim amid the clamor there:
Above, the iron roar of "L's,"
Below, the din of horns and bells.
He never starts, he never blinks;
He thinks and sits, he sits and thinks.

And I, who caught his mournful stare,
Pitied the man on Greeley Square.
"Here's grief," said I. "A sorry fate
Amid this noise to meditate."
And then—the lips moved, and I swear
Thus spoke the man on Greeley Square:

"The traffic thunder, shrieks and toots,
The honks, the bellows, frenzied hoots,
The clank and clatter, I don't mind;
I'm stone deaf—but I'm not stone blind!"
And with a baleful eye he froze
The news-stand right beneath his nose.

I looked—and, sure, on Greeley Square
Blossomed the pinkest tabloids there,
And girls disrobed proclaimed the Arts
'Mid betting sheets and racing charts,
While sin and carnage screamed their woe
In letters six feet high or so.

Alack! the man on Greeley Square!
I caught the horror in his glare;

And, as I watched, he seemed to shrink;
The statue crumbled chink by chink,
Until there sat (I rubbed my eyes)
A publicist of tabloid size.

THE COLUMNIST'S LOT

Capital Times *May 4, 1928*

With mingled pleasure and grief I note that John Culnan, inventor of Dr. Bazinook and other eccentrics, has started a daily column in the *Wisconsin State Journal.* The pleasure comes from reading his puns and poems, for John is a delightful punster, and a good pun, however much the intellectuals may grunt, is as entertaining as a good juggling act. And if you don't like good juggling acts, that's a misfortune, not a virtue; you are simply too grown up to live very long.

Yet the announcement that "Skipper" has joined the ranks of the column writers somehow depresses me. He is my friend, and I wished for him a kinder fate. He has an even temper, and a great fund of sunny enthusiasm, both of which are periled by the enterprise upon which he so lightly embarks. It may seem an act of malice to warn him of the snares and sorrows that lie in his road, but who could remain silent on seeing a friend set sail in an oyster shell in the teeth of a tempest?

For a month or two "Skipper" will cruise before the headwinds with bulging sails. Nothing easier than to write a blithe column a day; the world is lush with subjects for his pen—they will crowd his imagination like flowers in a June garden, and he will pluck here and there fastidiously, picking only the best. But then, alas, his eyesight will fail him a bit, or his fingers grow weary, and he will find now and then that what he thought was a rose is only a cabbage, slightly decayed. He will go home from writing his daily task pummeling his brains, wondering how in heaven's name he could have committed such a botchery. And the good people who yesterday applauded will curse him today as a dunderhead, and he will be a strong man, indeed, if he does not drown his sorrow in cherry phosphate and butterscotch sundaes.

For the most part, however, John will write his whimsies into vast and awful silence. He will rarely know from others whether what he composes is good or bad; he will console himself with the child-like faith that somewhere west of the cornfields some simple soul appreciated his pun about courses in husbandry frowning on bachelor degrees. Yet as a rule, for all he knows, the editor takes every issue of the newspaper and sinks them each night in the middle of Mendota, and goes home gloating.

But let him write seriously of something that is near to his heart, let him touch on some tender prejudice that people hug close, let him disturb even slightly the dust of old faiths—and the righteous thunder will rage around his head. Then, indeed, will he know that someone has read his writing; for he will be unveiled before the community as a monster who eats little babies, and his column will be held up as a corrupter of youth, and a menace to old ladies.

He will, if he is cautious, write only of cabbages and kings, always being careful to stipulate that the cabbages are moral and the kings monogamous.

* * *

John will discover, as the months unwind, that the fears and misgivings that assail a writer will stab him at his daily task. There is such a gulf between the conception of an idea and its expression in decent prose—as wide a gulf as between laying an egg and hatching it. It is so easy to have a vague and beautiful and clever notion about a subject, and so difficult to trap the notion in a hard and fast verbal cage. Thoughts are so fresh and fragmentary; yet their expression, once down on paper, looks stale as old buns. And fragmentary impressions, sober critics agree, have no place in formal prose. If you set down accurately what runs through your mind you will be looked at as a mad thing, and you will join James Joyce in limbo. So John will find himself hampered and harassed by the compulsions of conventional forms and passable diction. He will try to do a decent job of it, and will groan at the ineptitudes, the trash and the vapors that creep into his writings. He will hear people all about him declare that it is easy to write; that they could write wonders if they only gave their minds to it, and he will clench his teeth to keep from slaying them on the spot.

John will tell these boasters, I hope, to look at Stevenson, the novelist, who by common conspiracy is described as a master stylist. Yet Stevenson, at the height of his career, wrote a friend: "I sometimes wonder whether I shall ever learn to write."

John, too, will wonder if he will ever learn to write; it is the peculiar fate of literary hens to lay solid gold eggs at night, only to discover, in the cruel pallor of dawn, that the eggs have gilt shells and are as hollow as drums.

* * *

And John will point to Katherine Mansfield. She was one of the most brilliant stylists of our day, fresh, compact, clear and ringing as a bronze bell. And yet the *Journal of Katherine Mansfield,* recently published, reveals her pain and humility. "What dreadful, awful rot," she exclaims on reading over what she had written. "I am so frightened of writing mockery instead of satire that my pen hovers and won't settle." Again she writes in her diary: "Oh, I failed today. I turned back, looked over my shoulder, and felt as though I were struck down. I do not think I am a good writer: I realize my faults better than anyone else could realize them. I know exactly where I fail. And yet, when I have finished a story and before I have begun another, I catch myself preening my feathers; a root of pride puts out a thick shoot on the slightest provocation. . . . I have failed. I haven't been able to yield to the kind of contemplation that is necessary. I've decided to tear up everything that I've written and start again."

These are the terrors that lie in John's path. They will waylay him at every turn in the road, they will stride his chest at night, when he should be dreaming of a stream we once traveled together, a silver thread weaving through the hills, creeping under willows where the cardinal fluted his morning song.

And some day, when John is bending over the typewriter, enslaved by the imps and ogres that lay their snares for columnists, he may rebel, and curse his typewriter with a good, sufficient curse and walk down to the river again, a free man. And I, perhaps, will go with him.

A LAND WITHOUT ALIENS

New York Post *October 12, 1940*

[Orange, Tex. *Chairman Martin Dies of the House Committee Investigating un-American Activities said that if Congress permits the committee to continue its activities the investigation will result in deportation of 7,000,000 aliens.—News Item.]*

Now it happened that Martin Dies rubbed the magic lamp, and the genie appeared, and the genie said: "What is thy will, master?"

And Martin Dies answered: "It is my will that straightway all the aliens in America be exiled to some distant and most inhospitable spot, and there do sufferance for their sins."

And the genie said: "Truly I can grant thy wish, master, but there is a law in my land which says that whosoever is sent into exile shall be allowed to take with him whatever he has created by his own efforts. This is, I think, a just law, and if you abide by it, I can grant your desire."

And Martin Dies said: "Indeed, your law is quite just. Let the aliens be deported, and let them take with them what they have created, for surely they have fashioned nothing but dissent and plots and radical heresies and sins and sabotage, and to these they are welcome."

And the genie said: "So be it, master." And he uttered a few words of strange power, and a miracle happened.

It followed in that very instant that a vast fleet of barges and boats was fashioned, and into them, millions upon millions, flocked the aliens, and they took with them what they had created in America.

They took with them highways hewn out of the wilderness by Sicilians and Slavs, and great rafts of lumber felled in the forests by the Irish, Swedes and Norwegians, and many millions of square miles of earth made fertile by the Germans, the Danes and the Dutch, and billions of garments woven by the Jews, and mountainous masses of coal and iron and copper dug from the pits by Italians and Finns and Poles, and whole cities of skyscrapers and subways and railroads and mills and marts wrought by the sinews of many aliens from the earth's four quarters.

And they took with them also their alien culture, their music and their songs, their languages and their literature, their books and their Bibles and their cookery, their piety and their passions, their ideals and philosophy and folk-dances and fun which had been woven into the rich and multi-colored fabric of America.

Now all this happened when the genie granted the wish of Martin Dies, when the aliens left with all their works, and a great want followed, and a great and strange silence. And in that silence there was naught to be heard save the frightened whimpering of Martin Dies crying: "Genie, genie!" But there was no answer, for the genie, an alien, was on one of the boats to Bagdad, and after that there was nothing and the night.

E. B. White

E. B. White (1899–1983) flirted for years without succumbing to the column. At Cornell, he edited The Sun *and contributed poems and paragraphs to "The Berry Patch." Out of college and in Washington for fifteen months, he sought to emulate B.L.T. for the* Seattle Times; *neither he nor the editor was pleased, and the latter is now remembered for suggesting that young Andy White seek his literary fortune elsewhere. When he arrived in New York, as he later wrote:*

> *. . . my personal giants were a dozen or so columnists and critics and poets whose names appeared regularly in the papers. I burned with a low steady fever just because I was on the same island with Don Marquis, Heywood Broun, Christopher Morley, Franklin P. Adams, Robert C. Benchley, Frank Sullivan, Dorothy Parker, Alexander Woollcott, Ring Lardner, and Stephen Vincent Benét. I would hang around the corner of Chambers Street and Broadway, thinking: "Somewhere in that building is the typewriter that archy the cockroach jumps on at night."*

In 1925, he joined in founding The New Yorker, *where his unsigned contributions to "Talk of the Town," and "Note and Comments" helped fix the tone of the new weekly. When he resettled in Maine in 1938, the editor of* Harper's *suggested a monthly column: "It turned out to be one of the luckiest things that ever happened to me. I was a man in search of a first person singular, and lo here it was—handed to me on a platter before I ever left town." White's "One Man's Meat," when collected in a book of the same title, set a standard for prose that has chastened would-be emulators. Samples of his work follow; an informed commentary is in Scott Elledge's* E. B. White *(New York, 1984).*

PERSONALS

Seattle Daily Times *circa 1923*

[White's "Personal Column" was in the form of classified advertisements, published too confusingly in the actual classified section.]

June 21—The Longest Day

Tomorrow is the longest day of all,
The punctual sun rides brightly into port;
The charms of lengthy days will never pall—
But oh, they tend to make the nights so short!

There used to be an old family ritual about "the first drink." When a youth touched spiritous liquor for the first time, he marked his downfall, from which there was small hope of redemption.

But we would like to record another ritual. It concerns "the first scribbling." This seems to us far more abysmal and important than the first drink.

Up until the time when a youth first seizes a pencil and furtively scratches a thought on paper, he is secure. He rides the top of the world, a normal person.

But with the first inspired pencil mark, he signs his life away, unwittingly. The door slams shut. He has accepted words as his medium of exchange—and words, more than dollars, are elusive and painful to a degree which the youth yet knows nothing of.

We Answer Hard Questions

Sir:

How can a man determine how much sugar he is putting in his coffee when using one of those restaurant sugar shakers?

Rough Rollo

Rollo:

Pour your coffee into your empty soup bowl, put the desired amount of sugar into the empty coffee cup, and then pour the coffee back, straining the soup particles out through your napkin. Ask us anything at all.

HOT WEATHER

Harper's, "One Man's Meat" *July 1939*

The sound of victrola music right after breakfast gives the summer day a loose, footless feeling, the sort of inner sadness I have experienced on the outskirts of small towns on Sunday afternoon, or in the deserted city during a holiday, or on beaches where the bathhouses smelled of sour towels and yesterday's levity. Morning is so closely associated with brisk affairs, music with evening and day's end, that when I hear a three-year-old dance tune crooned upon the early air while shadows still point west and the day is erect in the saddle, I feel faintly decadent, at loose ends, as though I were in the South Seas—a beachcomber waiting for a piece of fruit to fall, or for a brown girl to appear naked from a pool.

* * *

Asterisks? So soon?

* * *

It is a hot-weather sign, the asterisk. The cicada of the typewriter, telling the long steaming noons. Don Marquis was one of the great exponents of the asterisk. The heavy pauses between his paragraphs, could they find a translator, would make a book for the ages.

* * *

Don knew how lonely everybody is. "Always the struggle of the human soul is to break through the barriers of silence and distance into companionship. Friendship, lust, love, art, religion—we rush into them pleading, fighting, clamoring for the touch of spirit laid against our spirit." Why else would you be reading this fragmentary page—you with the book in your lap? You're not out to learn anything, certainly. You just want the healing action of some chance corroboration, the soporific of spirit laid against spirit. Even if you read only to crab about everything I say, your letter of complaint is a dead give-away: you are unutterably lonely or you wouldn't have taken the trouble to write it.

* * *

How contagious hysteria and fear are! In my henhouse are two or three jumpy hens, who, at the slightest disturbance, incite the whole flock to sudden panic—to the great injury, nervously and sometimes physically, of the group. This panic is transmitted with great rapidity; in fact, it is almost instantaneous, like the wheeling of pigeons in air, which seem all to turn and swoop together as though controlled electrically by a remote fancier.

* * *

The cells of the body co-operate to make the man; the men co-operate to make the society. But there is a contradiction baffling to biologist and layman alike. On the day last spring that I saw a flight of geese passing over on their way to the lonely lakes of the north (a co-operative formation suggesting a tactical advantage imitated by our air corps)—on that same day cannibalism broke out among my baby chicks and I observed the brutality with which the group will turn upon an individual, literally picking his guts out. This is the antithesis of co-operation—a contrariness not unobserved in our own circles. (I recently read of a member of an actors' union biting another actor quite hard. I believe it was over some difference in the means of co-operation.)

* * *

"How are you going to keep from getting provincial?" asked one of our friends quite solemnly. It was such a sudden question, I couldn't think of any answer, so just let it go. But afterward I wondered how my friend, on his part, was going to keep from getting metropolitan.

As a matter of fact the provinces nowadays are every bit as lurid, in their own way, as the centers of culture. One of the farm owners here—a very rich man who up until quite recently owned herds and flocks for the sheer hell of associating with animals—sent his registered Guernseys on a tour of the fairs last fall. When the cows returned home heaped with glory they were met at the station by a trumpeter and led triumphantly through town in a pompous parade that conquerors of old would have envied.

All sorts of things go on in this provincial existence. To the north of us, photographers in airplanes have been making a vast aërial picture map of the country, showing every fence and lane. Eventually the whole nation will be so mapped. Individual maps are already available; a farmer can send in to Washington and they will send him a picture showing how his place looks from three miles up.

And I see by the paper that a hundred million parasites have been turned loose in the State this summer, to war on the spruce sawfly—a challenge to the balance of nature that seems rather alarming to a man who hardly dares shoot a crow for fear of upsetting the fine adjustment in the world of birds and insects, predator and prey. How could I become provincial, with parasites being loosed against the foe? I am in the very center of everything.

* * *

There is furthermore slight chance of my becoming provincial *this* summer, because I am raising a baby seagull and there isn't time. A young gull eats twice his own weight in food every ten minutes, and if he doesn't get it he screams.

The gull was a present from Mr. Dameron, who wore an odd look of guilt on his face as he approached, that evening, proffering the chick in a pint ice-cream container as tentatively as though it were a bill for labor. The occupant (about the size of a billiard ball) took one look at me, stretched out his stubby wings, and cried: "Daddy!" I must say I haven't failed him.

He was so tiny, so recently shell-girt, that I put him with a broody hen, thinking she might adopt him. Nothing ever came of that brief connection. The gull wanted me, not a hen. I imagine the nest seemed stuffy to him after the windblown, fog-drenched island of his nativity. I asked Mr. Dameron what to feed him. "I dunno," he replied, "but I don't think you can upset a gull's stomach."

I began cautiously with a tiny piece of hamburger. It was the merest beginning. In the last three weeks he has swallowed a mixture of foods that would sicken you to listen to. (His favorite dish is chicken gizzards chopped with clams, angle worms, and laying mash.) He has eaten ten thousand clams—of my own digging—and still screams accusingly every time I go by. He has drained my strength, yet somehow it all seems worth while. A mature gull in flight is simple beauty. Some day this child of mine is going to be stretching his wings and a gentle puff will come along and he will take off. The pleasure of seeing my worms and gizzards translated into perfect flight will be my strange reward. I just hope I live that long.

MAINE SPEECH*

Harper's, "One Man's Meat" *October 1940*

I find that, whether I will or no, my speech is gradually changing, to conform to the language of the country. The tongue spoken here in Maine is as different from the tongue spoken in New York as Dutch is from German. Part of this difference is in the meaning of words, part in the pronunciation, part in the grammar. But the difference is great. Sometimes when a child is talking it is all one can do to translate until one has mastered the language. Our boy came home from school the first day and said the school was peachy but he couldn't understand what anybody was saying. This lasted only a couple of days.

For the word "all" you use the phrase "the whole of." You ask, "Is that the whole of it?" And whole is pronounced hull. Is that the hull of it? It sounds as though you might mean a ship.

For lift, the word is heft. You heft a thing to see how much it weighs. When you are holding a wedge for somebody to tap with a hammer, you say: "Tunk it a little." I've never heard the word tap used. It is always tunk.

Baster (pronounced bayster) is a popular word with the boys. All the kids use it. He's an old baster, they say, when they pull an eel out of an

*Editor's Note: Later on, E. B. White felt that this description of Maine dialect was dated and inaccurate; it is reprinted here as a contemporary document, along with other columns describing the American language.

eel trap. It probably derives from bastard, but it sounds quite proper and innocent when you hear it, and rather descriptive. I regard lots of things now (and some people) as old basters.

A person who is sensitive to cold is spleeny. We have never put a heater in our car, for fear we might get spleeny. When a pasture is sparse and isn't providing enough feed for the stock, you say the pasture is pretty snug. And a man who walks and talks slowly or lazily is called mod'rate. He's a powerful mod'rate man, you say.

When you're prying something with a pole and put a rock under the pole as a fulcrum, the rock is called a bait. Few people use the word "difference." When they want to say it makes no difference, they say it doesn't make any odds.

If you have enough wood for winter but not enough to carry you beyond that, you need wood "to spring out on." And when a ewe shows an udder, she "bags out." Ewe is pronounced yo by old-timers like my friend Dameron.

This ewe and yo business had me licked at first. It seemed an affectation to say yo when I was talking about a female sheep. But that was when I was still thinking of them as yews. After a while I thought of them as yos, and then it seemed perfectly all right. In fact, yo is a better-sounding word, all in all, than yew. For a while I tried to pronounce it half way between yew and yo. This proved fatal. A man has to make up his mind and then go boldly ahead. A ewe can't stand an umlaut any more than she can a terrior.

Hunting or shooting is called gunning. Tamarack is always hackmatack. Tackle is pronounced taykle. You rig a block and taykle.

If one of your sheep is tamer than the others, and the others follow her, you say she will "toll" the others in. The chopped clams that you spread upon the waters to keep the mackerel schooling around your boat are called toll bait. Or chum bait. A windy day is a "rough" day, whether you are on land or sea. Mild weather is "soft." And there is a distinction between weather overhead and weather underfoot. Lots of times, in spring, when the ground is muddy, you will have a "nice day overhead."

Manure is dressing, not manure. I think, although I'm not sure, that manure is considered a nasty word, not fit for polite company. The word dung is used some but not as much as dressing. But a manure fork is always a dung fork.

Wood that hasn't properly seasoned is dozy. The lunch hour is one's nooning. A small cove full of mud and eelgrass is a gunkhole. When a pullet slips off and lays in the blackberry bushes she "steals a nest away." If you get through the winter without dying or starving you "wintered well."

Persons who are not native to this locality are "from away." We are from away ourselves, and always will be, even if we live here the rest of our lives. You've got to be born here—otherwise you're from away.

People get born, but lambs and calves get dropped. This is literally true of course. The lamb actually does get dropped. (It doesn't hurt it any —or at any rate it never complains.) When a sow has little ones, she "pigs." Mine pigged on a Sunday morning, the ol' baster.

The road is often called "the tar." And road is pronounced ro-ud. The other day I heard someone called President Roosevelt a "war mongrel." Statute is called statue. Lawyers are busy studying the statues. Library is liberry. Chimney is chimley.

Fish weir is pronounced fish ware. Right now they're not getting anything in the wares.

Hoist is pronounced hist. I heard a tall story the other day about a man who was histed up on the end of a derrick boom while his companions accused him of making free with another man's wife. "Come on, confess!" they shouted. "Isn't it true you went with her all last year?" For a while he swung at the end of the boom and denied the charges. But he got tired finally. "You did, didn't you?" they persisted. "Well, once boys," he replied. "Now hist me down."

The most difficult sound is the "a." I've been in Maine, off and on, all my life, but I still have to pause sometimes when somebody asks me something with an "a" in it. The other day a friend met me in front of the store, and asked, "How's the famine comin' along?" I had to think fast before I got the word "farming" out of his famine.

The word dear is pronounced dee-ah. Yet the word deer is pronounced deer. All children are called dee-ah, by men and women alike. Workmen often call each other dee-ah while on the job.

The final "y" of a word becomes "ay." Our boy used to call our dog Freddie. Now he calls him Fredday. Sometimes he calls him Fredday dee-ah; other times he calls him Fredday you ol' baster.

Country talk is alive and accurate, and contains more pictures and images than city talk. It usually has an unmistakable sincerity that gives it distinction. I think there is less talking merely for the sound that it makes. At any rate, I seldom tire of listening to even the most commonplace stuff, directly and sincerely spoken; and I still recall with dread the feeling that occasionally used to come over me at parties in town when the air was crowded with loud intellectual formations-the feeling that there wasn't a remark in the room that couldn't be brought down with a common pin.

* * *

A note from my garage this morning, saying that my oil was changed at 7839 and that it was time I came in to have the crankcase drained. "You've got enough to think about," the note said, "without trying to remember when your car needs its next Mobilubrication."

It is true, we all have much to think about. I used to try to remember about the oil, used to try to change it according to mileage on the car, but not any more. Now I change oil ritualistically, four times a year, on the summer and winter solstices and the spring and fall equinoxes. They are the dates I keep with my car. It seems to work all right; yet what a falling off the centuries have seen in men's customs. The first day of spring was once the time for taking the young virgins into the fields, there in dalliance to set an example in fertility for Nature to follow. Now we just set the clock an hour ahead and change the oil in the crankcase.

Damon Runyon

Damon Runyon (1884–1946), like the unnamed narrator of his Broadway fables, was a more or less legit guy who knew many citizens: miners, cowpokes, drifters, grifters, soldiers, hustlers, and scribes like Waldo Winchester, who falls hard for Miss Billy Perry in one of the earliest of the Broadway tales, ''Romance in the Roaring Forties.'' When Runyan took a job with Hearst's New York American *in 1910, he brought a gift for fast, easy prose and good doggerel learned in the Rockies, where he started at thirteen as a printer on his father's paper in Pueblo, Colorado. He was soon on the Hearst highwire, writing about baseball, boxing, racing, crime, and show business. Out of this came his Broadway tales in slick magazines; sixteen became movies, one a musical,* Guys and Dolls, *before he was visited by Death, ''a large and most distinguished looking figure in beautifully tailored soft white flannels.'' The visit is described in one of the last of his columns, collected in* Short Takes *(New York, 1946) and* Runyon from First to Last *(London, 1954). Other columns are gathered in* My Old Man *(New York, 1939) and* In Our Town *(New York, 1946). His biographer is Edwin P. Hoyt,* A Gentleman of Broadway *(New York, 1964).*

PRIVATE THOUGHTS OF A COLUMNIST

King Features *circa 1945*

My column is slightly irregular and loyal customers sometimes do me the honor of saying:

"I didn't see your stuff in the paper today. What happened?"

This indicates that they looked for it, which pleases me no little.

However, I am sometimes perturbed by the fact that when it does appear the customers make no comment. They notice when I am out but not when I am in. It is something to worry about.

Well, now, there are two reasons for the irregularity, one being physical. But the major reason is that I have nothing to say.

I hesitated a long time before revealing this. I realize that it may be a great shock to some of my customers and might create a precedent of far reaching consequences among newspaper columnists.

I mean it might change the present old established practice of columnists who have nothing to say, saying it.

Anyway, they might take to following my example in remaining out of the paper when they find themselves in the situation of having nothing to say. While this would save a vast amount of white paper, I fear it would be deplored by many editors.

When an editor contracts for a column to appear daily in public print, he naturally wants that contract fulfilled to the letter and thought the columnist often has nothing to say the editor is apt to feel that it is better to say it in the space and to the extent of the wordage allotted it than deprive his readers of the opportunity of guessing what the columnist means when he says nothing. Because quite often columns of this type arouse more discussion than those in which the columnist really says something.

The man never lived who had something to say every day of his life. By something to say, I mean of course, something worth listening to or reading. But if editors took to leaving out columnists when they have nothing to say, every columnist would be reduced to about three appearances per week and some less and that would be bad for the columning racket.

I expect no applause from my fellow columnists in this confession of the reason for my irregularity. They will probably say that I am non-union. Nor do I expect the commendation of editors. They are more likely to put me away as lazy. I can only hope for the approval of some of my customers for sparing them the boredom of reading my column when I had nothing to say.

The shock of many of them will probably be in their discovery of my lack of versatility on finding I had nothing to say. They will perhaps wonder why I did not fill in with a few letters from Vox Populi or other omnivorous readers. I thought of that. On reading a batch of the letters at hand I found they had nothing to say either.

I worry like the dickens when I find I have nothing to say, and never mind telling me I must be worrying all the time. I want to say something as much as any man alive, yet I have a feeling that whatever I say it should be interesting or entertaining. I have a number of stock subjects that I know I ride too hard, such as the case of the discharged soldiers and my national lottery, but they are not elastic enough to stretch over all the periods when I have nothing to say like those that some of my more facile contemporaries keep in store.

I envy those guys who when they have nothing to say, can always turn to labor and to Russia and to belting the administration and also to re-arranging the world after the war. This last, ladies and gentlemen, is a matter of which I know so little that I cannot even say nothing about it though certain of my contemporaries can do it to the extent of eight or nine hundred words daily.

NO LIFE

King Features *circa 1945*

You have been noticing an uneasy sensation in region of the Darby Kelly and the croaker says it looks to him like it might be—

Well, nothing serious, if you are careful about what you eat and take these here powders.

All right, Doc. Careful is the word from now on. Thanks.

Wait a minute. No orange juice.

What, no orange juice, Doc? Always have orange juice for breakfast.

No, no orange juice.

Okay, Doc. That's gonna be tough, but grapefruit is just as good.

No grapefruit, either. No acids.

No grapefruit? Say, what does a guy do for breakfast, Doc?

Cereals.

Don't like cereals, Doc.

No syrup.

You don't mean a little sorghum on wheat cakes, do you, Doc?

No sorghum. No wheat cakes. No sugar.

You don't mean no sugar in the coffee, Doc? Just a couple of spoons a cup?

Yes, and no coffee.

No look, Doc. You don't mean no coffee at all?

No coffee.

Say, Doc, that's all right about no sugar, but you must be kidding about no coffee at all.

No coffee.

Not even a coupla cups a meal, Doc? Why, that's just a taste.

No coffee.

Doc, that ain't human.

No candy.

Not even a little bitsy box of peppermints at the movies, Doc?

No, no peppermints. No ice.

Yo ain't talking about a tiny dab of banana ice cream, are you, Doc? The kind that goes down so slick?

Yes, no sweets at all. No highly spiced stuff. No herring.

What kind, Doc?

Any kind. No herring.

But you don't mean a little of that chef's special, Doc? The kind with the white sauce on it?

No herring.

Not even matjes, Doc?

No herring.

Well, all right, Doc. No herring. Gefüllte fish will have to do.

No gefüllte fish. No goulash.

What kind of goulash, Doc? Hungarian?

Any kind. No salami. No highly seasoned Italian food.

I never eat that more than a couple of times a week, anyway; I'll take a lobster Fra Diavolo now and then.

None of that.

Are you sure about the herring, Doc? There must be some kind that're all right.

No herring.

It's a conspiracy. Whoever heard of a little herring hurting anybody? Why, Doc, people have been eating herring for years and it never bothered them.

No herring.

Well, all right, no herring after to-night and to-morrow. What's this list, Doc?

It's your diet. Follow it closely.

But there ain't anything on it a guy can eat, Doc. It's terrible. You were just kidding about the coffee, weren't you, Doc? No coffee! Can you imagine a guy trying to live without coffee—what? You can't!

And no cigarettes.

Doc, a guy might as well be dead, hey?

DEATH PAYS A SOCIAL CALL

King Features *circa 1945*

Death came in and sat down beside me, a large and most distinguished-looking figure in beautifully-tailored soft, white flannels. His expensive face wore a big smile.

"Oh, hello," I said. "Hello, hello, hello. I was not expecting you. I have not looked at the red board lately and did not know my number was up. If you will just hand me my kady and my coat I will be with you in a jiffy."

"Tut-tut-tut," Death said. "Not so fast. I have not come for you. By no means."

"You haven't?" I said.

"No," Death said.

"Then what the hell are you doing here?" I demanded indignantly. "What do you mean by barging in here without even knocking and depositing your fat Francis in my easiest chair without so much as by-your-leave?"

"Excuse me," Death said, taken aback at my vehemence. "I was in your neighbourhood and all tired out after my day's work and I thought I would just drop in and sit around with you awhile and cut up old scores. It is merely a social call, but I guess I owe you an apology at that for my entrance."

"I should say you do," I said.

"Well, you see I am so accustomed to entering doors without knocking that I never thought," Death said. "If you like, I will go outside and knock and not come in until you answer."

"Look," I said. "You can get out of here and stay out of here. Screw, bum!"

Death burst out crying.

Huge tears rolled down both pudgy cheeks and splashed on his white silk-faced lapels.

"There it is again," he sobbed. "That same inhospitable note wherever I go. No one wants to chat with me. I am so terribly lonesome. I thought surely you would like to punch the bag with me awhile."

I declined to soften up.

"Another thing," I said sternly, "what are you doing in that get-up? You are supposed to be in black. You are supposed to look sombre, not like a Miami Beach Winter tourist."

"Why," Death said, "I got tired of wearing my old working clothes all the time. Besides, I thought these garments would be more cheerful and informal for a social call."

"Well, beat it," I said. "Just Duffy out of here."

"You need not fear me," Death said.

"I do not fear you Deathie, old boy," I said, "but you are a knock to me among my neighbours. Your visit is sure to get noised about and cause gossip. You know you are not considered a desirable character by many persons, although, mind you, I am not saying anything against you."

"Oh, go ahead," Death said. "Everybody else puts the zing on me so you might as well, too. But I did not think your neighbours would recognize me in white, although, come to think of it, I noticed everybody running to their front door and grabbing in their 'Welcome' mats as I went past. Why are you shivering if you do not fear me?'

"I am shivering because of that clammy chill you brought in with you," I said. "You lug the atmosphere of a Frigidaire around with you."

"You don't tell me?" Death said. "I must correct that. I must pack an electric pad with me. Do you think that is why I seem so unpopular wherever I go? Do you think I will ever be a social success?"

"I am inclined to doubt it," I said. "Your personality repels many persons. I do not find it as bad as that of some others I know, but you have undoubtedly developed considerable sales resistance to yourself in various quarters."

"Do you think it would do any good if I hired a publicity man?" Death asked. "I mean, to conduct a campaign to make me popular?"

"It might," I said. "The publicity men have worked wonders with even worse cases than yours. But see here, D., I am not going to waste my time giving you advice and permitting you to linger on in my quarters to get me talked about. Kindly do a scrammola, will you?"

Death had halted his tears for a moment, but now he turned on all faucets, crying boo-hoo-hoo-hoo.

"I am so lonesome," he said between lachrymose heaves.

"Git!" I said.

"Everybody is against me," Death said.

He slowly exited and, as I heard his tears falling plop-plop-plop to the floor as he passed down the hallway, I thought of the remark of Agag, the King of the Amalekites, to Samuel just before Samuel mowed him down: "Surely the bitterness of death is past."

Walter Winchell

Walter Winchell (1897–1972) wrote ''like a man honking in a traffic jam,'' said an awed Ben Hecht. Sunday evenings at nine Eastern time, Winchell was on the air declaiming, ''Good evening, Mr. and Mrs. North America and all the ships at sea. Let's go to press!'' In 1940, about 50 million out of 130 million Americans heard Winchell and/or read his syndicated column in 1,000 newspapers, writes John Mosedale in his entertaining The Men Who Invented Broadway *(New York, 1981). The co-inventor was Damon Runyon. The two Hearst columnists led the way in conjuring a Broadway spun from their own infectious slang. Runyanisms included ''dukes'' (fists), ''cock-eyed,'' ''drop dead,'' ''kisser,'' ''equalizer'' (gun), ''cheaters'' (glasses), ''monkey business,'' and ''hot spot'' (predicament). Winchell coined or popularized (in Mencken's reckoning) ''whoopee,'' ''blessed event'' (birth of a child) ''moompitcher,'' and ''the Hardened Artery'' (Broadway).*

Winchell's reign began in the 1920s when his column first appeared in the New York Graphic, *best remembered for its faked composite photographs. Winchell did not fabricate, but was disinclined to check interesting rumors. When he moved in 1929 to the new Hearst tabloid,* The Mirror, *''On Broadway'' became the paper's biggest draw. In the 1950s, Runyan was dead, Broadway crumbling, and Winchell a lapsed liberal turned bitter rightwinger; after the* Mirror *folded in 1963, the column ended its run in fringe newspapers. I have taken random samples from 1930 to convey its early vigor. See Winchell's memoir,* Winchell Exclusive *(New York, 1975). The classic hostile deposition remains St. Clair McKelway,* Gossip: The Life and Times of Walter Winchell *(New York, 1940).*

PORTRAIT OF A MAN TALKING TO HIMSELF

New York Daily Mirror *October 23, 1930*

Funny how some people react to various things . . . That old cynic, Sime of *Variety,* says that he encountered over thirty people one day who told him they intensely disliked me—until I started paragraphing about Walda and Gloria . . . Which is a helluva reason to switch your opinion of a guy you've intensely disliked personally . . . The big sentimentalists! . . . I told Sime how every time I wrote about my sugar pies, two morns later the mail was choked with affectionate letters from all types of readers—mothers, fathers, grandfolks, chorines, business executives and others, who applauded the items about the kids . . . But how some others would write and tell me those items made them sick and would I cease? . . . Too bad about those people . . . I will continue writing about Walda and Gloria, not only to keep making those readers ill but because it brings me closer to my pussycats . . . two of the most important reasons I have to keep working . . . when my desire is to do a Crater . . . For a week at least . . . Or maybe a month, maybe . . . Not that I am tired Of It All, either . . . I'm not tired . . . I want more than twenty minutes once a week—to play grocer boy with Walda, and hear her seriously say: "Please, grocer boy, bring me some noodle shoop for my chillun, and some bibs for my babies, an' some tahtoos," which happens to mean potatoes . . .

A mag wants an interview, Ruth tells me . . . They would like also to interview the bride . . . Huh-huh . . . They'd better not get anywhere near June . . . That's my biggest worry . . . She would like to spill plenty about "success" . . . "When you were making $25 a week and we had one room and no bath on 47th Street, it was heaven!" she said the other day . . . Of course it wasn't "heaven" at all . . . It was terrible, that's what it was . . . How the deuce can you be happy on $25 a week? . . . To her, however, it was plenty jolly . . . No children—just June and me . . . Delicatessen from the joint near the Somerset, and a large container of Coca-Cola and sweet pickles . . . I didn't know Broadway existed then—even though it is a half block away . . . Used to get up at 8 to be at *Vaude News* office by 9 . . . Some fun, eh, June? . . . Say, that was a pretty important job, at that, lemme tell you . . . Wasn't I working for Albee? . . . Albee! . . . What a power he was . . . He's six feet deep today, and show business keeps going on, anyhow . . . That's what June keeps saying . . . "If you really care anything about the three of us you'd do something about it!" . . . "But I can't do anything about it," I say back. If I do not hustle around for the next paragraph, then what? Don't you see, honey? Paragraphs! Things! Stuff! To keep them from saying things. If I start getting careless, where's the coin coming from to pay for those fancy skirts, and your dressy boots, and those combinations, and those baby coats with the "pussy feathers" on Walda's collar, and the very "smart" and fancy 19-dollar dresses for Gloria and so forth? . . . And now look at Ray Long, of Cosmopolitan. He's mad, I think, He's mad because I had an appointment with him the other day at 4 and couldn't make it. I swore a long time ago I'd never make appointments. Long probably thinks I was giving him some top-hat stuff. But I wasn't . . . I tried to go to sleep a little earlier. So what happened? I was just about to fall off when I realized my radio stuff was only half done, and those dopes up there probably would delete a few paragraphs at the last moment, and what would I do for material if I didn't think up some right then? Well, three hours later I played dead, and when I got up there was just enough time left to read proof and make the radio station. So Mr. Long probably thinks I didn't think he was important enough to keep an appointment with him . . . Oh, I am so weary of remembering what people not to offend. Keefe was saying how he, too, gets the longing to tell certain people who get on his nerves that *do* get on his nerves, and Keefe was saying, also, that being nice to people you really don't give a damn about is what makes you die young . . . But you mustn't be rude to people . . . Why mustn't you be rude to people? . . . What if you are rude to people? . . . So they quit talking to you! . . . Now isn't that a shame—they quit talking to you! . . . That is why Miss Guinan is in the hospital—that, and a misplaced vertebrae, because of a careless osteopath or chiro . . . Texas is always being nice to people, until she feels like choking . . . She'd better be careful herself and not do a Jack Donahue, that's what Tex better be careful about . . .

That maniac who nearly smashed a cabload of people to death at 49th and 7th at 1:15 yesterday morning with his crazy driving is lucky the cops on the beat were asleep at the time . . . A Cadillac sedan it was, numbered 2V176 . . . You can't drive machines in heavy traffic carefully, and try to pick up heavy-hipped things at the same time . . . Dot Parker would dismiss it with: "Aw, excuse his lust!" . . . Excuse nothing! . . . He probably hasn't any children or anyone depending on him for support . . . I don't want to shove off because of some bum's carelessness . . . I want a few more years' time yet—so I can get enough practice at columning to learn how to throw *anything* into type and be two weeks in advance like the experts do it . . . Then, perhaps, I will find time to play grocer boy with Walda, and let her ride horseback on my back the way I see fathers do it in the movies . . . I want a snatch of what Victor Hugo described when he said Paradise was a place where parents are always young, and the children always little.

THINGS I NEVER KNEW TILL NOW

New York Daily Mirror *October 11, 1930*

That in the old days Grand and Broome Streets (N.Y.) were known as "Murderers' Row." Since, all business has moved uptown.

That Shakespeare was buried 17 feet down.

That the phenomenon which causes a blotter to absorb ink is capillary attraction, which is the same thing that causes blood to circulate in the body. (How could you all exist without me?)

That humming birds can fly backward, you lovely people!

That Friday, the sixth day of the week, is the only day named for a woman. From a goddess of the character of the Roman Venus.

That modern whalers capture them by electrocuting them.

That five hundred people assembled daily to see Louis XIV get up and go to bed. A plaz-ure!

That John Van Eyck, of Bruges, was the inventor of oil painting; the ancient painters used wax.

That Sterne's dead body was peddled by his landlady to defray the lodgings. You don't know who Sterne was? Imagine!

That the first circulating library in America was the outgrowth of a club called the Junto, established by Franklin in Phila.

That Balzac wrote and published forty volumes before he could write one to which he was willing to put his name.

That Chaucer, Burns, Lamb and Hawthorne (you're gonna die at this one) were all customs house officers!

That as late as 1850 you could buy building lots on 7th avenue right here in New York City, for $500.

That Nat Hale, said: "I regret that I have but one life to give to my country, etc." at the corner of E. Broadway and Market Street, where the statue now in City Hall really should be.

That Greenwich Village used to be a big tobacco farm, and as for Hudson street, which runs through it, it got its name from old Hendrik Hudson, who used to trade with the Injuns there, of course.

That a guy named Adrian Block built the first street of houses in New York City, which is why they call them "Squares" in Phila.

That Meyerbeer, the composer of "La Prophète," suffered all his life with stomach ache, and his music sounds that way.

That there's a goldfish hospital at 86 Cortlandt St., N.Y.

That 10,000 people can all sit down in Central Park at once, there being that many seats in the place.

That those mystery books that sell for four-bits at the seegar stores come out in first editions of 200,000, which probably breaks the record for first printings of all time; they being regular $2.00 books in format. I merely write 'em, I haven't time to draw pictures.

That of the distinguished groups of Concord literary men, Thoreau was the only native of the place.

That the poet Longfellow, when in collitch, used to warble in the Brunswick Unitarian Choir. Probably tenor.

That Victor McLaglen of the phlickers who uses such "What Price Glory?" language is the son of a Church of Eng. Bishop.

That Sam Fraunces, founder and first proprietor of the Fraunces Tavern, still at Pearl and Broad Sts., was a crapshooter.

That the English artist, Hudson, instructor of Sir Joshua Reynolds, could paint a head, but needed help to arrange the shoulders.

That the Gin Rickey was born in 1870 in the St. James Hotel, which used to stand at the corner of Broadway and 26th Street. It was christened after Col. Joseph K. Rickey of Calloway County, Mo.

That some of us who sell gossip for a living would welcome one of those threatened "knockout blows" so we could relax.

That the Hamilton, Ontario, telephone directory reveals a Dr. Pain, a Dr. Wrong and a Dr. Deadman.

That the Packard Motor Co. of 1861 Grandest Canyon, N.Y., has a Mr. Ford, and a Miss Overland. What! No Mr. Austin?

That Simeon Stylites, who was an orginator of ideas, had his imitators, too. Shipwreck Kelly swiping his pole sitting theme. Simeon, you recall, at the age of 40 built a pillar 60 ft. high, climbed to the top and stayed there for 30 years till he croaked.

That Congressman Bloom spends $100 yearly feeding pigeons which flock to his window sill six stories above Dizzy Avenue.

That li'l ol' Fifi D'Orsay, Greta Garbo's pal and Mary Pickford's chum, is our guest star Chewsday eve WABC at 7:30. And that the following week the Duncan Sisters will warble and then Joe Cook.

That you're a dope if you miss the show at the Broadhurst Sunday night for Heywood Broun. What a bill!

That Shelley read the Bible through four times before he was 21.

That it was a fancy of John Stuart Mill that when the greater evils of life shall have been removed, the human race is to find its chief enjoyment in reading Wordsworth's poetry.

And that most of the grouches who grumble how much they hate the world never pause to realize the world Doesn't Care!

Dorothy Thompson

American journalism has long been a men's club, with women customarily relegated to soft features and the society page. Not so with Dorothy Thompson (1893–1961). The self-assured daughter of a Methodist minister, she grew up in upstate New York, where "God was everywhere, Jesus was father's personal friend . . . and the Thompsons under special protection." She became a foreign correspondent for the Philadelphia Ledger *in the 1920s, sweeping through Europe, as her friend John Gunther remarked, "like a blue-eyed tornado." At a dinner party in Berlin in 1927, she met Sinclair Lewis, who instantly proposed marriage. Not long after their wedding the following year, the hard-drinking novelist complained to Vincent Sheean, "If I ever divorce Dorothy, I'll name Adolf Hitler as correspondent."*

In 1934, Thompson was expelled from the Third Reich; two years later her "On the Record" column started in the New York Herald Tribune. *She was the scourge of appeasers, and an admirer of President Franklin Roosevelt. In supporting him for a third term in 1940, she lost her job on the traditionally Republican* Tribune, *and her column moved to the* New York Post. *A recent sympathetic biography is* Dorothy Thompson *(Boston, 1973) by Marion K. Sanders. See also* Dorothy and Red *(Boston, 1963), by Vincent Sheean. Her columns are collected in* On the Record *(Boston, 1939), from which the following are taken.*

WRITE IT DOWN

New York Herald Tribune *February 18, 1938*

Write it down. On Saturday, February 12, 1938, Germany won the world war, and dictated, in Berchtesgaden, a peace treaty to make the Treaty of Versailles look like one of the great humane documents of the ages.

Write it down. On Saturday, February 12, 1938, Naziism started on the march across all of Europe east of the Rhine.

Write it down that the world revolution began in earnest—and perhaps the world war.

Write it down that what not even the leaders of the German army could stomach—they protested, they resigned, they lost their posts—so-called Christian and democratic civilization accepted, without risking one drop of brave blood.

Write it down that the democratic world broke its promises and its oaths, and capitulated, not before strength, but before terrible weakness, armed only with ruthlessness and audacity.

What happened?

On February 4, Hitler made a purge of his army. He ousted his chief of staff and fourteen other generals. Why? Because the army leadership refused to undertake a brazen *coup d'état* against an unarmed friendly country—their German-speaking neighbor, Austria. Why did they refuse? Because of squeamishness? Hardly. Because they thought that Britain and France would interfere? Perhaps. Or because they themselves feared the ultimate catastrophe that would be precipitated for the future by this move? I think the latter is the best guess.

A week later, Hitler, with his reorganized army, made his move. How did he make it? He called in the Chancellor of Austria, Doctor von Schuschnigg, and gave him an ultimatum. Sixty-six million people against six million people. German troops massed before Passau, on the Danube, before Kufstein and Salzburg in the Alps. Hitler's generals stood behind him as he interviewed the Austrian chancellor. Hitler taunted his victim. "You know as well as I know that France and Britain will not move a hand to save you." Under such circumstances there emerges what Hitler, on Sunday, will doubtless hail as a friendly reconciliation between two German-speaking peoples and the consolidation of peace in eastern Europe.

What does the Chancellor of Austria really think about Naziism?

He expressed himself hardly more than a month ago, on January , in the *Morning Telegraph* of London.

This is what he said:

"There is no question of ever accepting Nazi representatives in the Austrian cabinet. An absolute abyss separates Austria from Naziism. We do not like arbitrary power, we want law to rule our freedom. We reject uniformity and centralization. . . . Christendom is anchored in our very soil, and we know but one God: and that is not the State, or the Nation, or that elusive thing, Race. Our children are God's children, not to be abused by the State. We abhor terror; Austria has always been a humanitarian state. As a people, we are tolerant by predisposition. Any change now, in our *status quo,* could only be for the worse."

And he spoke in a room where hangs the death mask of his predecessor, Chancellor Engelbert Dollfuss, his own greatest friend, who was assassinated by the Nazis in 1934. And in the adjoining room, a lamp burns continually before a shrine, which belonged to Dollfuss, and is set on the spot where he fell.

In 1933, to please another despot, Mussolini, Dollfuss himself dissolved the Social Democratic Party and shot workmen in their own homes. Not to please Austria. To please Mussolini. And the little daughter of Dollfuss said to the child of a friend of mine: "Does your father cry all the time? Mine does."

Why does Germany want Austria? For raw materials? It has none of consequence. To add to German prosperity? It inherits a poor country with serious problems. But strategically, it is the key to the whole of central Europe. *Czechoslovakia is now surrounded. The wheat fields of Hungary and the oil fields of Rumania are now open. Not one of them will be able to stand the pressure of German domination.* One of them, and one only, might fight: Czechoslovakia. And that would mean: either another Spain or, immediately, a world war.

It is horror walking. Not that "Germany" joins with Austria. We are not talking of "Germany." We see a new Crusade, under a pagan totem, worshiping "blood" and "soil," preaching the holiness of the sword, glorifying conquest, despising the Slavs, whom it conceives to be its historic "mission" to rule; subjecting all of life to a collectivist militarized state; persecuting men and women of Jewish blood, however diluted it may be; moving now into the historic stronghold of Catholic Christianity, into an area of mixed races and mixed nationalities, which a thousand years of Austro-Hungarian Empire could only rule tolerably with tolerance. And led by a patricide. For Adolf Hitler's first hatred was not Communism, but Austria-Hungary. Read "Mein Kampf." And he loathed it for what? For its tolerance! He wanted eighty million Germans to rule with an iron hand an empire of eight million "inferiors"—Czechs, Slovaks, Magyars, Jews, Serbs, Poles and Croats.

Today, all of Europe east of the Rhine is cut off completely from the western world. The swastika banner, we are told, is the crusader's flag against Bolshevism! Madness! Only the signs of the flags divide them. Oswald Spengler wrote, in our times, "We shall see the era of world wars and of Caesarism." Ortega y Gasset wrote, "We shall see the rise of barbarism."

Both are here recorded—in the morning newspapers.

And it never needed to have happened. One strong voice of one strong power could have stopped it.

Tomorrow, one of two things can happen. Despotism can settle into horrible stagnation, through the lack of real leadership and creative brains. For the law of despotisms is that they decapitate the good, and the brave, and the wise. The Danubian Basin, into which Hitler now moves, has ruined many. And a wiser man than Hitler, Bismarck, fought his first war to break Germany loose from its headaches. Perhaps, then, all of Europe east of the Rhine will become, eventually, a no-man's land of poverty, militarism and futility. But none the less a plague spot.

More likely the other law of despotism's nature—the law of perpetual aggressiveness—will cause it to move always, farther and onward, emboldened, and strengthened, by each success.

To the point where civilization will take a last stand. For take a stand it will. Of that there is not the slightest doubt.

Too bad that it did not take it this week.

OBITUARY FOR EUROPE

New York Herald Tribune *September 21, 1938*

To write with calm about the news that broke on Monday in London is almost impossible, but calm is needed to consider that news and to review as many of its implications as it is possible to envisage.*

One can, of course, make a general statement no less true because it is sweeping.

On Monday the prime minister of Great Britain and the prime minister of France announced to the world that they had decided to throw the last democratic republic east of the Rhine into the jaws of the Nazis.

In doing so France broke a solemnly given pledge, reiterated time and again within the very last few weeks. In doing so Great Britain cut herself loose from a republic of which she herself was one of the prime architects.

But this is not by any means the worst aspect of what has happened. Great Britain could, from the beginning, have withheld herself from the whole matter.

She could silently have decided that she stood for isolation from the problems of Europe east of the Rhine. She could have left the matter to her ally, France—let France, who was definitely committed, take the lead—and abided by the results.

*France and Britain announced acceptance of the partition of Czechosolovakia, as demanded by Hitler, so that its German-speaking minority could establish a pro-Nazi ministate.

Instead, Great Britain first entertained Mr. Henlein; then took the lead and sent the Runciman commission to Czechoslovakia. Czechoslovakia, when that commission arrived to "make peace," was an extremely strong state, with a centralized government.

Lord Runciman persuaded the Czechoslovak government greatly to weaken that state by making concessions to the minority groups which would decentralize the government.

The Czechs made those concessions, all of them except one, for the sake of "conciliating Hitler" and keeping the peace of Europe.

The one which they refused to make they were advised against even by the *Times* of London—namely, the right of the Germans to set up a Nazi state inside Czechoslovakia.

They were even prepared to negotiate further until Henlein, on German soil, offered them an ultimatum.

All the concessions were made to show good will toward Great Britain and to prevent Czechoslovakia's allies from having to go to war to save her. It was known, it was announced, that Lord Runciman believed these were all the concessions which any national state could decently be asked to make.

A week ago these British recommendations were rejected by Hitler —that is to say, by the head of an outside state.

And now Great Britain took up the ultimatum from Hitler, without even consulting Czechoslovakia, and France, her ally, permitted Great Britain to do so, and even collaborated.

And Great Britain and France, on behalf of Czechoslovakia though by no means delegated to do so by her, accepted the ultimatum.

And with this goes the last vestige of law in Europe, the last shred of prestige of either Great Britain or France, the last hope of settling anything by negotiation, compromise, treaty and law, or by anything whatsoever except sheer brute force.

And with it goes the last shred of respect for what calls itself 'democracy,' but what has plainly degenerated into sheer personal government inside the so-called democratic great powers.

For who instructed Mr. Chamberlain to go to Berchtesgaden? Who instructed him to embark upon a course which would prove to be irrevocable? Under whose orders was he turning all Europe either over to Fascism or over to revolution and war, or a combination of the last two?

Under whose instructions was he personally dissolving the League of Nations?

Not only did France and Britain desert Czechoslovakia, but they weakened her for defense in advance; they encouraged her to delay; they undermined the authority of her leaders at home—undermined it, because what Czech today believes that Mr. Beneš and Mr. Hodza were wise in trusting to the good will and honorable intentions of France and Britain?

And now, finally, they have even broken down her moral case! They have put her in the position where, in the minds of hundreds of thousands of people naturally ignorant of Czechoslovakia and of central Europe, she, and not Germany, will be considered the aggressor!

Not only have they assassinated her, but they have besmirched her character!

Czechoslovakia never had a propaganda ministry until last week. In all the years that I have been a journalist, I have never received a single piece of unsolicited information from the Czech government! But I know Czechoslovakia as I know every country in Europe where there is a German minority.

But this is not all, nor is it even, in the wide view, the most important thing.

Great Britain and France, by their action, have aligned themselves on the side of the racial theories of Hitler, against the conception of national sovereignty for which Britain has consistently stood and for which she must stand for sheer self-preservation.

The implications are appalling! First of all, Hitler's doctrine threatens with disintegration every small state where there is a German population, or any other racial minority—Switzerland, Belgium, Denmark, Poland, among the first; Rumania, Yugoslavia, among the second.

And the British and French empires? The doctrine applied to Czechoslovakia by the British will flare in Arabia and in Morocco, and throughout all the colonies of both empires!

Germany will rule from the North to the Black Sea—rule with her mobilized "Nation in Arms," her concentration camps, her arbitrary law, her barter policy, which makes her a parasite upon the whole world economy and its chief underminer. And that is not all.

The very thing that Britain has tried to stem—the awful ideological war, with all its accompaniments of civil strife and revolution—has now been let loose! The one thing that could have prevented a horrible conflict between Fascism and Communism inside all countries was the unquestioned honor, prestige, wisdom and courage of the great democracies. The great democracies today are worth nothing whatsoever in the minds of the common people of the world.

If it is possible to halt the internal disintegration of France, that will be a miracle.

Nor shall we escape the inexorable effects of what has happened.

The United States this morning is isolated. But it has Fascism rampant on its southern frontiers, and that Fascism, enormously strengthened by victory, will become more virulent.

A totally victorious European Fascism, or European civil strife between ideologies, will have repercussions here.

And if we are to remain a democracy we shall have to make the most heroic, intelligent and united efforts, to achieve the fullest sense of responsibility, the greatest democratic discipline and inner collaboration of all elements and the deepest sense and appreciation of our communal existence.

NIHILISM EAST OF THE RHINE

New York Hearld Tribune *March 15, 1939*

There have been those who have believed that a free hand for Germany in eastern Europe would mean the efficient organization of that great complex of states held between the Rhine and the Russian frontier.

Ever since the war and the break-up of the Austro-Hungarian Empire publicists have been accustomed to speak of the "Balkanization" of eastern Europe, and some have looked to Germany to bring a greater amount of order into this territory. Some people believed that this would be the function of the Nazis.

On the belief that Nazi Germany is first of all a legitimate government and, secondly, represents a conservative and integrated order of society, much of the western world has gone astray.

As anyone who reads "Mein Kampf" carefully, following its main threads through the miasma of disorganized material, must see, Naziism does not represent an organized state at all. It is nothing more or less than the kernel of a movement which will theoretically come to rest nowhere until it has established the supremacy of the Teutonic and allied races over the earth, reducing the rest to vassalage.

Such phrases as "the union of all Germans within the Reich" are totally misleading as descriptions of Nazi policy. They have become more and more misleading since the conquest of Austria.

For it now appears that Hitler lays claim to the whole territory of the former Austro-Hungarian Empire, in which non-Germans outnumber Germans in the ratio of four to one.

Hitler's attitude toward the Austro-Hungarian monarchy is made crystal-clear in his book. He believes that this territory consists of inferior Slavs, Magyars and other not quite white people who should be ruled by the "Herrenvolk"—the Germans.

It is now possible to see, first, that he really means this and, second, how he intends to accomplish it. He intends to accomplish it by the pulverization, the atomization, of the existing states to the point where they will be utterly helpless and can be enslaved and plundered.

There was Czechoslovakia: a well-organized, on the whole prosperous, well-governed, liberal state, its finances in excellent order, its defenses admirable, its people enjoying the freedoms of western civilization, its education progressive, its population industrious and peaceable.

Since last September this state has been pulverized.

The lopping-off of the Sudetenland was accomplished with the collaboration of Britain and France on the theory that the Sudetens should "go home to the Reich"—a Reich to which they had never in all history belonged.

The lopping-off of Slovakia can hardly thus be explained. It is the result of a systematically engineered *coup d'état,* effected by a complete travesty of the democratic process, by revolutionary agents, treasonable Slovaks and unexampled high-pressure propaganda in a country where the press had been silenced in advance by Nazi threats.

An independent Slovakia is a monstrosity. It is incapable of living by itself as a state, and will simply become a vassal.

What is left of a progressive state—Bohemia and Moravia—is now an island completely surrounded by Germany. These, too, will be pulverized.

When this is accomplished Poland will be cut off from Hungary, and will be, probably, the next state to be smashed.

Germany will claim the provinces of Pomorze, Poznan and Silesia as being German. These are the "Sudetenlands" of Poland. They are not German and they never have been German, as far as the population is concerned.

The so-called Polish Corridor was always a land predominantly populated by Poles. Since 1918 it has become almost wholly populated by Poles.

But Germany may also claim the former Austrian provinces of Poland, to which, ethnically, she has not a shred of claim.

The technique will be exactly the same as that used against Czechoslovakia. There will be a continuance of the provocation of the Polish minority in Danzig in the hope that this will bring retaliations against Germans in the Corridor. We shall then hear that a great power of eighty millions cannot tolerate the barbarous treatment of its people; there will be a mobilization on the Polish border; another ultimatum; Poland will be reminded that her ally, France, can do nothing whatever to help her; Stalin will sit tight, and Germany will accomplish her purpose 'without spilling a drop of blood.'

Probably Germany will try to make a deal with Poland and will offer her an outlet to the sea in Libau, in Latvia. This will accomplish a double purpose. It may persuade the Poles to take national dismemberment more gracefully, and it will break up Latvia, thus carrying the pulverization process a step farther.

But the acquisition of these territories will incorporate in Germany nearly a million Poles, and the incorporation into Poland of that section of Latvia necessary for an outlet to the sea will put nearly a million Latvians into Poland, and the whole process will change Poland from being a national state into being a conglomeration of nationalities, pulverized, atomized and ripe for complete vassalage.

Of course, as the Nazi empire makes its way, taking new territories, it will make no compensation whatsoever to the states which it is robbing.

The modern port of Gdynia, which has been built with millions of Polish money, will be simply taken by the Germans, just as what they have done in Czechoslovakia is a huge act of expropriations.

The whole of southern and eastern Europe is in for this pulverization and plunder process.

Nazi Germany does not intend to allow a single strong or numerous state to exist between her frontiers and the Russian border.

The end result of all this is revolution and chaos. It can have no other result. Everything that Germany is acquiring now is an ultimate liability unless she terrorizes the whole world. It cannot be administered in any orderly fashion, for the Nazi Reich is incapable of tolerable administration of peoples who have once enjoyed national freedom. It can only enslave them.

Most significant is Stalin's speech in which he proclaims the isolation of Russia. Not a word is being said now in Germany about the Soviet Ukraine. On the contrary, feelers are being constantly put out between Russia and Germany. The hope of certain French and British conservatives that Russia and Germany would spring at each other's throats and bleed each other white in a war is not at all likely to be fulfilled. Naziism will not make war. Naziism is making a great nihilist revolution.

Westbrook Pegler

Publisher E. W. Scripps called himself a "damned old crank," which also describes Westbrook Pegler (1894–1969). Like his friend and later adversary Heywood Broun, he came to column writing out of sports and honed a direct, graphic style. This he displayed to advantage in "Fair Enough" during a dozen years with Scripps-Howard, then with diminishing effect for Hearst. Pegler initially took on crooks and dictators, winning a Pulitzer Prize for exposing a gangster union in Hollywood. But the column declined into a shooting gallery for pet hates, notably "La Boca Grande" (Eleanor Roosevelt) and "Old Bleeding Heart" (Broun). He published three collections, T'Aint Right *(New York, 1936),* Dissenting Opinions *(New York, 1941), and* George Spelvin, American *(New York, 1942). Finis Farr's* Fair Enough *(New Rochellle, N.Y., 1975) is a balanced biography. To a pair of early columns on fascism, I append Alistair Cooke's memorial recollections as broadcast in his "Letter from America" for BBC radio. These letters are themselves a unique long-running aural column, beginning in March 1946 and continuing weekly after Cooke's retirement from the* Manchester Guardian, *a marathon feat too little known by his admiring public television audience.*

MUSSOLINI AND HIS TOUGHS

New York World-Telegram *circa 1935*

One rainy afternoon in Rome I went out in a taxi to circle the huge park, surrounded by a high wall, where Mussolini's house stands alone amid a grove of trees. The house was hidden from view, and the trip would not have paid out in results except for the presence along the sidewalk beneath the wall of a detail of cutthroats the like of which has not been assembled under one command since the days of His Majesty's British Black and Tans.

Mussolini's household guards were stationed at intervals of about fifty yards all around the park, each under his umbrella in a drizzle, all keeping up a nominal pretense of waiting for a street car, of which there was none in that vicinity, or for a friend, of which there can be none on earth. There were other secret policeman on the far side, one carrying a suitcase which could have contained the conventional Tommy gun. This one was strolling up the block the first time I passed. When I came around again he was strolling the other way with his suitcase.

They were a homey touch, in a way, for they were the first of that type that I had seen in several weeks in Italy, and for a moment I thought I was back in Miami Beach, Fla., in the winter colony of the artichoke and slot-machine aristocracy of our Northern cities.

At a distance, and in print, the castor-oil treatment with which the Black Shirts overcame political doubt seems not very brutal and even slightly humorous. But at close quarters, as the boys peered out from underneath their dripping umbrellas, it was not so easy to see the joke.

The customary procedure of these missionaries is to call at night in a body on the victim of a political error, twist his arms and legs until the bones crack, then pour down his neck a quart of castor oil which may be mixed with kerosene. It is inartistic to use kerosene, however, for enough castor oil taken all at once is enough to rupture the human plumbing and to bring about death from natural causes within a few days. There is always plenty of castor oil.

A few days later, on a trip to Pontinia, the tailor-made village which Mussolini built on a reclaimed swamp somewhat resembling the Florida glades, the strong-arm detail was out again, this time in greater numbers, and the scene was pleasantly reminiscent of those Kentucky Derbies at Louisville in the days of prohibition when all the Italian alky cookers and bootleggers east of the Rocky Mountains gathered for their annual convention.

They were everywhere. They mingled with the crowd and climbed the steps to overlook the press of people. They sidled up and looked earnestly into your eyes and gazed at your camera and studied your overcoat for bulges.

I felt nervous, for we had left Rome in the dark at 7 A.M. without breakfast, and I had in my pocket an eating apple whose contour might have suggested the hoodlums' pineapple.

This trip was organized after the fashion created by the late Tex Rickard when he was conducting ballyhoo parties for sports writers to the training camps of his fighters. Mussolini has a special preference for the foreign press on such occasions, for he has the natural desire that his achievements in this great reclamation work shall be known in other lands.

He has salvaged the swamps, built several towns, and lifted whole communities out of their old surroundings, to set them down in a new country where everything is ready for life. It is the kind of work that we in America are always going to do. Consequently, the petty Mussolinis of Mussolini's foreign press bureau, so lazy and stupid most of the time, round up all the foreign representation they can find when the Duce is about to put on one of his shows. If they don't turn up with enough foreign press Mussolini gets sore and gives them hell.

So there were about fifty of us on this trip, and we all had a long look at the Duce putting on his No. 3 routine. This is the one in which he is a kindly man of the people, and the living spirit of a proud and mighty but friendly race demanding only justice for Italy. It is said to be the best act in his repertoire, and I will say that I have never seen anything of its kind, although Red Mike Hylan, when he was Mayor of New York, resembled Mussolini around the ears.

After the act Mussolini received us in the council room of the new town hall. He is not as tall as I had expected, being about 5 feet 10 and thick in the barrel, but not fat. He moves around with the quick, jerky motions of a major general tearing along a company street on an inspection trip, and his yes men, like the humble shavetail, are frightened speechless in his presence except to throw back their heads and yell, "Doo-chay, Doo-chay."

His face surprised me, for I had expected a popeyed glare, an undershot jaw and the generally bloodthirsty mien, whereas there was a soft expression in Mussolini's eyes, he spoke in a quiet, civil tone, and his lips parted in a smile three or four times. His eyelids were red, as though he had been working nights. But the Black Shirts told me Mussolini never gets tired. His eyelids are naturally red.

A Frenchwoman journalist stepped out and dropped her wedding ring into his hand for his gold collection. Mussolini said he was touched, and went away in a pleasant mood. It was strange to feel drawn to this man, knowing that a word from him might cause another world war any minute, and that down in the village street and all over the place were those plain-clothes missionaries of his who might rub against that apple any minute and take impulsive steps.

THE AMERICAN DUCE

New York World-Telegram *Circa 1935*

Within a few months I have seen three dictatorships in operation, and the experience confirms my belief that the death of Huey Long removed a serious menace to the liberty of the United States. This may sound high-powered, but I have been seeing some grim demonstrations of the condition which we narrowly escaped. It is alarming to realize that a majority of the people of Louisiana regarded Huey as their friend, and were not only willing but eager to sign over their citizenship to him.

That was not so shocking when it happened in Italy, Germany and Austria, because those people had always lived under the authority of kings and armies. But the citizenship of the American was supposed to be a possession worth fighting for. Huey himself denied that he had dictatorial ambitions, insisting, on the contrary, that he was the protector of the common man, and I do not deny that he liked poor people better than businessmen and the rich, although he hated poor people who opposed him.

But Mussolini and Hitler both say the same thing, and, like Huey, they proceed on the theory that the common man is too dumb to know what is best for him.

By trick and stratagem Huey Long had abolished the Legislature in Louisiana, and the two houses which were jumping through hoops for him at the time of his death were just as farcical as the so-called legislative bodies which answer the commands of the dictators in Rome and Berlin.

Elections in Louisiana also had been reduced to absurdity by the application of various laws which Huey had enacted, particularly the law permitting him to engage unlimited numbers of poll inspectors at $5.00 each to protect the purity of the ballot.

Huey himself snickered as he pronounced the words "purity of the ballot" in explaining the purpose of this bill to the committee in Baton Rouge. For his Legislature knew that it really granted him the power to buy votes at $5.00 each, sufficient to control the result in any doubtful precinct and charge the cost to the public treasury.

Like Mussolini and Hitler, Huey was organizing a terrorist organization under the authority of the State Department of Criminal Investigation, with no limit on the number of operatives or the amount to be spent meeting the pay roll. He already had the courts and the lawyers under control, and the Department of Criminal Investigation, if it had reached its full development, would have been a terrible force.

In Germany the Brown Shirts have now been superseded by the army and have lost most of their old power, but they were a bad lot when they were running wild. They could drag a man out of his bed at night and put him away in a concentration camp without the merest pretense of a trial. And often when a Brown Shirt happened to owe money to the prisoner, the accused was executed for treason or shot in the back while attempting to escape.

It was the same in Italy in the early years of Fascism, and it would have been the same in our country under Huey Long if he had lived and enjoyed a few more of the political breaks which were bringing him along so alarmingly up to the day of his death.

Another characteristic which Huey shared with Il Duce and Der Fuehrer was his cruelty toward people who wouldn't quit to him. He enjoyed seeing unarmed men beaten by his armed guards. He enjoyed feeling that a machine gun was mounted at his hotel in Baton Rouge, ready to pour it into anyone who opposed him. And if he had lived to develop his dictatorship he would not have hesitated to fire on the common man.

For Huey had the heart and soul of a demagogue and dictator, and his closest associates in the development of his power were men of the same brutal type as Goebbels, Streicher, Roehm and the rest of Hitler's handy men.

Huey used to boast of the mileage of paved roads and the bridges which he had built and of the grandeur of his state capital and the buildings of the state university. But Mussolini and Hitler also boast of their roads and bridges, railroad stations and model tenements. But the state capitol of Louisiana had ceased to be a Legislature, and the university, like Heidelberg and Munich, was no longer free.

Huey lynched freedom in the university when he kicked out of school half a dozen students of journalism for criticizing him in print, and his university president was a promising candidate for the office of director of education under the national dictatorship. He also had a clergyman lined up for the job of national bishop, and his system of persecuting his opponents by raising their taxes was a method of confiscation adjusted to the early stage of an elastic plan. He refused to permit any inspection of the public treasury accounts, and his appeal to enthusiasm in public appearance was as magnetic as Mussolini's or Hitler's, with the additional menace of good-natured hilarity.

He could make the suckers laugh. There is no denying that he was a good fighter; all dictators are. But Huey was stealing the freedom of the very people he claimed to love, and he might have had them completely at his mercy in four years more.

Probably his method was impulsive rather than studied, but he was following exactly the system that had made slaves of the Italians and Germans. It is no laughing matter that such a man with such a plan was hell bent for national power when mere chance cut him down.

AN OUTRAGEOUS MAN

by Alistair Cooke

British Broadcasting Corporation, "Letter from America" *June 29, 1969*

Westbrook Pegler died suddenly the other day and I was saddened. I am puzzled to say why this should be so. I was never close to him, he was seventy-four, he'd been sick in more than a physical way for years, and he had not been seen in print much in a decade or so.

Scurrility was his trade, you might say, and in the 1930s and '40s it was a breathtaking thing to see how close he could sail to the wind of the libel laws. Long before the end he lost all skill in coming about when the wind was raging. He grossly libeled an old newspaper friend by describing him, among other repulsive things, as a coward, a war profiteer, a fugitive from the London blitz in the bowels of comfortable hotels, "an absentee war correspondent," and he wrote also that the man had once gone "nuding along the road with a wench in the raw." This spasm of whimsy and malice cost Pegler's employers $175,000 in the biggest libel settlement awarded up to that time.

He was always a scornful man, but after that his scorn turned rancid and he babbled on and on about old enemies, about both President and Mrs. Roosevelt long after they were dead. His last years seemed to have been spent in fuming total recall of all the Presidents he had watched, from the first Roosevelt to John F. Kennedy, whom he called "a mean, ratty, dough-heavy Boston gang politician." Toward the end even the monthly magazine of the John Birch Society found his last piece, on Chief Justice Warren, too raw to print.

You would think that here was the case of great talent gone to seed. And, as a man certainly, he was for years simply thrashing in deep water and making incoherent sounds before he went under. Why, then, should I have felt sad at his going? It is because, I think, of a fact of relativity not mentioned in Einstein's theory. Some people are so bristling with life that when they have been dead for years they still seem to be on call, whereas many sweet but pallid people who are up and about have nothing more to offer.

Pegler, newspaperman, is the man I am talking about, and he had a lion's share of the vitality that outlasts its time and place. Picture him first. A big, hulking man but erect as a grandee, with glimmering blue eyes that flashed an ultimatum to all simple believers and all secret slobs. A grim, mischievous Irish mouth. Two shaggy, forked eyebrows to stress that the message—via Western Union—came from nobody but Mephistopheles. Plainly not the kind of reporter to be brushed off by a handout, or a telephone call, or a presidential "no comment."

He was what they used to call a muckraker, and in his middle years he was the best. Once, over a drink in Denver, when I asked him what he was doing so far from his den in Connecticut, he said he'd heard that a couple of insurance companies had suddenly shown alarming fat profits. "The trail led out here," he said, "and I thought I might come and—sink a pick. Could be pay dirt." I imagined every insurance man in the West out of bed and doctoring the books.

When, soon after, he appeared in Hollywood to look into the way the motion-picture unions worked, his hosts should have been warned. At that time a man named Willie Bioff was the labor boss of the movie unions and a man highly thought of by the bigwigs of the New Deal.

Pegler threatened nothing and nobody. He mooched around the studios and the houses in Beverly Hills and dingy offices downtown, and picked up a private scent that led him to Chicago and on to other, obscurer towns in the hinterland. He pored over police blotters and old newspapers and tramped off to interview this anonymous old man and that forgotten old madam. And after six months of pick-sinking he wrote a series of searing columns. The first one began with the firecracker of a sentence: "Willie Bioff is a convicted pimp." Period. By the time Pegler was through, so was Bioff, whose shady past Pegler had reconstructed with the tedious accuracy of one of those picture puzzles in a hundred bits and pieces that emerge at last as the Taj Mahal. Bioff went to jail. So did the national president of a building employees' union, and the prison gates closed behind him on the whining phrase: "I've been Peglerized."

Pegler was born in Minneapolis, a skinny little runt, irascible at his first gasp. He delivered newspapers as a boy in the paralyzing northern winters, and one day he was shoving his little wagon along in Chicago when the arctic wind whistling in across the lake blew his papers away, and while he was chasing them his route book took off in another direction. He was disgusted. He tore off home and when the route boss called him up and said, "You're fined three dollars," he shouted back, "Oh, shut your face." And that was the end of his first job.

He was the son of a newspaperman and he never thought there was any other trade to follow. He started, at sixteen, as a $10-a-week cub reporter and described himself as "a raw kid as freckled as a guinea egg." He was transferred to St. Louis and at twenty-two was in London as a fledgling foreign correspondent. Then he went into the Navy in the First World War and after that turned sports reporter and in no time showed that he had a rowdy, biting style that was to make him a star. He moved up to a sports column, and when nothing much was happening he wrote about this and that and found himself sounding off about the cost of living, the gangsters, and the man in the White House. In 1933 he set up shop as a national columnist with no illusions about the pretensions of the breed. He called all columnists "myriad-minded us . . . experts on the budget who can't balance an expense account, pundits on the technological age who can't put a fresh ribbon in their own typewriters, and resounding authorities on the problems of the farmer who never grew a geranium in a pot."

The first fifteen years of his column were the great years. He wrote about everything with the unsleeping skepticism of a man in the bleachers who had watched many a dumbbell turn into a national hero on the baseball mound, who had seen many a horserace fixed and who was therefore quite ready to believe that a labor leader, a Governor, a Secretary of the Treasury or even a President might be no better than he should be.

He refused to be snowed by the vocabulary of the sociologists or distracted from his own horse-sense by the hushed pronouncements of the Walter Lippmanns and the Arthur Krocks, what he used to call "double-dome" commentators. . . .

Well, in time Pegler's meanness got to be a national scandal, and the obituary writers made much—too much, I think—of his almost psychotic hatred of the Roosevelt family, and his ornery conviction that practically all Americans of Slavic origin were probably communists. In life, in fact, he was most of the time, and in a private room, an amiable and surprisingly softy-spoken companion. But the path to his prose led through the bile duct. He had a perverse love of demoting all current heroes in a single phrase. Vice President Henry Wallace was "old bubble-head." J. Edgar Hoover, when everybody thought of him as the national scourge of all evil men, was put down as "a nightclub fly-cop." Mayor Fiorello La Guardia, the plucky Little Flower to the citizens of New York, was to Pegler "the little padrone of the Bolsheviki." And though it may seem odd, his devotion to Franklin Roosevelt till America got into the war was an expression of this same contempt for people in power, for he saw Roosevelt as the champion of "the hired help" against "the meanness of a complacent upper class."

He was brought up in the tough and talented school of Chicago reporters when Chicago was the best newspaper town in the country. And when he came East, he carried with him this air of being a prairie lad permanently unfooled by the rich, the genteel, the powerful, and all foreigners. He must have adopted the manner early on, possibly as a small fry's defense against the jeering reminders of his gang that his father (a diligent and respectable Cockney immigrant) was English and his mother was Canadian: Irish-Canadian but still Canadian. Like many other Midwesterners, Pegler was specially on his guard against any beguilements that came from England and the English. . . .

It's true that Pegler often got angry about many foolish things and never forgot a grudge. But just as often his indignation was nobly directed against unfashionable targets, and sometimes his scorn made Dean Swift read like Lewis Carroll. He bucked the Ku Klux Klan when it was dangerously powerful, and he belabored the ruthlessness of union leaders when their power was sacrosanct. He wrote withering pieces out of Nazi Germany, which was more than many resident correspondents did, but when Hitler was everybody's Evil Eye and Stalin was his benign counterpart, Pegler saw Stalin as at least an equal monster. . . .

If the spiritualists are right, and Pegler is somewhere within the sound of these words, he is certainly tearing at his robes and bashing in his harp.

Ernie Pyle

Ernest Taylor Pyle (1900–1945) was well cast for reporting World War II from the private's vantage, as a roving Scripps-Howard columnist. Pyle had made a name for himself as a chronicler of back-country America, so self-deprecating he could not muster the nerve to introduce himself to another Scripps-Howard columnist, Eleanor Roosevelt. His home paper was the old Washington Daily News, *for which he had started out as an aviation reporter, coming from Indiana University and the* Laporte *(Ind.)* Herald. *He was killed on Okinawa, April 18, 1945. Pyle's wartime dispatches have been newly collected in* Ernie's War *(New York, 1986), edited by David Nichols with a foreword by Studs Terkel. His biographer is Lee G. Miller,* The Story of Ernie Pyle *(New York, 1950). What follows also draws from a compilation of his at-large columns,* Home Country *(New York, 1947).*

THE MIDWEST WIND

Washington Daily News *September 23, 1935*

I don't know whether you know that long, sad wind that blows so steadily across the thousands of miles of Midwest flatlands in the summertime. If you don't, it will be hard for you to understand the feeling I have about it. Even if you do know it, you may not understand.

To me the summer wind in the Midwest is one of the most melancholy things in all life. It comes from so far and blows so gently and yet so relentlessly; it rustles the leaves and the branches of the maple trees in a sort of symphony of sadness, and it doesn't pass on and leave them still. It just keeps coming, like the infinite flow of Old Man River. You could—and you do—wear out your lifetime on the dusty plains with that wind of futility blowing in your face. And when you are worn out and gone, the wind—still saying nothing, still so gentle and sad and timeless—is still blowing across the prairies, and will blow in the faces of the little men who follow you, forever.

One time in 1935, when I was driving across Iowa, I became conscious of the wind and instantly I was back in character as an Indiana farm boy again. Like dreams came the memories the wind brought. I lay again on the ground under the shade trees at noontime, with my half hour for rest before going back to the fields, and the wind and the sun and the hot country silence made me sleepy, and yet I couldn't sleep for the wind in the trees. The wind was like the afternoon ahead that would never end, and the days and the summers and even the lifetimes that would flow on forever, tiredly, patiently.

Maybe it's a bad job, my trying to make you see something that only I can ever feel. It is just one of those small impressions that form in a child's mind, and grow and stay with him through a lifetime, even shaping a part of his character and manner of thinking, and he can never explain it.

MY DAY WITH MRS. ROOSEVELT

Washington Daily News *circa 1937*

I don't know that Mrs. Roosevelt and I had anything especially in common, except that we both wrote columns and saw a lot of country. But I'd been in her camp for a long time; I considered her a mighty fine woman. And I thought it would be nice if our wandering reportorial paths should cross some time. I hoped it might be out in the country somewhere, where she wouldn't be so busy, and we could sit down and chat —just one old columnist to another.

Well, our paths crossed. In San Antonio, Texas. We two Washingtonians, meeting so far away from home.

I'd had it all planned how I'd sent a note to her secretary and ask if I might come up privately and say hello. They say she's awfully nice about such things. And after all, our two columns had run side by side for years in many newspapers. But do you know what I did? I lost my nerve. I fought with it a long time, and finally lost. I just couldn't send a note. So I got blue and lay down on the bed and said to myself. "Well, some other time, maybe."

But I had a friend on a San Antonio newspaper, and he said, "Why, you damn fool, you mean two travelers like you and Mrs. Roosevelt have never met? Why, this is news. Go on up and see her. If you're afraid to send a note, you can at least go to her press conference with our local reporters."

So, under his goading, I went. There were four local reporters, four photographers, and four high-school journalism students. And me—the thirteenth. She stood at the door, smiling, and shook hands with each of us as we introduced ourselves. I guess I really went because I hoped vaguely that Mrs. Roosevelt might recognize the almost unforgettable mug that ran in the column alongside hers. Or that she might catch the name as we introduced ourselves. But she didn't. She smiled upon me as upon all the others. I blushed, and sat in the corner with the two high-school boys. Mrs. Roosevelt looked at me frequently. I'm sure she must have thought, "How admirable it is for a bald-headed man to keep on trying to get through high school."

While the other reporters took notes, I just sat and watched how Mrs. Roosevelt handled things. Her graciousness had been mentioned many times, and I could see it had not been exaggerated. There wasn't a question she declined to answer, and she did practically no hedging; I've never seen anyone treat strange reporters with more intellectual honesty. And she was kind, even about the dumbest questions. During the embarrassing silences when we country reporters couldn't think of anything to say, she would fill in by elaborating on the last answer. She had a beautiful enunciation. She also split an infinitive now and then, which further warmed me toward her.

During her lecture tours Mrs. Roosevelt would be on the go from about six in the morning till late at night. Every minute was full; everybody tried to see her. She had almost no privacy.

Thinking of my own seemingly full days, I said to her, "Don't you get terribly worn out on these tours?"

She said, "Oh, no. They're much less strenuous than when I'm in Washington and have to keep up with the social routine." She looked right at me when she spoke, and I thought to myself, she knows I'm from Washington. She's trying to place me.

Then one of the reporters asked her what her program was for the day. She said she was to go out riding with some friends shortly, and after that she had to get back because "I've got a daily column to write, you know." She laughed when she said that, one of those "between us newspaperfolks" laughs. She started her smile at one end of the group, and bestowed it clear around the circle to me on the other end. It seemed a sort of knowing smile when it settled on me. Ah, she's got it, I thought. Now she knows who I am. She'll say something later.

We talked with her for about half an hour. One very intelligent and pretty girl reporter did most of the questioning. The high-school boys were so scared they never once opened their mouths. One high-school girl asked for an autograph, and got it.

"Is there anything else?" asked Mrs. Roosevelt. There wasn't. So we all stood up, and once again filed past, to smile and shake hands and say goodbye. I was the last in line. I guess maybe I hung back on purpose so I could be last—I don't know. I still had one faint hope that she really knew. I thought she might lean over and whisper, "Stay behind a minute, Ernie. Let's talk about our columns." But it was not to be. In that brief second of her smile and handshake I couldn't have been any more anonymous if I'd been a fish in the sea. And so—onward, and out the door.

That's how I met Mrs. Roosevelt.

HOW IT WAS ON ANZIO

Scripps-Howard Syndicate *March 28, 1944*

With the Allied Beachhead Forces in Italy,—When you get to Anzio you waste no time getting off the boat, for you have been feeling pretty much like a clay pigeon in a shooting gallery. But after a few hours in Anzio you wish you were back on the boat, for you could hardly describe being ashore as any haven of peacefulness.

As we came into the harbor, shells skipped the water within a hundred yards of us.

In our first day ashore, a bomb exploded so close to the place where I was sitting that it almost knocked us down with fright. It smacked into the trees a short distance away.

And on the third day ashore, an 88 went off within twenty yards of us.

I wished I was in New York.

* * *

When I write about my own occasional association with shells and bombs, there is one thing I want you folks at home to be sure to get straight. And that is that the other correspondents are in the same boat —many of them much more so. You know about my own small experiences, because it's my job to write about how these things sound and feel. But you don't know what the other reporters go through, because it usually isn't their job to write about themselves.

There are correspondents here on the beachhead, and on the Cassino front also, who have had dozens of close shaves. I know of one correspondent who was knocked down four times by near misses on his first day here.

Two correspondents, Reynolds Packer of the United Press and Homer Bigart of the *New York Herald Tribune,* have been on the beachhead since D-day without a moment's respite. They've become so veteran that they don't even mention a shell striking twenty yards away.

* * *

On this beachhead every inch of our territory is under German artillery fire. There is no rear area that is immune, as in most battle zones. They can reach us with their 88's, and they use everything from that on up.

I don't mean to suggest that they keep every foot of our territory drenched with shells all the time, for they certainly don't. They are short of ammunition, for one thing. But they can reach us, and you never know where they'll shoot next. You're just as liable to get hit standing in the doorway of the villa where you sleep at night, as you are in a command post five miles out in the field.

Some days they shell us hard, and some days hours will go by without a single shell coming over. Yet nobody is wholly safe, and anybody who says he has been around Anzio two days without ever having a shell hit within a hundred yards of him is just bragging.

* * *

People who know the sounds of warfare intimately are puzzled and irritated by the sounds up here. For some reason, you can't tell anything about anything.

The Germans shoot shells of half a dozen sizes, each of which makes a different sound of explosion. You can't gauge distance at all. One shell may land within your block and sound not much louder than a shotgun. Another landing a quarter mile away makes the earth tremble as in an earthquake, and starts your heart to pounding.

You can't gauge direction, either. The 88 that hit within twenty yards of us didn't make so much noise. I would have sworn it was two hundred yards away and in the opposite direction.

Sometimes you hear them coming, and sometimes you don't. Sometimes you hear the shell whine after you've heard it explode. Sometimes you hear it whine and it never explodes. Sometimes the house trembles and shakes and you hear no explosion at all.

But I've found out one thing here that's just the same as anywhere else—and that's that old weakness in the joints when they get to landing close. I've been weak all over Tunisia and Sicily, and in parts of Italy, and I get weaker than ever up here.

When the German raiders come over at night, and the sky lights up bright as day with flares, and ack-ack guns set up a turmoil and pretty soon you hear and feel that terrible power of exploding bombs—well, your elbows get flabby and you breathe in little short jerks, and your chest feels empty, and you're too excited to do anything but hope.

THEY ALMOST GOT ME

Scripps-Howard Syndicate *March 21, 1944*

With the Fifth Army Beachhead Forces in Italy,—When our bombing was over, my room was a shambles. It was the sort of thing you see only in the movies.

More than half the room was knee-deep with broken brick and tiles and mortar. The other half was a disarray all covered with plaster dust and broken glass. My typewriter was full of mortar and broken glass, but was not damaged.

My pants had been lying on the chair that went through the door so I dug them out from under the debris, put them on and started down to the other half of the house.

Down below everything was a mess. The ceilings had come down upon men still in bed. Some beds were a foot deep in debris. That nobody was killed was a pure miracle.

Bill Strand of the *Chicago Tribune* was out in the littered hallway in his underwear, holding his left arm. Maj. Jay Vessels of Duluth, Minnesota, was running around without a stitch of clothing. We checked rapidly and found that everybody was still alive.

The boys couldn't believe it when they saw me coming in. Wick Fowler of the *Dallas News* had thought the bombs had made direct hits on the upper part of the house. He had just said to George Tucker of the Associated Press, "Well, they got Ernie."

But after they saw I was all right they began to laugh and called me "Old Indestructible." I guess I was the luckiest man in the house, at that, although Old Dame Fortune was certainly riding with all of us that morning.

* * *

The German raiders had dropped a whole stick of bombs right across our area. They were apparently five-hundred-pounders, and they hit within thirty feet of our house.

Many odd things happened, as they do in all bombings. Truthfully, I don't remember my walls coming down at all, though I must have been looking at them when they fell.

Oddly, the wall that fell on my bed was across the room from where the bomb hit. In other words, it fell toward the bomb. That is caused by the bomb's terrific blast creating a vacuum; when air rushed back to the center of that vacuum, its power is as great as the original rush of air outward.

When I went to put on my boots there was broken glass clear up into the toes of them. My mackinaw had been lying on the foot of the bed and was covered with hundreds of pounds of debris, yet my goggles in the pocket were unbroken.

At night I always put a pack of cigarets on the floor beside my bed. When I went to get a cigaret after the bombing, I found they'd all been blown out of the pack.

The cot occupied by Bob Vermillion of the United Press was covered a foot deep with broken tile and plaster. When it was all over somebody heard him call out plaintively, "Will somebody come and take this stuff off of me?"

* * *

After seeing the other correspondents, I went back to my shattered room to look around again, and in came Sgt. Bob Geake of Fort Wayne, Indiana, the first sergeant of our outfit. He had some iodine, and was going around painting up those who had been scratched.

Bob took out a dirty handkerchief, spit on it two or three times, then washed the blood off my face before putting on the iodine, which could hardly be called the last word in sterilization.

Three of the other boys were rushed off to the tent hospital. After an hour or so, five of us drove out to the hospital in a jeep to see how they were.

We found them not in bad shape, and then we sat around a stove in one of the tents and drank coffee and talked with some of the officers.

By now my head and ears had started to ache from the concussion blasts, and several of the others were feeling the same, so the doctors gave us codeine and aspirin.

Much to my surprise, I wasn't weak or shaky after it was all over. In fact I felt fine—partly buoyed up by elation over still being alive, I suppose. But by noon I was starting to get jumpy, and by mid-afternoon I felt very old and "beat up," as they say, and the passage of the afternoon shells over our house gave me the willies.

We got Italian workmen in to clean up the debris, and by evening all the rooms had been cleared, shaky walls knocked down, and blankets hung at the windows for blackout.

All except my room. It was so bad they decided it wasn't worth cleaning up, so we dug out my sleeping bag, gathered up my scattered stuff, and I moved to another room.

The hospital has invited Wick Fowler and me to move out with them, saying they'd put up a tent for us, and I wouldn't be surprised if we took them up on it. There's such a thing as pressing your luck too far in one spot.

A NEW KIND OF WARFARE

Scripps-Howard Syndicate *March 30, 1944*

With the 5th Army Beachhead Forces in Italy,—You've heard how flat the land of the Anzio beachhead is. You've heard how strange and naked our soldiers feel with no rocks to take cover behind, no mountains to provide slopes for protection.

This is a new kind of warfare for us. Here distances are short, and space is confined. The whole beachhead is the front line. The beachhead is so small that you can stand on high ground in the middle of it and see clear around the thing. That's the truth, and it ain't no picnic feeling either.

I remember back in the days of desert fighting around Tébessa more than a year ago, when the forward echelons of the corps staff and most of the hospitals were usually more than eighty miles back of the fighting. But here everybody is right in it together. You can drive from the rear to the front in less than half an hour, and often you'll find the front quieter than the rear.

Hospitals are not immune from shellfire and bombing. The unromantic finance officer counting out his money in a requisitioned building is hardly more safe that the company commander ten miles ahead of him. And the table waiter in the rear echelon mess gets blown off his feet in a manner quite contrary to the Hoyle rules of warfare.

* * *

It's true that the beachhead land is flat, but it does have some rise and fall to it. It's flat in a western Indiana way, not in the billiard-table flatness of the country around Amarillo, Texas, for example.

You have to go halfway across the beachhead area from the sea before the other half of it comes into view. There are general rises of a few score feet, and little mounds and gulleys, and there are groves of trees to cup up the land.

There are a lot of little places where a few individuals can take cover from fire. The point is that the generalized flatness forbids whole armies taking cover.

Several main roads—quite good macadam roads—run in wagon-spoke fashion out through the beachhead area. A few smaller gravel roads branch off from them.

In addition, our engineers have bulldozed miles of road across the fields. The longest of these "quickie" roads is named after the commanding general here, whose name is still withheld from publication. A painted sign at one end says "Blank Boulevard," and everybody calls it that. It's such a super-boulevard that you have to travel over it in super-low gear with mud above your hubcaps, but still you do travel.

* * *

Space is at a premium on the beachhead. Never have I seen a war zone so crowded. Of course, men aren't standing shoulder to shoulder, but I do suppose the most indiscriminate shell dropped at any point in the beachhead would land not more than two hundred yards from somebody. And the average shell finds thousands within hearing distance of its explosion. If a plane goes down in No-Man's Land, more than half the troops on the beachhead can see it fall.

New units in the fighting, or old units wishing to change position, have great difficulty in finding a place. The "already spoken for" sign covers practically all the land in the beachhead. The space problem is almost as bad as in Washington.

Because of the extreme susceptibility to shelling, our army had moved underground. At Youkous and Thélepte and Biskra, in Africa a year ago, our Air Forces lived underground. But this is the first time our entire ground force has had to burrow beneath the surface.

Around the outside perimeter line, where the infantry lie facing the Germans a few hundred yards away, the soldiers lie in open foxholes devoid of all comfort. But everywhere back of that the men have dug underground and built themselves homes. Here on this beachhead the dugouts, housing from two to half a dozen men each, will surely run into the tens of thousands.

As a result of this, our losses from shelling and bombing are small. It's only the first shell after a lull that gets many casualties. After the first one, all the men are in their dugouts. And you should see how fast they can get there when a shell whines.

In addition to safety, these dugouts provide two other comforts our troops have not always had—warmth and dryness.

A dugout is a wonderful place to sleep. In our Anzio-Nettuno sector a whole night's sleep is as rare as January sun in sunny Italy. But for the last three nights I've slept in various dugouts at the front, and slept soundly. The last two nights I've slept in a grove which was both bombed and shelled, and in which men were killed each night, and yet I never even woke up. That's what the combination of warmth, insulation against sound, and the sense of underground security can do for you.

Walter Lippmann

Walter Lippmann (1889–1974) possessed, a British colleague wrote, ''the name that opened every door.'' While at Harvard ('10), he studied with Santayana and was sought out by William James; after his first book, A Preface to Politics *(1913), Theodore Roosevelt acclaimed him as the ''most brilliant young man of his age in the United States.'' He worked for Lincoln Steffens, then joined in founding* The New Republic, *advised on drafting President Wilson's Fourteen Points, and in 1924 was put in charge of the editorial page of the* New York Evening World, *the best in the country.*

After the World*'s demise in 1931, he moved to Washington to write ''Today and Tomorrow,'' initially for the* New York Herald Tribune, *then for the* Washington Post *until his retirement in 1967. In striving for reason and civility, Lippmann could seem bloodlessly remote; he assailed book-burnings in Germany but never discussed Nazi extermination camps. He was a better judge of policies than politicians; he once famously dismissed Franklin Roosevelt as ''a pleasant man who, without any important qualifications for the office, would very much like to be President.'' Yet Lippmann left an Olympian void that no successor has managed to occupy. Ronald Steel's* Walter Lippmann and the American Century *(Boston, 1980) is a superlative biography. The following selections are from* The Essential Lippmann *(New York, 1963), edited by Clinton Rossiter; and* Interpretations, 1933–1935 *(Boston, 1936), edited by Allan Nevins.*

THE VINDICATION OF DEMOCRACY

New York Herald Tribune *July 5, 1934*

To those who for one reason or another have been losing faith in constitutional government and democratic methods, the course of events in Germany must bring a sharp realization of what the alternatives really are* "The National Socialist State," says Dr. Frick, the Minister of the Interior, "is built up on unconditional obedience to the orders of the Führer and his deputies." How has this doctrine worked out? To express disagreement with the Führer is forbidden. Yet many Germans disagreed with him. The result was that they plotted underground. In place of free discussion there was conspiracy and intrigue. Furthermore, since there exists no orderly way of settling differences of opinion, the conspirators were presumably contemplating massacres and assassinations. To defeat them, to avert their treachery, the Führer then felt compelled to resort to massacre and assassination. And where does this leave him? It leaves him sitting on top of a people who have been taught that killing is politics.

* * *

A great deal of scorn has been poured out upon the endless talking done in representative parliaments. It is often tiresome. In great emergencies it may be dangerous. But this endless talking marks a very great advance in civilization. It required about five hundred years of constitutional development among the English-speaking peoples to turn the pugnacity and the predatory impulses of men into the channels of talk, rhetoric, bombast, reason and persuasion. Deride the talk as much as you like; it is the civilized substitute for street brawls, gangs, conspiracies, assassination, private armies. No other substitute has as yet been discovered.

The doctrine preached by the fascists that a nation can think and feel with one mind and one heart, except on details, is contrary to all human experience. For a short time, in a mood of exaltation or under the crushing power of terrorism, a nation may appear to be of one mind. But that cannot possibly last, and among highly civilized people it has neither been expected nor been desired. It is assumed that people will think differently and will have opposing interests and views. Unanimity is not desired, because people have learned that no man is omniscient and that therefore no man should be omnipotent. An opposition is just as much a part of the government as the party in power. Since unanimity is neither desirable nor for the long run possible, instead of suppressing the opposition, civilized countries guarantee it representation and opportunity to express its views.

*At the end of June, 1934, in a so-called blood-purge, a number of Nazi leaders and other prominent Germans were summarily executed.

* * *

Once a nation abandons these principles it is inevitably driven to the disorders which constitutional government gradually overcame. The idea that these armies of Black Shirts and Brown Shirts represent some great new twentieth century invention can be entertained only by those who have never read any history. They are unmistakably a reversion to political practices which prevailed in Western Europe up to about the seventeenth century. That they often have patriotic ideals, that they are often inspired by great zeal is nothing new. Cesare Borgia also saw visions of national greatness, and the armies of the Pretender felt they were saving England.

But the progress of civilization has required that they be suppressed. Our democratic principles are the product of this experience in overcoming the disorders of government by plot, intrigue, assassination, and partisan armies. Democracy is not the creation of abstract theorists. It is the creation of men who step by step through centuries of disorder established a regime of order. The forms of representative government may vary. They may be amended. But for its essential principle, that opposition is legal, and that it may win control of the government, there is among modern people no alternative except terrorism, assassination, and the continual threat of civil war.

SHOULD I TELL YOU HOW TO VOTE?

New York Herald Tribune *October 1940*

*To Alexander Woollcott**

My dear Alec:

Not only are you entitled to an answer to your letter, but I am glad of the chance to explain to you personally why, unless something develops which I do not now foresee, I am taking no part in this campaign. I am making this explanation to you personally as an old friend but I do not think I shall make any public statement during the campaign. For while I want my friends to understand what I do, I have an almost fanatical conviction that columnists who undertake to interpret events should not regard themselves as public personages with a constituency to which they are responsible.

It seems to me that once the columnist thinks of himself as a public somebody over and above the intrinsic value and integrity of what is published under his name, he ceases to think as clearly and as disinterestedly as his readers have a right to expect him to think. Like a politi-

*Lippmann used most of this letter to his former *World* colleague in "Today and Tomorrow."

cian, he acquires a public character, which he comes to admire and to worry about preserving and improving; his personal life, his self-esteem, his allegiances, his interests and ambitions become indistinguishable from his judgment of events. In thirty years of journalism I think I have learned to know the pitfalls of the profession and, leaving aside the gross forms of corruption, such as profiting by inside knowledge and currying favor with those who have favors to give, and following the fashions, the most insidious of all the temptations is to think of oneself as engaged in a public career on the stage of the world rather than as an observant writer of newspaper articles about some of the things that are happening in the world.

So I take the view that I write of matters about which I think I have something to say but that as a person I am nobody of any public importance, that I am not adviser-at-large to mankind or even to those who read occasionally or often what I write. This is the code which I follow. I learned it from Frank Cobb who practiced it, and abjured me again and again during the long year when he was dying that more newspaper men had been ruined by self-importance than by liquor. You will remember that he had had opportunity to observe the effects of both kinds of intoxication.

So I do not feel called upon to make a public explanation why I do not take a position in this campaign. Now I wish to tell you why I am not taking a position. For reasons which seem to me good, I have not wanted to take a position ever since Willkie was nominated and Roosevelt was re-nominated. By accident I do not have to take a position, even privately, because being a resident of the District of Columbia I have no vote. If I had a vote, I should have to choose, and having chosen I should not wish to take a political position privately which I did not explain publicly. But since I cannot vote, I have felt free not to choose and, therefore, to devote all my attention to what I feel are the transcendent problems posed by the calamitous fact that we are compelled to have an election during what may be the culminating crisis of modern history.

The first problem was how the campaign could be conducted so that it would not irreparably divide the nation on the issue of foreign policy, paralyze the action of the government, prevent rearmament, demoralize the free peoples abroad, and cause the American democracy to destroy men's faith in democracy. The second problem was how, once the election was over, to reunite the nation promptly because, obviously without unity, we shall go to a disaster.

Having written so much about the war and the national defense, I made up my mind before the conventions that if there was any way of keeping these questions detached from the party struggle and from candidates, it was my duty to do so. When Willkie was nominated, and after I had satisfied myself personally of the sincerity of his convictions, I felt that foreign policy could be kept out of the campaign, and that second only to the successful defense of Britain, this was the most important and fortunate event that had happened in the midst of an unbroken series of horrors and catastrophes. So I decided that if Willkie stood fast on his commitments and convictions, the best thing I could do in this campaign was to stand aside, try to deflate the foolish and embittering arguments of both sides, and to keep reminding people that they were going to have to live after November fifth.

Though the campaign is very bitter, it might have been infinitely more bitter if the two candidates had not been Willkie and Roosevelt. The bitterness which exists is class bitterness, a most dangerous thing, but it would have been a thousand times worse if in addition the country had been torn by a struggle between "appeasers" and "warmongers." We have escaped that, thanks to the nomination of Willkie and his fairly firm resistance to the isolationist politicians in his own party. But the ominous class conflict remains and, whoever is elected, the liquidation of that conflict is our immediate and our paramount task.

If Roosevelt is re-elected, a very large part of the most energetic part of the population will be disaffected, and it will require a supreme effort, I think, to produce in Roosevelt and in the defeated opposition a genuine spirit of conciliation and unity. If Willkie is elected, he will have a divided and incoherent party and there will be profound distrust among millions of people. To overcome this he will need from Roosevelt and the Democrats a kind of magnanimity and patriotism which is rare in politics.

There are in both camps men close to Willkie and to Roosevelt who see this and they are the men we must count upon to bridge the chasm which has been opened up. A few journalists who have kept themselves detached from the charges and counter-charges may, then, come in handy.

You asked me how I was going to vote. You have gotten back a little treatise. But I wanted to answer your question because it was you who asked it, and if I have been long-winded it is because I haven't the time to be more concise. . . .

REFLECTIONS ON GANDHI

New York Herald Tribune *February 3, 1948*

In the life and death of Mahatma Gandhi we have seen re-enacted in our time the supreme drama of humanity. Gandhi was a political leader and he was a seer, and perhaps never before on so grand a scale has anyone sought to shape the course of events in the world as it is by the example of a spirit which was not of the world as it is.

Gandhi was, as St. Paul said, transformed in the renewing of his mind, he was not "conformed to this world." Yet he sought to govern turbulent masses of men who were still very much conformed to this world, and have not been transformed. He died by violence as he was staking his life in order to set the example of non-violence.

Thus he posed again the perennial question of how the insight of the seers and saints is related to the work of legislators, rulers, and statesmen. That they are in conflict is only too plain, and yet it is impossible to admit, as Gandhi refused to admit, that the conflict can never be resolved. For it is necessary to govern mankind and it is necessary to transform men.

Perhaps we may say that the insight of the governors of men is, as it were, horizontal: They act in the present, with men as they are, with the knowledge they possess, with what they can now understand, with the mixture of their passions and desires and instincts. They must work with concrete and with the plainly and generally intelligible things.

The insight of the seers, on the contrary, is vertical: They deal, however wide their appeal, with each person potentially, as he might be transformed, renewed, and regenerated. And because they appeal to experience which men have not yet had, with things that are not at hand and are out of their immediate reach, with the invisible and the unattained, they speak and act, as Gandhi did, obscurely, appealing to the imagination by symbolic evocation and subtle example.

The ideals of human life which the seers teach—non-resistance, humility, and poverty and chastity—have never been and can never be the laws of a secular society. Chastity, consistently and habitually observed, would annihilate it. Poverty, if universally pursued, would plunge it into misery and disease. Humility and non-resistance, if they were the rule, would mean the triumph of predatory force.

Is it possible that the greatest seers and teachers did not know this, and that what they enjoined upon men was a kind of suicide and self-annihilation? Obviously not. Their wisdom was not naïve, and it can be understood if we approach it not as rules of conduct but as an insight into the economy and the order and the quality of the passions.

At the summit of their wisdom what they teach is, I think, not how in the practical issues of daily life men in society can and should behave but to what ultimate values they should give their allegiance. Thus the injunction to render unto Caesar the things that are Caesar's is not a definite political principle which can be applied to define the relation of Church and State. It is an injunction as to where men shall have their ultimate obligations, that in rendering to Caesar the things that are Caesar's, they should not give to Caesar their ultimate loyalty, but should reserve it.

In the same manner, to have humility is to have, in the last reaches of conviction, a saving doubt. To embrace poverty is to be without possessiveness and a total attachment to things and to honors. To be non-resistant is to be at last non-competitive.

What the seer points toward is best described in the language of St. Paul as the creation of the new man. "And that ye put on the new man, which after God is created in righteousness and true holiness." What is this new man? He is the man who has been renewed and is "no longer under a schoolmaster," whose passions have been altered, as Gandhi sought to alter the passions of his countrymen, so that they need no discipline from without because they have been transformed from within. Such regenerated men can, as Confucius said, follow what their hearts desire without transgressing what is right. They are "led of the spirit" in the Pauline language, and therefore they "are not under the law."

It is not for such men as them that governments are instituted and laws enacted and enforced. These are for the old Adam. It is for the aggressive, possessive, carnal appetites of the old Adam that there are punishments and rewards, and for his violence a superior force.

It is only for the regenerate man, whose passions have been transformed, that the discipline of the law and of power are no longer needed, nor any incentive or reward beyond the exquisite and exhilarating wholesomeness and unity and freedom of his own passions.

TO OURSELVES BE TRUE

Washington Post *May 9, 1961*

[Written in the wake of the Bay of Pigs debacle]

We have been forced to ask ourselves recently how a free and open society can compete with a totalitarian state. This is a crucial question. Can our Western society survive and flourish if it remains true to its own faith and principles? Or must it abandon them in order to fight fire with fire?

There are those who believe that in Cuba the attempt to fight fire with fire would have succeeded if only the President had been more ruthless and had had no scruples about using American forces. I think they are wrong. I think that success for the Cuban adventure was impossible. In a free society like ours a policy is bound to fail which deliberately violates our pledges and our principles, our treaties and our laws. It is not possible for a free and open society to organize successfully a spectacular conspiracy.

The United States, like every other government, must employ secret agents. But the United States cannot successfully conduct large secret conspiracies. It is impossible to keep them secret. It is impossible for everybody concerned, beginning with the President himself, to be sufficiently ruthless and unscrupulous. The American conscience is a reality. It will make hesitant and ineffectual, even if it does not prevent, an un-American policy. The ultimate reason why the Cuban affair was incompetent is that it was out of character, like the cow that tried to fly or a fish that tried to walk.

It follows that in the great struggle with communism, we must find our strength by developing and applying our own principles, not in abandoning them. Before anyone tells me that this is sissy, I should like to say why I believe it. Especially after listening carefully and at some lengths to Mr. Khrushchev, I am very certain that we shall have the answer to Mr. Khrushchev if, but only if, we stop being fascinated by the cloak and dagger business and, being true to ourselves, take our own principles seriously.

Mr. K. is a true believer that communism is destined to supplant capitalism as capitalism supplanted feudalism. For him this is an absolute dogma, and he will tell you that while he intends to do what he can to assist the inevitable, knowing that we will do what we can to oppose the inevitable, what he does and what we do will not be decisive. Destiny will be realized no matter what men do.

The dogma of inevitability not only gives him the self-assurance of a man who has no doubts but is a most powerful ingredient of the Communist propaganda. What do we say to him, we who believe in a certain freedom of the human will and in the capacity of men to affect the course of history by their discoveries, their wisdom, and their courage?

We can say that in Mr. K.'s dogma there is an unexamined premise. It is that the capitalist society is static, that it is and always will be what it was when Marx described it a hundred years ago, that—to use Mr. K.'s own lingo—there is no difference between Governor Rockefeller and his grandfather. Because a capitalist society cannot change, in its dealings with the underdeveloped countries it can only dominate and exploit. It cannot emancipate and help. If it could emancipate and help, the inevitability of communism would evaporate.

I venture to argue from this analysis that the reason we are on the defensive in so many places is that for some ten years we have been doing exactly what Mr. K. expects us to do. We have used money and arms in a long losing attempt to stabilize native governments which, in the name of anti-communism, are opposed to all important social change. This has been exactly what Mr. K.'s dogma calls for—that communism should be the only alternative to the status quo with its immemorial poverty and privilege.

We cannot compete with communism in Asia, Africa, or Latin America if we go on doing what we have done so often and so widely—which is to place the weak countries in a dilemma where they must stand still with us and our client rulers, or start moving with the Communists. This dilemma cannot be dissolved unless it is our central and persistent and unswerving policy to offer these unhappy countries a third option, which is economic development and social improvement without the totalitarian discipline of communism.

For the only real alternative to communism is a liberal and progressive society.

James Reston

So equable and reassuring is the pipe-smoking James (Scotty) Reston (1909–) that he sometimes surprised New York Times *readers with his gloomy decline-and-fall ruminations. His occasional sermons owed something to his origins in Clydebank, Scotland, just as his ''Uniquack'' machine for decoding political bombast has something to do with his early years as an Associated Press sportswriter. He joined the* Times *in London in 1939, moving in five years to Washington, where he won two Pulitzer prizes and presided as ''the Washington correspondent,'' or bureau chief. A graceful writer perhaps too addicted to symmetrical fair-mindedness, Reston continued as an emeritus columnist for two years after his formal retirement in 1987. His columns are gathered in* Sketches in the Sand *(New York, 1967) and* Washington *(New York, 1986); I have drawn on the latter for three interrelated pieces.*

FREEDOM VERSUS SECURITY

New York Times June 19, 1971

> *Here various news we tell, of love and strife,*
> *Of peace and war, health, sickness, death and life . . .*
> *Of turns of fortune, changes in the State,*
> *The falls of favorites, projects of the great,*
> *Of old mismanagements, taxations news,*
> *All neither wholly false, nor wholly true.*
>
> New London (Conn.) *Bee* March 26, 1800

Great court cases are made by the clash of great principles, each formidable standing alone, but in conflict limited, ''all neither wholly false nor wholly true.''

The latest legal battle, *The United States* v. *The New York Times,* is such a case: The Government's principle of privacy, and the newspaper's principle of publishing without Government approval.

This is not essentially a fight between Attorney General Mitchell and Arthur Ochs Sulzberger, publisher of the *New York Times.* They are merely incidental figures in an ancient drama. This is the old cat-and-dog conflict between security and freedom.

It goes back to John Milton's pamphlet, *Areopagitica,* in the seventeenth century against Government censorship or, as he called it: "for the liberty of unlicenc'd printing." That is still the heart of it: the Government's claim to prevent, in effect to license, what is published ahead of publication, rather than merely to exercise its right to prosecute after publication.

Put another way, even the title of this case in the U.S. District Court is misleading, for the real issue is not the *New York Times* versus the United States, but whether publishing the Government's own analysis of the Vietnam tragedy or suppressing that story is a service to the Republic.

It is an awkward thing for a reporter to comment on the battles of his own newspaper, and the reader will make his own allowances for the reporter's bias, but after all allowances are made, it is hard to believe that publishing these historical documents is a greater threat to the security of the United States than suppressing them or, on the record, as the Government implies, that the *Times* is a frivolous or reckless paper.

The usual charge against the *New York Times,* not without some validity, is that it is a tedious bore, always saying on the one hand and the other and defending, like *The Times* of London in the thirties, "the Government and commercial establishment."

During the last decade, it has been attacked vigorously for "playing the Government game." It refused to print a story that the Cuban freedom fighters were going to land at the Bay of Pigs "tomorrow morning." It agreed with President Kennedy during the Cuban missile crisis that reporting the Soviet missiles on that island while Kennedy was deploying the fleet to blockade the Russians was not in the national interest.

Beyond that, it was condemned for not printing what it knew about the U.S. U-2 flights over the Soviet Union and, paradoxically, for printing the Yalta Papers and the Dumbarton Oaks Papers on the organization of the United Nations.

All of which suggests that there is no general principle which governs all specific cases and that, in the world of newspapering, where men have to read almost two million words a day and select a hundred thousand to print, it comes down to human judgments where "all [is] neither wholly false nor wholly true."

So a judgment has to be made when the Government argues for security, even over historical documents, and the *Times* argues for freedom to publish. That is what is before the court today. It is not a black-

and-white case—as it was in the Cuban Missile crisis when the Soviet ships were approaching President Kennedy's blockade in the Caribbean.

It is a conflict between printing or suppressing, not military information affecting the lives of men on the battlefield, but historical documents about a tragic and controversial war; not between what is right and what is wrong, but between two honest but violently conflicting views about what best serves the national interest and the enduring principles of the First Amendment.

WHO ELECTED THE *TIMES?*

New York Times *June 24, 1971*

The public reaction to the publication of the Pentagon Papers has been overwhelmingly on the side of the newspapers, but there is a strong and vehement view that it is wrong, dangerous, and even criminal for a newspaper to assume responsibility for publishing private official documents without the consent of the Government.

Who, it is asked, elected the *New York Times?* How can outsiders judge better than the official insiders what damage may be done by publication of secret documents? By what right do newspapers presume to print official information which may embarrass the Government and give comfort to the enemy?

These are serious questions which deserve serious answers, for it is clear that the publication of the Pentagon Papers *has* embarrassed the Government, disclosed evidence of official deception, and in the process provided Hanoi, Moscow, and Peking with material for anti-American propaganda.

At first glance, it is a devastating indictment, but should documents not be published because they embarrass the Government? Nobody is arguing that newspapers have the right to publish the nation's war plans or troop movements, or anything else that would endanger the lives of the men in the American expeditionary force. But historical documents? Evidence that the Congress and the people were misled years ago—even if this embarrasses the Government and provides propaganda for the enemy? This is clearly another matter.

After all, every time Mike Mansfield, the opposition leader in the Senate, calls on the Government to end the war by a date certain, or any newspaper or preacher or group of citizens condemns the bombing or questions the loss of life or the diversion of resources or what the war is doing to divide and weaken the nation—all this is picked up by our adversaries and used against the United States.

Should we then suppress the documents because they "embarrass" the Government? Deceive the people about the record of the war? Submit to the Government's argument that publication will cause "irreparable injury" to the national defense rather than "irreparable injury" to the nation's reputation for fair dealing and plain and honest speaking to the Congress and the people? Confuse "embarrassment" to the Government and its officials with the security of the Republic?

In the absence of clear evidence that publication of these old documents is truly a threat to the defenses of the nation—which the Government has not provided—these are good political but bad philosophical and historical questions. Still, they are being raised by influential men and they come closer to the Marxist view of the press—that it should be a servant of the government—than to the American view of the press as defined in the First Amendment.

It is not good enough to suppress facts relating to the past, as distinguished from dangerous military information affecting the present or future, on the ground that this may be awkward. This resembles Nikolai Lenin's view of the press.

"Why should freedom of speech and freedom of press be allowed?" he asked in 1920. "Why should a government which is doing what it believes to be right allow itself to be criticized? It would not allow opposition by lethal weapons. Ideas are much more lethal than guns. Why should any man be allowed to buy a printing press and disseminate pernicious opinions calculated to embarrass the government?"

Well, many men who oppose publication of the Pentagon Papers don't go this far, but the violent opponents of publication, like Herbert Rainwater, the national commander of the Veterans of Foreign Wars, who is crying "treason," come very close to the Lenin thesis that opposition to the Government is unpatriotic or worse.

It is true that newspaper editors, raised in the American tradition of "publish and be damned," do not always know what damage they may do to the diplomatic process by publishing official documents. Their information is limited, and no doubt the official insiders know more than the outsiders, but even this is a dubious argument.

As Walter Lippmann wrote many years ago, you had better be careful not to go too far with the "insiders" argument. "For if you go on," he told the National Press Club in Washington on his seventieth birthday in 1960,

> . . . you will be showing how ridiculous it is that we live in a republic under a democratic system, and that anyone should be allowed to vote.
>
> You will be denouncing the principle of democracy itself, which asserts that the outsiders shall be sovereign over the insiders. For you will be showing that the people, since they are ignoramuses, because they are outsiders, are therefore incapable of governing themselves.

> If the country is to be governed with the consent of the governed, then the governed must arrive at opinions about what their governors want them to consent to. . . . Here we correspondents perform an essential service. In some field of interest, we make it our business to find out what is going on under the surface and beyond the horizon. . . .
>
> In this we do what every sovereign citizen is supposed to do, but has not the time or the interest to do for himself. This is our job. It is no mean calling. We have a right to be proud of it, and to be glad that it is our work.

WHY PUBLISH SECRETS?

New York Times *June 27, 1971*

A troubled friend wants to know why the newspapers don't leave the questions of secret documents and national security to the President. Let us suppose that we did.

Presidential power is now greater than at any other time in the history of the Republic. Ever since the invention of atomic weapons and intercontinental ballistic missiles, it is clear that the nation could be mortally wounded before the Congress could ever be assembled on Capitol Hill.

Accordingly, the balance of decisive power in the foreign field—but not over internal policy—has passed from the Congress, where it lay before the two world wars, to the White House. This may or may not have been what we wanted, but it was clearly what we had to do.

Other inventions tipped the balance of political power toward the President, especially nationwide television. It is at his disposal whenever he likes, with a studio in the White House. He has instant communications with the people and the world, all of which is necessary. The Congress cannot compete with him in the use of these modern instruments in the conduct of public policy.

But these unavoidable facts raise serious questions. Should such power not be subject to review by the representatives of the people? Should the Congress not know what is going on? Should the executive be free to use the power it needs to deal with the threat of nuclear war in undeclared wars like Vietnam? Should the press shut its eyes to any documents, even old historical documents, the executives chooses to mark top secret?

The fuss over the Pentagon Papers is only a symbol of a much larger problem. It is true that these papers raise questions of "national security," but the greatest threat to national security in this time is the division of the people over a war they have had to fight in accordance with decisions of governments that didn't tell them the truth. The nation is seething with distrust, not only of the Government but of the press, and the issue of the Pentagon Papers is merely whether we should get at the facts and try to correct our mistakes or suppress the whole painful story.

Fundamentally, this is not a fight between the Government and the press. It is not even a fight over the President's decisive power to defend the nation in an age of nuclear missiles. Congress has submitted to the scientific facts on the ultimate questions of nuclear war.

But now it has been asked, in the name of "security," not even to look at a historical analysis of a war it has financed but not declared, not to question the unelected members of the White House staff, who had access to the papers Congress could not see, and to respect the Administration's right to stamp "secret" on any documents it likes, and to keep them secret years after the event, when officials long out of office are writing their own versions of history out of the "secret documents."

My "troubled friend" has good cause for anxiety. He is right to wonder whether the press knows enough and is responsible enough to publish things the Attorney General wants suppressed. He is right to concern himself with the security of the nation.

But what is being exposed here is not primarily some Government documents that might cause "irreparable damage" to the defense of the nation, but a system of secrecy, of Presidential presumption, of influential staff advice by men who cannot be questioned, of concealment and manipulation, all no doubt with the best motives, but nevertheless, a system which has got out of hand and could really cause "irreparable damage" to the Republic.

No doubt the press itself is often poorly informed and clumsy in its efforts to expose the dangers of this system, but the greater the power in the hands of the executive, the greater the need for information and skepticism on the part of the Congress and the press.

My anxious friend might be careful about weakening the instruments of information and review at such a time. No doubt they are blunt instruments, often misused, but in this case of the Pentagon Papers, or so it seems here, the greater danger is the system of executive secrecy; and the greater danger to the security of the nation is the mistrust this system of secrecy and contrived television propaganda has caused.

James Madison summed up the problem at the beginning of the Republic:

> Among these principles deemed sacred in America, among those sacred rights considered as forming the bulwark of their liberty, which the Government contemplates with awful reverence and would approach only with the most cautious circumspection, there is no one of which the importance is more deeply impressed on the public mind than the liberty of the press.

That this liberty is often carried to excess; that it has sometimes degenerated into licentiousness, is seen and lamented, but the remedy has not yet been discovered.

Perhaps it is an evil inseparable from the good with which it is allied; perhaps it is a shoot which cannot be stripped from the stock without wounding vitally the plant from which it is torn. However desirable those measures might be which correct without enslaving the press, they have never yet been devised in America.

Joseph and Stewart Alsop

Joseph Alsop (1910–89) was once a familiar Washington landmark, known for his bullying certitude and eloquent ''huhs.'' For thirty-seven years he conducted or co-conducted a Washington column, collaborating from 1946 until 1958 with his brother Stewart. The Alsops had the advantage of caste ties to Washington's ruling elite. While that opened doors, what gave ''Matter of Fact'' its impact in the New York Herald Tribune, *besides its able reportage, was its passionate hostility to Communism abroad and to anti-Communist demogogy at home. The column's urgency during the Eisenhower years pointed the way to John Kennedy, but also to Vietnam. When the partnership ended, Joseph carried on until 1974 in the* Washington Post *while Stewart from 1968 was a columnist for* Newsweek. *The brothers collected their columns in* The Reporter's Trade *(New York, 1960). Joseph Alsop was also author of* The Rare Art Tradition *(Princeton, 1983), a 250,000-word ''essay'' on the cultural origins of art collecting.*

THE MAD-HATTER LOYALTY PURGE

(Joseph and Stewart Alsop)

New York Herald Tribune *August 22, 1948*

There is not the slightest doubt that in these times the American government must have the right to protect itself from attack from within. The immensely delicate and difficult problem of national security exists and it must somehow be solved. No sensible man believes that J. Parnell Thomas's headline hunts can solve it. And any one who believes that the solution is to be found in the present loyalty program should consider certain questions which the conduct of the loyalty program has posed.

First, there is the whole troubling question of guilt by association. Obviously, a man who has constantly associated with Communists and promoted Communist fronts should not have access to state secrets. But how far should this principle be carried?

Take the recent case of a State Department employee. He was charged with having associated with ten persons, all presumably suspect. He had never heard of five of them and had had only the most casual contacts with four. But he had known one of the ten intimately for several years.

This man was a banker, and stood high in the banking community, a community not conspicuous for its radical tendencies. The State Department man could only defend himself by defending the banker, and neither he nor the banker had any notion why the banker was suspect. It finally developed that the banker had lived briefly in Albuquerque, N.M., several years before, and that an anonymous landlady had reported to the F.B.I. that he kept Communist literature in the basement. On further inquiry, it was established that the Communist literature consisted of accumulated copies of *The New Republic* which the banker had discarded.

This sort of thing is making the loyalty program a pretty sour joke. Moreover, the case of the State Department man is typical in that in the great majority of cases of guilt by association, at least half the supposed associates are wholly unknown to the accused.

Second, it seems obvious enough that an intellectual pro-Communist, who accepts unhesitatingly the Communist dogma, should hold no sensitive position—this was the sort of individual who figured conspicuously in the Canadian spy case. But, again, how far should this principle be carried?

Here are a few verbatim quotes from recent loyalty hearings: "What type of books did your associates have—any on political or social economy?" "What kind of books, by title, did you purchase, what kind of literature?" "How many copies of Howard Fast's novels have you read?" "Do you read 'The Newspaper PM'?" "Have you a book by John Reed?"

All government employees know that such questions are asked in loyalty cases, and the assumption that any kind of intellectual or political curiosity provides grounds for suspicion is inevitable. If this country wants a government service in which ignorance is at a premium, it will doubtless get what it deserves.

Third, does it really matter in America whether a man's maternal grandparents came over from Poland, or whether all his ancestors landed decorously on Plymouth Rock? There have been countless instances when, during loyalty hearings, individuals have been asked where their parents and grandparents were born. An entirely American response would be that it was none of the loyalty board's damn business where a man's forebears came from.

Finally, there is the question of confrontation. Take the case of a man who has one item of "derogatory information" in his dossier. His file notes that F.B.I. source X-32 has said that the accused is known to have attended a meeting of high Communist leaders in Seattle. The accused man hotly denies it. The loyalty board has no way whatsoever of knowing whether X-32 or the accused man is lying, and the board knows nothing at all about X-32.

It is the F.B.I. position that to allow those charged to confront their accusers might compromise F.B.I. sources. This seems logical enough. It is also the F.B.I. position that F.B.I. agents should not be allowed to make estimates of the trustworthiness of their sources, and this too seems logical. Yet surely loyalty boards which must decide cases certain to affect the whole future of government employees should know the identity of the sources of their information. They should also, in case of necessity, be able to confront and cross-question such sources in secret.

Obviously there is no easy solution to the problem of national security. But a government of drones and boneheads and toadies hardly contributes to national security. And if the sort of thing outlined above is allowed to continue indefinitely, that is the kind of government we shall get.

A MIRACLE IN POLAND

(Joseph Alsop)

New York Herald Tribune *September 6, 1957*

Properly speaking, the miracle of Jasna Gora, the great fortified monastery of "The Bright Mountain" took place 301 years ago.

On that occasion, the monks and a few score of Polish men-at-arms held and hurled back 10,000 Swedish troops who besieged the monastery's bastioned walls through a long, bitter winter month. The victory was credited to Jasna Gora's precious Virgin image, long ago chance-brought out of late Byzantium into these wild Polish marches.

I do not think there was any higher intervention; but I, too, witnessed a miracle at Jasna Gora just the other day. It was not easy to define, being the peculiar combination of a theme, a ceremony, a crowd and a man. Yet it seemed to me decidedly miraculous.

Imagine, then, a high hill swelling upward from the suburbs of a dreary little industrial town. A broad way leads through trees to the hill's summit. And here there is the beginning of the miracle.

The summit has been leveled, to make a gigantic plaza three times larger than the great plaza of St. Peter's. Only a single column, bearing Christ with his thorny crown, interrupts this vast, flat expanse. And the whole expanse is filled, as though by a fantastic human inundation, by a single continuous sea of people. There they stand in silent patience, men and women, young people and children, literally hundreds of thousands of them. And all gaze toward the towering church, rising in a surge of baroque pinnacles from the same tall bastions that the monks held against the Swedes.

Suddenly, along the battlements, the long procession of the Eucharist winds its way, banner after banner, choir after choir. A monk stationed by the high temporary altar erected on the church front gives a signal through a loud speaker. The crowd bursts, full-throated, into the hymn recounting Jasna Gora's miracle, "On the Heights of Czestochowa."

For a long hour the people stand, singing hymn after hymn and watching the platform around the altar gradually fill with the churchmen of the procession. Last come the Bishops of Poland, splendid in their vestments and mitres. Alas, a closer view reveals chasubles sadly confected of old lace curtains and capes made of the shoddy silk of Communist Peiping. The Church in Poland today is not rich in this world's goods.

Suddenly there is a hush. Stefan Cardinal Wyszynski quietly takes his place under the scarlet canopy that stands by the altar's sides. The wise eyes in the pale, ascetic face briefly survey the scene. Music breaks the silence and Baranjak, Archbishop of Poznan—that place name heavy with memories—begins the celebration of the mass. The crowd joins the responses as though this were a single parish church holding hundreds insteads of hundreds of thousands; and after the consecration of the Host the whole multitude sinks to its knees.

When the mass is ended the Cardinal enters the pulpit, and again there is a moment of silence while he stands, vividly outlined against white draperies, a commanding figure in brilliant scarlet. Then he speaks, telling the story of Poland's dedication to the Virgin Mary by King Jan Kazimierz, and saying that this is a time to renew the dedication with an oath. And slowly, in a strong masculine voice, he repeats the oath.

It is a curious oath, resembling a set of New Year's resolutions on a national scale. One catches echoes of certain exhortations to the people by Poland's Communist government in Warsaw—exhortations against the absenteeism that afflicts industry, the alcoholism that is a curse in this country, the disorderliness and lack of discipline that have appeared here since freedom returned. But whether or not the crowd also catches these echoes, all follow the Cardinal when he asks them to repeat after him: "We swear to thee, Mary, Queen of Poland, we swear to thee!"

Then the slender hand is raised in blessing. At a signal, with a passion that fills the hilltop air, the whole multitude breaks into Poland's battle hymn, "Great God Through Ages Protector of This Polish Land."

And so the morning comes to an end; and the sea of people flows away again, to picnic in the sunshine and queue up by thousands to say their prayers before Jasna Gora's Virgin image.

But in the dusk, when the enormous bulk of the fortress monastery shows black against the evening sky, the sea of people flows back into the plaza, more numerous than ever. Here and there candles shielded by workworn hands make points of winking light. Once again the old hymns sound out while the procession winds along the battlements. Once again, when the service begins, all these hundreds of thousands join together in chanting the litany to the Virgin. And this time, the Cardinal himself preaches to his people. The sermon is not unlike the morning oath, pressing the government a little on such contentious points as the difficult situation of Poland's Catholic press, but above all exhorting the people to be good citizens, even exhorting the miners to bring up more coal.

"You must understand," the Cardinal tells the listening thousands, "that what could not be destroyed by erroneous political doctrine can still be ruined by national demoralization. You are a generation of heroes, and to you God has given a serious duty—not to lose what was saved by the blood of your fathers."

So the sermon ends. Once more, with the same strange, resonant passion, the whole multitude sings Poland's battle hymn. And then all is over and it is time to journey homeward through the night.

Such was this modern miracle of Jasna Gora. If you think about it, it says a good deal about this new Poland whose two leaders, linked in unlikely partnership, are the brave veteran Communist, Wladyslaw Gomulka and the brave Prince of the Church, Stefan Cardinal Wyszynski.

GO, STRANGER, AND TELL . . .

(Joseph Alsop)

New York Herald Tribune *November 25, 1963*

Of all the men in public life in his time, John Fitzgerald Kennedy was the most ideally formed to lead the United States of America.

Such, at any rate, is this reporter's judgment, perhaps biased, but at any rate based on long experience and close observation, and no longer possible to suspect as self-serving. To be sure, judging Kennedy was never easy, for he was no common man, to be judged by common standards.

Courage, intelligence, and practicality; a passion for excellence, and a longing to excel; above all, a deep love of this country, a burning pride in its past, and an unremitting confidence in the American future—these were the qualities which acted, so to say, as the mainsprings of Kennedy the President.

Kennedy the man, Kennedy the private face, was half the enemy and half the reinforcement of Kennedy the President. He had an enviable grace of manner and person. He enjoyed pleasure. After Theodore Roosevelt, he was the first American President to care for learning for its own sake. After Abraham Lincoln, he was the first American President with a rich vein of personal humor—which is a very different thing from the capacity to make jokes.

This strange, dry, detached, self-mocking humor no doubt aided him to assess men and events; but in his public role, it was a handicap. Certainly it was not the same sort of handicap as Lincoln's humor, which actually prevented great numbers of otherwise intelligent persons from taking Lincoln seriously.

President Kennedy's humor instead inhibited him from showing the depth of his feelings. Any public exhibition of emotion gave him gooseflesh. So foolish people said he was a cold, unfeeling man, although few men in our time have had stronger feelings about those things that mattered to him.

After his country, what mattered most to him was to live intensely, with purpose and effect. He was in some sense the ultimate personification of the observation of Justice Holmes: "Man is born to act; to act is to affirm the worth of an end; and to affirm the worth of an end is to create an ideal."

The ideal that Mr. Kennedy affirmed in action was singularly simple; for no man was ever more contemptuous of the theological complexities of ideology. (It was hard to know, indeed, whether he held a more sovereign contempt for the doctrinaire mushiness of the extreme American left or for the doctrinaire hate-preachings of the extreme American right. He was slow to anger, but these made his gorge rise.)

His ideal could be completely summed up in only a score or so of words—a nation conceived in liberty and dedicated to the proposition that all men are created equal; the proud stronghold of a new birth of freedom; and the standing promise to all men that Government of the people, by the people and for the people shall not perish from the earth. The noble, ancient phrases, the pieced-together tags from the finest of all American utterances, are as well-worn by now as antique coins, whose legend is illegible. But *he* could read the legend still. *He* still took this definition of our Nation's purpose with perfect literalness; and this was the ideal that his actions sought to affirm.

Whereas Franklin Delano Roosevelt took office when the Nation was clamoring for leadership and crying out to be shown a new course, John Fitzgerald Kennedy took office in a time of violent—yet hardly comprehensible—change.

Too many, then as now, confronted the vast revolutionary processes of our time either with fatty complacency or with shrill, embittered indignation. His task was therefore a hard task, and he was untimely cut off before his task could be half done.

Yet if we look at our country and the world in which we live—if we honestly compare the prospects now opening before us with the prospects as they seemed when Mr. Kennedy's presidency began—we can see that there has been a new birth of hope.

It is perhaps pardonable, at this moment, to be personal. Speaking for myself, I have not dared to hope as I do now since those first months of the Korean War, when such overly high hopes were born from a strong sense that America was grandly accomplishing a high, historic service. That service had its heavy price.

I still remember watching the wolfhound regiment through a long, hard fight, and how the bodies of the fallen were carried in when the fight was won, and how I suddenly could think only of Simonides' epitaph that was inscribed, for all to read, on the tomb of the dead Spartans at Thermopylae. The dead speak:

Go, stranger, and in Lacadaemon tell
That here obedient to the laws we fell.

But the President who is lost to us, like those men who were lost in Korea so many years ago, was no drilled, unthinking Spartiate. He was the worthy citizen of a Nation great and free—a Nation, as he liked to think, that is great because it is free. This was the thought that always inspired his too brief leadership of our republic.

Mary McGrory

Mary McGrory (1918–   ), too rarely celebrated by prize-givers and media critics, has, in the more exacting estimate of her peers, been in the first rank since 1953. That is when the Washington Star *assigned her to the Army-McCarthy hearings (she was then a book reviewer). Here she describes Joseph Welch, the courtly Bostonian who was the Army counsel:*

> *In the floodlighted jungle of the hearing room Mr. Welch, who might have stepped out of the "Pickwick Papers," does not appear entirely in his element. His habitual expression is dubious. . . . A tall man, he has a long face and owlish eyes. He beams rather than smiles, and sometimes when he is listening to a witness he puts the tips of his fingers together and looks rapt, as one might who is listening to the fine strains from the Boston Symphony Orchestra. Mr. Welch proceeds at the measured pace of the minuet, with frequent, courtly bows. Senator McCarthy favors the tarantella, moving almost faster than the human eye can follow.*

In 1960, she launched her thrice-weekly column in the Star, *where it appeared until the once dominant afternoon paper folded in 1979. On moving to the* Washington Post, *she stipulated that her column appear, as before, in the news pages. Her minimalist syntax suggests a devotion to Jane Austen and a debt to four years of Latin at the Girls' Latin High School in her native Boston. Her columns on Watergate received a Pulitzer Prize in 1974. There is no published collection.*

A DAY OF ENDLESS FITNESS

Washington Star *November 26, 1963*

Of John Fitzgerald Kennedy's funeral it can be said he would have liked it.

It had that decorum and dash that were in his special style. It was both splendid and spontaneous. It was full of children and princes, of gardeners and governors.

Everyone measured up to New Frontier standards.

A million people lined every inch of his last journey. Enough heads of state filed into St. Matthew's Cathedral to change the shape of the world.

The weather was superb, as crisp and clear as one of his own instructions.

His wife's gallantry became a legend. His two children behaved like Kennedys. His 3-year-old son saluted his coffin. His 6-year-old daughter comforted her mother. Looking up and seeing tears, she reached over and gave her mother's hand a consoling squeeze.

The procession from the White House would have delighted him. It was a marvelous eye-filling jumble of the mighty and the obscure, all walking behind his wife and his two brothers.

There was no cadence or order, but the presence of Gen. de Gaulle alone in the ragged line of march was enough to give it grandeur. He stalked splendidly up Connecticut Avenue, more or less beside Queen Frederika of Greece and King Baudouin of Belgium.

The sounds of the day were smashingly appropriate. The tolling of the bells gave way to the skirling of the Black Watch Pipers whose lament blended with the organ music inside the Cathedral.

At the graveside there was the thunder of jets overhead, a 21-gun salute, taps, and finally the strains of the Navy hymn, "Eternal Father Strong to Save."

He would have seen every politician he ever knew, two ex-Presidents, Truman and Eisenhower, and a foe or two. Gov. Wallace of Alabama had trouble finding a place to sit in the Cathedral.

His old friend, Cardinal Cushing of Boston, who married him, baptized his children and prayed over him in the icy air of his Inaugural, said a low mass. At the final prayers, after the last blessing, he suddenly added, "Dear Jack."

There was no eulogy. Instead, Bishop Philip M. Hannan mounted the pulpit and read passages from the President's speeches and evoked him so vividly that tears splashed on the red carpets and the benches of the Cathedral. Nobody cried out, nobody broke down.

And the Bishop read a passage the President had often noted in the

Scriptures: "There is a time to be born and a time to die." He made no reference to the fact that no one had thought last Friday was a time for John Fitzgerald Kennedy to die—a martyr's death—in Dallas. The President himself had spent no time in trying to express the inexpressible. Excess was alien to his nature.

The funeral cortege stretched for miles. An old campaigner would have loved the crowd. Children sat on the curbstones. Old ladies wrapped their furs around them.

The site of the grave, at the top of one slope, commands all of Washington. Prince Philip used his sword as a walking stick to negotiate the incline.

His brother, Robert, his face a study in desolation, stood beside the President's widow. The children of the fabulous family were all around.

Jacqueline Kennedy received the flag from his coffin, bent over and with a torch lit a flame that is to burn forever on his grave—against the day that anyone might forget that her husband had been a President and a martyr.

It was a day of such endless fitness, with so much pathos and panoply, so much grief nobly borne that it may extinguish that unseemly hour in Dallas, where all that was alien to him—savagery, violence, irrationality—struck down the 35th President of the United States.

HALDEMAN HEARS THE TAPES

Washington Star *November 12, 1974*

On the 30th day of the trial, H.R. Haldeman's eyes are like two burnt holes in a blanket. He coaches his lawyers constantly, mouthing instructions as they object to "hearsay evidence" and "leading questions."

Haldeman is the only man in America in this generation who let his hair grow for a courtroom appearance. He is the most embattled and alert of the five Watergate defendants. He faces out at the spectators at his separate table. He scribbles bold black notes with a large pen on yellow legal pad.

Two eager careerists are telling their familiar stories of his connivance. L. Patrick Gray, the torpedoed submarine skipper, the short-term acting director of the FBI; portly Vernon Walters, still the deputy director of the CIA, who glides through his well-worn recital of how the agency was used as the rug under which Watergate was to be swept. There are no surprises.

But Haldeman can do nothing about the tapes. Without the tapes, the trial is a coherent, orderly rerun of the Ervin hearings. But the "little Sony" is the witness that cannot be cross-examined, cannot be cautioned against giving hearsay and cannot keep Richard Nixon out of the courtroom.

Right up until the moment the reels are played, his lawyers fight. Did Haldeman use "Gemstone," the code word for the break-in in his June 23 conversation with Nixon? The judge, John J. Sirica, as anxious for irreversibility as the Watergate principals once were for "deniability," says "whether it's gemstone or rhinestone," he's not going to argue and orders the word deleted.

It is a small victory. Haldeman's face shows a small tight grin. It is to be the last of the day.

Chief Prosecutor James F. Neal announces "government exhibit No. One A." He gives the date Southern style, "June the twenty-third, nineteen-and seventy-two."

There is a bustle in the courtroom. Transcripts are handed out. Everyone reaches for earphones, huge plastic discs like giant earmuffs. The spectators who have waited since midnight, the bored reporters, the weary jurors, lawyers, defendants begin fumbling with the volume control box that is under every other seat. The judge's earphones are different, white, lighter in weight. He looks like a medieval poet with the white headbands favored by Dante.

Haldeman, looking around to see that nothing else is afoot, clamps on his headset. John Mitchell does not. He holds one disc against his left ear. It serves as a shield, covering most of his gray face.

The sound begins. It is, despite Richard Nixon's famous put-down, fine. In one segment, to be sure, there is a whirring noise, rather like that of the dentist's drill warming up; in another, a sound which is identified as writing, but which could be the panting of King Timahoe.

There is no question, however, of what is being said. Haldeman comes on first, his tones low, unaccented, with blurred consonants. He is hastening through distressing developments, the money, the photographs. "We are back to the—in the problem area, because the FBI is not under control, because Pat Gray doesn't exactly know how to control them."

John Mitchell, he goes on, has come up with "the only way to solve this. . . . Have Walters call Pat Gray and just say 'Stay the hell out of this.' . . ."

A faint edge of pink appears over the top of the gray disc Mitchell is holding against his face. Haldeman slumps down in his seat.

The voice of the unindicted, pardoned coconspirator now in Long Beach Memorial Hospital is heard. "Um'hum," says the former president of the United States, calm, dispassionate.

They both knew the reels were running. Did Haldeman, whose inspiration it was to record every syllable of Richard Nixon, forget? It is hard to imagine. He remembers always to wear his American-flag lapel pin. He was an advance man, the breed that lives by detail.

Yet here he is, laying it all out, digging his own grave. By mid-day, in the second exchange, the two voices are stronger and more confident, as they flesh out their crazy, doomed scenario. Nixon likes it better every minute. He pronounces: "It's going to make the FBI and CIA, look bad, it's going to make Hung look bad, and it's likely to blow the whole, uh, Bay of Pigs thing which we think would be very unfortunate for the CIA and for the country, at this time, and for American foreign policy. . . ."

The release of the June 23 transcript last Aug. 5 sank Richard Nixon. The last clinging congressional fingers let go that day. What will it do to H. R. Haldeman, the prompter, the scribbler? His lawyers can confuse the witnesses and the jury. The transcripts, however, make everything "perfectly clear."

WISDOM, NOT HEAVING BOSOMS

Washington Post   *December 16, 1984*

Today is her birthday, and it offers an alibi—any will serve—to talk about Jane Austen and her mystery. Why does this novelist, who was born 209 years ago, and who wrote about the English gentry on, as she said, "two inches of ivory," continue to hold readers in thrall?

I can tell you that one account of a Jane Austen Society meeting brought more mail than any other topic discussed in this space.

The Jane Austen cult was the subject of a recent front-page story in the *Wall Street Journal.* Five hundred and sixty thousand copies of her novels have been printed since 1981. This might well amuse the daughter of the rector of Steventon, who offered her first novel, *Sense and Sensibility,* to a publisher with such diffidence that a refusal came back by return of post.

She had a few contemporary admirers. The prince regent of England, for instance, kept a complete set of her novels at all his estates. He never knew when he would require a quick fix of wit or principle. Today, my friend Linda Wertheimer has three sets of Jane Austen, one bound in leather, a second, more pedestrian hardback and a third in pocket books, of a size suitable for being slipped under a notebook at a press conference dealing with say, "revenue-enhancement" or possibly the choice of a new Democratic Party chairman.

Sir Walter Scott was smitten by her. So is Nora Ephron, an utterly contemporary young woman, who reads *Pride and Prejudice* at least once a year, because she is entranced, as any sensible person would be, by Elizabeth Bennett.

Volumes have been written about Jane Austen since she succumbed to an unknown illness in 1817 at the age of 41. I have gone through some of them, enjoying them most when she is quoted. Her critics point out that she dodged the hurly-burly of life, averting her eyes from passion, never writing about childbirth, death, poverty, war or any of life's more wrenching moments. Louis Kronenberger observes that some people fine her "tea-tablish." Charlotte Bronte laments the want of the "heaving bosom" in any of her novels.

And yet, new Jane Austen societies are formed every day, and once a year they meet in convention to devote two days to impassioned discussion of one of her six novels. Emma and Anne Elliot and Fanny Price seemed to be in the room.

Maybe it is because she deals with one subject, what is called today, "interpersonal relationships" and the eternal theme of young women in search of husbands. Jane Austen recorded the pursuit with fidelity and clarity that has never been matched. The young woman in the singles bar recognizes the truth, if not the circumstances, of her account.

She also describes loneliness, mostly through her delineations of old maids (she was one herself) who must be ingratiating, obliging, never revealing their own feelings in order to be tolerated. Miss Bates of *Emma* is her masterpiece in this line. Miss Bates is forever praising, doing, offering, hoping to be included. Emma makes fun of her, in one of the most memorable scenes of that much-praised novel, and when Mr. Knightley, her mentor—and her future husband—takes her to task, she gives way to bitter tears and turns a corner in her life. The thing about Jane Austen is that she is as decent as she is perceptive.

That's a stab at the "why?" The "how" remains.

She never left the south of England for any length of time, never traveled abroad, never met a writer or even corresponded with one. She had no friends outside her warm family circle, to whom her novels were read aloud with great delight and pride. She had no other sources of encouragement, except the prince regent's librarian, who suggested she write a romance about Saxe-Coburg, and some readers who marveled that she had taken them into another world.

She created those compelling characters and wrote those incomparably fine-grained books on a portable writing desk in the parlor, subject to frequent interruptions from family members, servants, visiting nieces and nephews wanting counsel on courtships and later on novels they were writing. She was warned of their coming by a squeaky door which she forbade to be fixed, and simply thrust the pages out of sight as she rose to greet them.

By today's standard, she had nothing: no word processor, no creative-writing courses, no workshops on the novel, no prospects, if successful, of being sent on publicity tours, autographing parties and appearances on late-night television shows with Joan Rivers.

All she had was her sharp eye, her true pen and her sense of what men and women are—and what they ought to be.

I. F. Stone

The reputation of I. F. Stone (1907–89) confirms a remark by Lippmann: "It is safer to be wrong before it is fashionable to be right." Ostracized as a fellow traveler by the Washington press establishment and having outraged the National Press Club by inviting a black (a Federal judge) to its dining room, Stone uttered in the 1950s what later became commonplace: that McCarthyism was a scandal and racism a disgrace, that Vietnam was a swamp and the missile gap a mirage. Stone was eventually rehabilitated, awarded honorary degrees, and guardedly praised by the mainstream press.

He came to Washington as correspondent for the leftish New York daily PM, *and its short-lived successor,* The Compass. *In 1953 he started* I.F. Stone's Weekly, *which became a biweekly in 1968 and closed up three years later. Stone's gifts were amply visible in his newsletter: erudition, the patience to mine official documents, and the courage to admit error. Stone published six collections of his pieces, and Neil Middleton has edited a selection,* The I.F. Stone Weekly Reader *(New York, 1973). Here is Stone's judgment on John F. Kennedy.*

AFTER DALLAS: WE ALL HAD A FINGER ON THAT TRIGGER

I.F. Stone's Weekly *December 9, 1963*

There was a fairy tale quality about the inaugural and there was a fairy tale quality about the funeral rites. One half expected that when the lovely princess knelt to kiss the casket for the last time, some winged godmother would wave her wand and restore the hero whole again in a final triumph over the dark forces which had slain him. There never was such a shining pageant of a Presidency before. We watched it as children do, raptly determined to believe but knowing all the time that it wasn't really true.

Of all the Presidents, this was the first to be a Prince Charming. To watch the President at press conference or at a private press briefing was to be delighted by his wit, his intelligence, his capacity and his youth. These made the terrible flash from Dallas incredible and painful. But perhaps the truth is that in some ways John Fitzgerald Kennedy died just in time. He died in time to be remembered as he would like to be remembered, as ever young, still victorious, struck down undefeated, with almost all the potentates and rulers of mankind, friend and foe, come to mourn at his bier.

For somehow one has the feeling that in the tangled dramaturgy of events, this sudden assassination was for the author the only satisfactory way out. The Kennedy Administration was approaching an impasse, certainly at home, quite possibly abroad, from which there seemed no escape. In Congress the President was faced with something worse than a filibuster. He was confronted with a shrewdly conceived and quietly staged sitdown strike by Southern committee chairmen determined to block civil rights even if it meant stopping the wheels of government altogether. The measure of their success is that we entered this final month of 1963 with nine of the thirteen basic appropriation bills as yet unpassed, though the fiscal year for which they were written began last July 1 and most of the government has been forced to live hand-to-mouth since. Never before in our history has the Senate so dragged its heels as this year; never before has the Southern oligarchy dared go so far in demonstrating its power in Washington. The President was caught between these old men, their faces set stubbornly toward their white supremacist past, and the advancing Negro masses, explosively demanding "freedom now." Mr. Kennedy's death, like those of the Birmingham children and of Medgar Evers, may some day seem the first drops portending a new storm which it was beyond his power to stay.

In foreign policy, the outlook was as unpromising. It was proving difficult to move toward co-existence a country so long conditioned to cold war. Even when Moscow offered gold for surplus wheat, it was hard to make a deal. The revolt in Congress against foreign aid illustrated how hard it was to carry on policy once tense fears of communism slackened even slightly. The President recognized the dangers of an unlimited arms race and the need for a modus vivendi if humanity was to survive but was afraid, even when the Sino-Soviet break offered the opportunity, to move at more than snail's pace toward agreement with Moscow. The word was that there could be no follow up to the nuclear test ban pact at least until after the next election; even so minor a step as a commercial airline agreement with the Soviets was in abeyance. The quarrel with

Argentina over oil concessions lit up the dilemma of the Alliance for Progress; however much the President might speak of encouraging diversity, when it came to a showdown, Congress and the moneyed powers of our society insisted on "free enterprise." The anti-Castro movement our CIA covertly supports was still a spluttering fuse, and in Vietnam the stepping up of the war by the rebels was deflating all the romantic Kennedy notions about counter guerrillas, while in Europe the Germans still blocked every constructive move toward a settlement in Berlin.

Abroad, as at home, the problems were becoming too great for conventional leadership, and Kennedy, when the tinsel was stripped away, was a conventional leader, no more than an enlightened conservative, cautious as an old man for all his youth, with a basic distrust of the people and an astringent view of the evangelical as a tool of leadership. It is as well not to lose sight of these realities in the excitement of the funeral; funerals are always occasions for pious lying. A deep vein of superstition and a sudden touch of kindness always lead people to give the departed credit for more virtues than he possessed. This is particularly true when the dead man was the head of the richest and most powerful country in the world, its friendship courted, its enmity feared. Everybody is anxious to celebrate the dead leader and to court his successor. In the clouds of incense thus generated, it is easy to lose one's way, just when it becomes more important than ever to see where we really are.

The first problem that has to be faced is the murder itself: Whether it was done by a crackpot leftist on his own, or as the tool of some rightist plot, Van Der Lubbe style, the fact is that there are hundreds of thousands in the South who had murder in their hearts for the Kennedys, the President and his brother the Attorney General, because they sought in some degree to help the Negro. This potential for murder, which the Negro community has felt for a long time, has become a national problem. But there are deeper realities to be faced.

Let us ask ourselves honest questions. How many Americans have not assumed—with approval—that the CIA was probably trying to find a way to assassinate Castro? How many would not applaud if the CIA succeeded? How many applauded when Lumumba was killed in the Congo, because they assumed that he was dangerously neutralist or perhaps procommunist? Have we not become conditioned to the notion that we should have a secret agency of government—the CIA—with secret funds, to wield the dagger beneath the cloak against leaders we dislike? Even some of our best young liberal intellectuals can see nothing wrong in this picture except that the "operational" functions of CIA should be kept separate from its intelligence evaluations! How many of us—on the left now—did not welcome the assassination of Diem and his brother Nhu in South Vietnam? We all reach for the dagger, or the gun, in our thinking when it suits our political view to do so. We all believe the end justifies the means. We all favor murder, when it reaches our own hated opponents. In this sense we share the guilt with Oswald and Ruby and the rightist crackpots. Where the right to kill is so universally accepted, we should not be surprised if our young President was slain. It is not just the ease in obtaining guns, it is the ease in obtaining excuses, that fosters assassination. This is more urgently in need of examination than who pulled the trigger. In this sense, as in that multi-lateral nuclear monstrosity we are trying to sell Europe, we all had a finger on the trigger.

But if we are to dig out the evil, we must dig deeper yet, into the way we have grown to accept the idea of murder on the widest scale as the arbiter of controversy between nations. In this connection, it would be wise to take a clear-sighted view of the Kennedy Administration because it was the first U.S. government in the nuclear age which acted on the belief that it was possible to use war, or the threat of war, as an instrument of politics despite the possibility of annihilation. It was in some ways a warlike administration. It seems to have been ready, soon after taking office, to send troops to Vietnam to crush the rebellion against Diem; fortunately both Diem and our nearest Asian allies, notably the Filipinos, were against our sending combat troops into the area. The Kennedy Administration, in violation of our own laws and international law, permitted that invasion from our shores which ended so ingloriously in the Bay of Pigs. It was the Kennedy Administration which met Khrushchev's demands for negotiations on Berlin by a partial mobilization and an alarming invitation to the country to dig backyard shelters against cataclysm.

Finally we come to the October crisis of a year ago. This set a bad precedent for his successors, who may not be as skillful as he was in finding a way out. What if the Russians had refused to back down and remove their missiles from Cuba? What if they had called our bluff and war had begun, and escalated? How would the historians of mankind, if a fragment survived, have regarded the events of October? Would they have thought us justified in blowing most of mankind to smithereens rather than negotiate, or appeal to the UN, or even to leave in Cuba the medium range missiles which were no different after all from those we had long aimed at the Russians from Turkey and England? When a whole people is in a state of mind where it is ready to risk extinction—its own and everybody else's—as a means of having its own way is an international dispute, the readiness for murder has become a way of life and a world menace. Since this is the kind of bluff that can easily be played once too often, and that his successors may feel urged to imitate, it would be well to think it over carefully before canonizing Kennedy as an apostle of peace.

Murray Kempton

The morning before his Pulitzer Prize was announced in 1985, Murray Kempton was chaining his five-speed bicycle along a Manhattan street to attend a press conference on the day's political scandal. He came early, hoping for surprises and remarking to a Village Voice *colleague: "I am as addicted to surprises as Don Giovanni was to love's illusions," a quintessential Kempton-ism by a craftsman as revered by colleagues as he is uncelebrated at large. Born in Baltimore of Confederate ancestry, Kempton was a page at the 1936 Democratic convention when he first encountered his model, H. L. Mencken: "He was like a sportswriter. After finishing a page, he would pull the paper and carbons out of the upright typewriter, and hold it up like a flag and a banner" for the copy boys to carry away. Kempton started in 1946 as a labor columnist for the* New York Post, *and under the tutelage of editor James Wechsler was soon writing about all else. Since 1981, his column has adorned* Newsday. *When he finally received a Pulitzer, Kempton was gracious, "Believe me, late is better." His books include a selection of his pieces,* America Comes of Middle Age *(New York, 1963).*

THE U.S. AS PICKPOCKET

New York Post *December 15, 1955*

The United States government, in all its awe and majesty, has now descended in malignity to picking the pocketbooks of poor old women.

Alexander Bittelman is a sixty-six-year-old Communist Party official now serving a prison term under the Smith Act. Early this fall, someone discovered that Bittelman was receiving social security payments from the United States government.

Bittelman was getting $88.10 a month mailed directly to his cell in Atlanta. He used to take $10 out for cigarets and other sundries from the commissary and send the rest back to his wife, Eva, living without other income in Croton, New York.

Old Mrs. Bittelman appears to be the menace from which the Social Security Board rescued us yesterday when it announced that persons employed by the Communist Party and its affiliates are no longer eligible for old-age retirement benefits.

Her husband is a professional Communist and has had no other employer in the seventeen years he and the party have been paying social security taxes. The government is only the custodian of the social fund, which is supported by equal payments by employer and employe. When the government stops Bittelman's retirement checks, it is not saving its own money; it is stealing his.

A few months ago, the Communist Party of the State of New York applied to the Health Insurance Plan of Greater New York for medical insurance for its thirteen employes. H.I.P. replied that the party's application has been rejected by decision of its underwriting committee. An H.I.P. spokesman said that it does not have to give reasons for rejecting applications, and I suppose it hasn't.

I only know that, if a child is born to a secretary in the Communist Party office some time in the future, it will be a child born under free private enterprise, because its mother is debarred from health insurance, a benefit taken for granted by most workers in this town.

In the last decade we have developed a sort of common law of justice for Communists. That common law is now in its full flower; the Communist is taxed as an American citizen and denied any other blessing of citizenship except taxation. From the Communist Party's typist's child to old Mrs. Bittelman, these people have total insecurity, from the cradle to the grave.

There is always a regulation somewhere that will allow a government officer to do a malignant thing. That regulation serves two purposes; it gives an excuse to him and it gives an excuse to such bystanders as should be outraged and need to justify their silence.

Bittelman's experience summons to memory the whole Republican campaign of 1936. Social security was a big factor in that campaign; the Republicans argued that it was no insurance at all, and that the government would steal it from prospective beneficiaries whenever the whim touched it. The Republicans could not have dreamed then that their case would be proved nineteen years later by a worn-out Communist hack and with not one word of protest from them.

Eva Bittelman is an old woman; and I would not believe the government of the United States if it argued without further proof that she is a wicked or dangerous one. That is irrelevant; virtue is irrelevant to any concept of a basic American right. As Calvin Coolidge might say, we hired her husband's money and the Communist Party's, didn't we? And, when we refuse to pay it back, we are welshers on a debt.

There are very few people in America who would stand up and say without shame that they favor the suppression of ideas. Yet are ideas so much more important than people that our government can without shame make its wars on old women? We have a government, which without protest from any liberal Senator, has taken social security away from an old man and a disability pension away from an ex-soldier solely because they were Communists. Is there any better measure of a nation than the way it treats the helpless?

. . . BUT WE ARE A COUNTRY OF LAWS

New York Post *December 21, 1955*

Maxwell Berman wrote the *Post* yesterday to dissent from my view that the government of the United States of America had stooped to picking pockets because it denied social security benefits to Alexander Bittelman, a convicted Communist, and his wife.

"We are," said Brother Berman, "a country of laws."

This is a splendid notion and one to be commended to the Department of Justice, the Social Security Board, the Bureau of Internal Revenue, and the Veterans Administration.

Brother Berman says that I should stop and reflect and not let emotion govern my thinking. It certainly plays hell with my writing. I got so worked up about poor Bittelman that I cheated the customers of the facts of his case.

The Social Security Board did not cut off Bittelman's old age pension because he was convicted under the Smith Act. Its order applied to all employes of the Communist Party. The Social Security Board found that, under the McCarran Act, the Communist Party has been found to be an agent of a foreign power, and, the board says, employes of a foreign government are ineligible for social security.

There are, it appears, two Alexander Bittelmans. There is Mr. Berman's Alexander Bittelman who tried to murder our government. That one is all right; he can get his social security. But there is another Alexander Bittelman; he is an employe of a foreign government and thus presumably enjoys a kind of diplomatic immunity from its taxes.

We are a government of laws; am I mistaken in the emotional notion that there is the same law for the Social Security Board and for other agencies of the government? Apparently I am.

George Blake Charnay is the state chairman of the Communist Party of New York. He makes $75 a week, all of it from the party. Under the social security law, he is an employe of a foreign government and exempt from social security tax. Charnay also served two years under the Smith Act and then got sprung for a new trial. During his stretch, he fell behind in his income tax payments, all on earnings paid him by the Communist Party. Does his diplomatic immunity extend to the Bureau of Internal Revenue? Of course not; Charnay is being harassed for taxes by its agents right now.

We are a government of laws. Saul Wellman is a Communist official convicted under the Smith Act for conspiring to overthrow the government of the United States. The Veterans Administration was paying him a disability pension for wounds suffered in Europe in World War II. Now that has been cut off. The excuse is that the statute provides that any veteran who gives aid and comfort to an enemy of the United States thereby loses his pension. The Wellman who tried to murder the government can be paid for his wounds; it's just this other fellow who gave aid and comfort to an enemy of the United States. And this is not, by the way, a charge for which Wellman has ever been tried.

We are a government of laws. A number of Communists convicted under the Smith Act have run enough of their course to be eligible for parole. The board asks them, as it asks all convicts, whether they repent of their crimes. These people, of course, deny they ever committed a crime; they propose to go on doing what they felt innocent of doing in the first place.

There is, of course, a relevant statute to deal with their attitude. Parole boards are abjured to make sure that a prisoner repents of his crime before considering his parole. This applies alike to bank robbers and sick old Communists. A sick old Communist does not repent of the crime he says he did not commit, and he—under the law, of course—is ineligible for parole.

We are a government of laws. Any laws some government hack can find to louse up a man who's down.

CAMPAIGN VULGARITIES (1960)

New York Post *January 9, 1962*

Comes now Public Document 75452, from the Subcommittee of the Subcommittee of the Senate Committee on Commerce, the sober record of the fall of 1960 when America was deciding whether to move again:

> *Vice President Nixon:* Could I ask you one favor, Jack?
>
> *Jack Paar:* Yes, sir; you can ask any favor you'd like.
>
> *Vice President Nixon:* Could we have your autograph for our girls?

The notes on that particular meeting at the summit (Paar: I can't tell you how much this means to our show. It gives us "class.") are the opening exhibit in a Senate report labeled, "The Joint Appearances of Senator John F. Kennedy and Vice President Richard M. Nixon and other 1960 Campaign Presentations."

That was Sept. 11, 1960, and Nixon had packed. The Kennedys rallied two weeks later.

> *Charles Collingwood:* Hello, Caroline.
>
> *Mrs. Kennedy:* Can you say hello?
>
> *Caroline:* Hello.
>
> *Mrs. Kennedy:* Here, do you want to sit up in bed with me?
>
> *Mr. Collingwood:* Oh, isn't she a darling?
>
> *Mrs. Kennedy:* Now, look at the three bears.
>
> *Caroline:* What is the dolly's name?
>
> *Mrs. Kennedy:* All right, what is the dolly's name?
>
> *Caroline:* I didn't name her yet.

The issues were, of course, sometimes met more decisively, so decisively in fact that both her elder campaigners may sometimes wish they had adopted Caroline's law.

For example, on the Paar show:

> *Question:* Well, Mr. Vice President, I was wondering if the Congolese Premier sends his troops in Katanga . . . would the United States back up the United Nations in this . . . ?
>
> *Vice President Nixon:* The United States does support the United Nations and must support the United Nations in the Congo . . .

Or:

> *Mr. Cronkite:* Would you feel any restrictions against naming a member of the family to the Cabinet, for instance.
>
> *Senator Kennedy:* I think it would be unwise.

This has to be an incomplete record of all the wonderful nonsense we were embracing in that lost time of crisis, but it is still a delectable sample. Where else could we have the text of the fifth campaign broadcast of the International Ladies Garment Workers Union, that citadel of adult education ("During the series, we've heard from Adlai Stevenson, Mrs. Eleanor Roosevelt, George Meany, Tony Curtis, Janet Leigh . . . telling you why they, too, are voting for Kennedy and Johnson.")?

> *Tallulah Bankhead:* I am here to introduce John F. Kennedy . . . who as of January 20 will be a resident of 1600 Pennsylvania Avenue, Washington, D.C. That's the White House, darlings.

Here are the debates and here are all the joint statements of Kennedy and Nixon to the American Bar Association (judges should be qualified lawyers), to the American Jewish Committee (there is no Jewish vote), to the Oklahoma Oil Producers (the depletion allowance is a shield of the republic), to the Inland Waterway Association (the inland waterway is a lifeline of the republic), to the small businessmen (the small businessman is the sentinel of the Republic).

Most of all there is Nixon, doom upon him, from the confident beginning, "Oh, I like sports," through the depressing middle, "Charles, I am not a natural politician," to the desperate end, "It's the millions of people that are buying new cars that have faith in America."

This painful, vulgar record evokes him again, but the mystery of his collapse taunts us yet. Still it was a terribly close election and who can say what small mistake cost him it?

There is one clue:

> *Bill Henry, of NBC:* I am so fascinated with that little kitten. Does the kitten have a name?
>
> *Julie Nixon:* Yes, its name is Bitsy Blue Eyes.

Maybe Caroline saved the package when she held off naming the doll.

ANENT THE BAY OF PIGS

New York Post *April 19, 1961*

All the civilized world rejoiced today in our Cuban adventure.

The French, for instance, rejoiced that we now have an Algeria. The British were glad to know we had found a Kenya. Only Mr. Tshombe, a representative of civilization by *grace belgique,* must as Tuesday wore on have felt a certain aesthetic contempt; they do these things more tidily in Katanga.

The debate in the United States followed the lines customary in our debate: (1) it did not exist; (2) what faint quivers there were appear to have been resolved by the reflection that Castro had betrayed the revolution. Ask no more why we do not give BARs to people attempting to overthrow Trujillo; Trujillo never betrayed a revolution.

We live in the year 1961 and America has made much progress since the 1850s. In the 1850s, Abraham Lincoln damaged his political future and Henry David Thoreau went to jail because they opposed the Mexican War. I do not propose to go to jail over this matter—assuming that the rulers of the earth would condescend to send me—and thus our grandchildren will miss the catch in the throat which would otherwise dampen their eyes at the news that a man I revere had said to me, "What are you doing in jail, uh, Murray?" and I firmly and bravely gave reply: "And what are you doing out there, Arthur Schlesinger, Jr., and what would your father think if he could see you?"

Poor Thoreau! He lived in a world before the invention of the Americans for Democratic Action and the syndicated columnist; he was thus pathetically uninformed. But, if Thoreau could stand and say a word for such as Santa Ana, can anyone do less for Castro?

The reports yesterday were not encouraging to those who invoked the ancient, queasy prayer that if it were done when 'tis done then 'twere well it were done quickly. The Latin Americans, none of whom had the loins to speak out for Castro in public, were somewhat derisive toward us liberators in the corridors. They were as usual wrong; we will, I fear, get the business done, sloppily but on the record.

One thought all day of Adlai Stevenson and of Raul Roa, Castro's Foreign Minister.

There was the feeling that the administration had handed Stevenson that cup and that he would drink it like the gentleman he is. There was also the feeling that he knew no more than the rest of us precisely what had happened.

Roa was something else, a free and happy man. "Today," he said, "I am fifty-four years old. It is the spring. With the revolution, it is always the spring."

It cannot always have been the spring. Roa knows the Cuban revolution. He knows the lies Castro has told himself and through himself the people. But yesterday it was spring again; he was only a Cuban patriot, standing against the alien.

We had a bad day in this rock pool. It left us only with the consolation that no truly great power can ever be called to book for its sins in the UN. But it was a day that belonged to the Poles, the Czechs, and the Rumanians—the enslaved given the floor, to cry out at slavery elsewhere.

In all I have to say about this dirty business, I hope that I would speak nothing harsh about those Cubans we sent on this errand. I would not be caught speaking against a soldier lying on his stomach on swampy ground being shot at. Under ordinary circumstances, being no Fidelista, I should wish them well; as it is, I could not wish them ill. There is every excuse for them; but there is no excuse for Allen Dulles or John F. Kennedy. If you think there is, you are, believe me, wrong.

Jimmy Cannon

The showman Billy Rose, once a columnist himself, said of James Joseph Victor Cannon (1910–1973): "With nothing more to write about than the same old muscle bustle, Jimmy sends ordinary words through the typewriter and makes them come out dancing." Born in New York, Cannon started as a copy boy on the Daily News. *His coverage of the Lindbergh trial won him an offer from Damon Runyon to join the rival* Mirror, *and by 1936 he was a sports columnists. Cannon soon headed out of an exceptional pack: Joe Williams, Dan Parker, Frank Graham, and the erudite John Kieran. After serving as a combat correspondent for* Stars and Stripes *in the European theater, Cannon started a six-day-a-week column for the* New York Post *in 1946. Addicted readers, as Wilfrid Sheed observed, took to "ripping their way blindly past Mary Worth and Murray Kempton" to reach "Jimmy Cannon Says." A favored running feature were his nobody-asked-me opinions about everything,* e.g., *"If Howard Cosell was a sport, it would be Roller Derby." His sports columns have been collected in* Who Struck John? *(New York, 1956) and* Nobody Asked Me, But . . . *(New York, 1978), edited by Jack and Tom Cannon.*

NOBODY ASKED ME, BUT

New York Post *1951–1954*

I don't fetch mambo music.

When I see a guy with lapels on his vest, I figure he's a big shot.

You're blowing the duke when she ridicules your best friend's wife after the first double date.

Girls who wore the longest pony tails now favor the shortest idiot boy's hair cuts.

Nanette Fabray improved television just by showing up.

Giant fans knock other clubs as much as they boost their own.

Sit with a guy with a mustache for five minutes and I'll bet he touches it at least once.

The sauted smelts at Sweet's in Fulton Fish Market are a treat.

I've never seen a better umpire than Larry Goetz.

Did you ever see a girl sitting on the back seat of a motorcycle who appeared to be enjoying it?

A man described as "a sportsman" is generally a bookmaker who takes actresses to night clubs.

Most bus drivers answer the simplest questions as though they were afraid of committing perjury.

Does Salvador Dali conceive the patterns of all imported neckties?

Be prepared to be insulted when your pal warns you his new petit four has an "off beat sense of humor."

Do those beer-shilling TV bartenders ever buy a round?

Maybe, the Senators should try soccer next season.

Beware if your dearly beloved suddenly develops an interest in chess and you've never played the game.

I'd like to see Fred Allen do a one-man show.

Guys who squire dames who lug those satchel-sized handbags have to wind up with cauliflower knees.

When I see a young girl with a gold tooth, I figure she just arrived in this country.

I wonder what makes parking attendants so sour-faced.

Guys who beg horse owners for tips squawk the loudest when they lose.

Steve Allen's intelligently casual humor is proof that comedians don't have to be violently vulgar to succeed.

If baseball goes for pay television, shouldn't the viewers be given a bonus for watching a ball game between Baltimore and Kansas City?

A fat woman's enemy is a sales lady who recommends bullfighter's shorts.

I'd take my business around the corner if a guy had one of those white tobacco urns resembling a human skull on his desk.

Has there ever been a tea room with a TV set?

I can't ever remember a barber smoking a pipe in the shop.

Not even flat girls look good in the flat look.

I figure most teenage girls spend more time mussing up their blue jeans than they do primping for a party.

Some girls can pick a thread off a man's suit and make it seem as if they were handing him a rose.

My idea of class if Eleanor Powell dancing the soft shoe.

When I hear a rich man described as a colorful character I figure he's a bum with money.

You're betting better than a draw if she's a bad cook and goes to cooking school after your first couple of dates.

Jackie Robinson still generates more excitement than any player in the game when he gets on base.

Burlesque lost its appeal for me when comedians such as Red Buttons, Rags Ragland and Phil Silvers quit it.

No matter how tense the game, the sound of a plane motor will attract the interest of a baseball crowd.

Chances are the next jay-walker who almost gets run over will bawl out the driver of the car.

More people dine single-handed in health food restaurants than any other type of eating place.

When an actress is described as a Hollywood starlet, I figure she's a pretty girl with no talent.

What is the difference between a busted marriage and a trial separation?

YOU'RE WILLIE MAYS

New York Post *April 1954*

You're Willie Mays, of Fairfield, Alabama, who is part of the small talk of New York. You follow your trade through the seasons of work and then return to the hamlet you call your hometown. But this shall be your city as long as your talent lasts. You're Willie Mays of the Giants and you get your mail at the Polo Grounds. Be sure you leave a forwarding address.

Cherish what occurs but dilute the joy with cynicism, too. This is a prize for being young and strong. Hold onto it and drain the fun out of it. It happens only to a few. But all ballplayers are transients.

Big men, important in the affairs of the world, discuss you as though they were your friends. Famous people such as Gen. Douglas MacArthur, Jim Farley, Robert Wagner, the Mayor of the City of New York, and Jerry Lewis, of Martin and Lewis, watched you yesterday as your home run in the sixth inning beat the Dodgers, 4–3.

Kids forget the squalor of their childhood as they emulate the shambling urgency of your gait. They speak as though you lived on the same block with them. You entertained 32,397 spectators at the Polo Grounds yesterday. Strangers, aching with loneliness, spoke to those who sat alongside of them. And they mentioned your name.

There were hoots from the grouches among the Brooklyn partisans. There was applause from those who support the Giants. But you brought people together in the bantering arguments of sports. You made time pass for the bored with a bright rush. It is a fine accomplishment in a terrible age. The pay is good, too.

Your frantic image dashed across the screens of television sets in living rooms decorated with a splendor you wouldn't believe. You've become a metropolitan fable, told in saloons and pool rooms and related on street corners, in home and playground. You have the gift of excitement.

Here come the dreamers from all the Fairfields of the land. They arrive, like medieval entertainers flocking to the court of a bored old king who demands to be amused. Here is the greatest community of jugglers and acrobats and magicians in the world. They perform the tricks in all the arts and the professions.

But how many pugs push hand trucks on the docks? How many writers lug bus boys' trays? How many painters work behind counters? How many actors drive cabs? How many poets feed a machine? How many ballplayers, settling in middle age for what they are, lost a day's pay to watch you hit the home run yesterday?

Didn't playing with the Birmingham Barons make you a celebrity in Fairfield? You were still in high school when you played shortstop for them. They knew you in Birmingham, which isn't the smallest place in the world, either. It's not a bus-stop when you're Willie Mays, a high school kid, from Fairfield. So, except when you were very little, you were big stuff in a lot of towns.

It was Eddie Montague, scouting for the Giants, who visited the ball park to see a first baseman. He was looking at a guy for Sioux City but he made you. That started you. But he couldn't get it on paper because you were still a high school student. So they waited until June. But what about the guy Montague was supposed to collar for Sioux City? Did he make it? What's his name? What does he do now? Nobody remembers. It matters only to him if he traveled to Sioux City.

So you broke in with Trenton and hit .353. Few become famous in Trenton except the guy who is elected Governor. But people you never met shook your hand when you walked those streets after a ball game. You're better known nationally than most Senators and Governors.

You really turned it on at Minneapolis in the spring of '51. One time you were hitting .500. Your average was .477 when the Giants reached down for you. You came to the National League after being in only 116 minor league games. You played your first game in Philadelphia with the Giants. You had a miserable road trip. You got up 25 times and didn't get a hit.

You wept in the clubhouse. You told Leo Durocher to send you back. You shook inside. What, you asked yourself, was this? But what else could they do? They had to go with you. Your reputation forced them to. And the first time up in the Polo Grounds you hit a home run.

It changed. You became the pet of the ball club. They kidded you.

They made you the butt of practical jokes. What a time you had. You're innocent and full of kid's mischief and laughter. You made catches that amazed the big leaguers.

You're no better than Terry Moore or the DiMaggios, Joe and Dom. But you did it by instinct alone. You were thrilling to watch. Sometimes they astonished you when they told you how great you were. You make the play without thinking about it. And the Giants won the pennant in '51.

But after you had played 34 games in '52 and hit .236 you were drafted. The duty wasn't tough. Now you're back playing. And what a load you're carrying. You read where you'll guarantee the Giants another pennant. They can't miss, you hear, with you. You like to hear that kind of talk but sometimes you wonder.

But yesterday you did what they anticipated. You hit one and that was the ball game. You're Willie Mays and you got 153 more games to go.

NOBODY ASKED ME, BUT

New York Post *circa 1955*

I have more faith in brusque doctors than oily-mannered ones.

You're middle-aged if you remember Larry Semon, the movie comic.

Laurette, a biography of Laurette Taylor by her daughter, Marguerite Courtney, is a sadly moving book.

El Morocco is still the most exciting night club in the country.

One of the greatest of all impossible matches to make would be Tony Canzoneri versus Henry Armstrong.

I have only pity for those legitimate business men who boast about their friendship with gangsters.

Do women ever wear those outfits exhibited by models on TV shows?

I always feel a guy who wears a black homburg and unshined shoes is in financial trouble.

Doesn't Marty Glickman, the sports announcer, sound like an Atlantic City boardwalk auctioneer?

A talkative bartender is usually a sloppy one.

I'm always surprised when a thin, pinch-faced blond has a sense of humor.

Guys who use other people's coffee saucers as ash trays should be barred from public places.

I'd like to see a movie revival of my favorite melodrama, *The Front Page,* with Frank Sinatra playing Hildy Johnson.

I'm depressed when I see a homely young girl drinking a midnight soda alone at a fountain on a Saturday night.

When a pug picks street fights, I figure there's dog in him and he has to show how tough he is.

Look out for the stranger who beats you to the match every time your girl needs a light at a cocktail party.

No one is more sensitive about his game than a weekend tennis player.

I still get a wallop out of watching the lights go on as the dusk sifts into Broadway.

I always feel a wife is subtly knocking her husband when she lavishly praises her sister's old man.

I suspect guys who say, "I just send out for a sandwich for lunch," as lazy men trying to impress me.

Most guys who discuss *War and Peace* as a great novel are odds-on not to have read it.

No one is more generous than a guy spending a rich wife's dough.

Go out and buy Ella Fitzgerald's record, "Taking a Chance on Love."

John K. M. McCaffery is one of the few news commentators who appears to understand the events he is reporting.

Too few people realize what a fine race rider Conn McCreary is.

Why do teenagers stage wrestling matches in the aisles of crowded subway trains but generally sit still in half-empty ones?

Will everybody be trying in this year's Derby Trial since Andy Crevolin was suspended for admitting Determine wasn't pushed in '54?

I'm partial to shrimps but they must be drenched in Russian dressing.

Why do laundries puncture socks with sharp-pronged tags?

I fell a wife is a sloppy housekeeper when she has a strong opinion on bridge.

Even a beautiful woman looks ugly when she's rooting for her money at a race track.

Bet me the next miniature hot dog you eat at a cocktail party will be cold.

Surliest drinkers I know are fat women who are aggressively cheerful when sober.

No restaurant is a good one for me if the coffee is stale.

When a panhandler gives me an elaborate story, I pass because I figure he's a professional moocher.

Chances are the next person you see feeding pigeons in the park will be shabbily dressed.

Why do people who run race tracks think they should be regarded as benefactors of mankind?

Even bad piano music in a cocktail lounge provokes a torch carrier to discuss his anguish.

Women with untidy finger nails try to hide them. Guys seem to make defiantly unnecessary gestures so you won't miss them.

Red Smith

Red Smith (1905–82) violated the adage that aging sportswriters lose their fastball. Born Walter Wellesley Smith in Green Bay, he was dubbed ''Brick'' at Notre Dame and became ''Red'' thereafter. Hired as a reporter by the Milwaukee Sentinel, *he broke into sports on the* St. Louis Star-Times *and was soon a columnist on the* Philadelphia Record. *His work impressed Stanley Woodward, sports editors of the* New York Herald Tribune, *in whose pages Smith offered his ''Views of Sports'' daily from 1945. When the* Trib *died, Smith in due course joined the* Times, *where he won a Pultizer Prize for Distinguished Criticism in 1976. His* Times *colleague, Dave Anderson, has edited* The Red Smith Reader *(New York, 1982); another anthology,* To Absent Friends *(New York 1982) collects his evocative eulogies. Ira Berkow, also of the* Times, *has written* Red: The Life and Times of a Great American Writer *(New York, 1986).*

ON GRANTLAND RICE

New York Herald Tribune *July 15, 1954*

Coming home from vacation is different this time. New York isn't the same town at all. Grantland Rice isn't here. This isn't going to be maudlin but it has to be personal. Grantland Rice was the greatest man I have known, the greatest talent, the greatest gentleman. The most treasured privilege I have had in this world was knowing him and going about with him as his friend. I shall be grateful all my life.

I do not mourn for him, who welcomed peace. I mourn for us.

Granny was a restless sleeper. Sometimes he thrashed about and muttered in his sleep and sometimes he cried out in the dark. "No!" he would shout. "No, dammit, no! Frankie, help me! No, I say!"

Does that seem a curious thing to tell about him now? It isn't, really, because it was so characteristic. It required no dream book by Freud to help interpret those cries. All through his waking hours Granny was saying *yes, surely, glad to, of course, no trouble at all, certainly, don't mention it.* Not only to his friends, but to all the others who imposed on his limitless generosity. And so, when he slept . . .

The only thing greater than his talent was his heart—his gentle courtesy, his all-embracing kindness. And as great as that was his humility.

Once his working press ticket for the Army-Notre Dame football game went astray in the mail. This was Grantland Rice, who did as much for American football as any other man who ever lived; who practically invented the Army-Notre Dame game, who made it a part of American literature with his "Four Horsemen" story of 1924.

He went down Broadway and bought a ticket from a scalper and watched the game from the stands, with his typewriter between his knees. When it was over he made his way apologetically to the press box to do his work.

A friend who heard of this was aghast. "Why didn't you throw some weight around?" he demanded.

"Tell you the truth," Granny said, "I don't weigh much."

In 1944, when the whole World Series was played in St. Louis, the working press had one ticket for the Cardinals' home games and another for the Browns'. On the day of the last game Granny arrived at Sportsman's Park with the wrong ticket.

Nobody crashes the World Series. Granny was going to catch another cab, fight the traffic three or four miles to the Chase Hotel and return with his proper credentials. "You'd miss the first six innings," Frank Graham said. "Come with me."

Leading his reluctant friend through the press gate, Frank whispered to the man at the turnstile, "This is Grantland Rice behind me. He has the wrong ticket."

"Where?" the gateman said, his face lighting. "Come in, Mr. Rice, come in."

Now they were inside, but at the entrance to the press box proper another guard held the pass. Again the conspiratorial whisper, the thumb gesturing back over the shoulder.

"Grantland Rice!" the guard said. "Mr. Rice, how are you, sir. I've always wanted to meet you."

"Frankie, you are marvelous," Granny said as they took their seats. "How did you manage that?"

Perhaps it is not literally true that Grantland Rice put a white collar upon the men of his profession, but not all sportswriters before him were cap-and-sweater guys. He was, however, the sportswriter whose company was sought by presidents and kings.

When Warren Harding was president he asked Granny down to Washington for a round of golf and Granny invited his friend, Ring Lardner.

"This is an unexpected pleasure, Mr. Lardner," Harding said as they hacked around. "I only knew Granny was coming. How'd did you happen to to make it, too?"

"I want to be ambassador to Greece," Lardner said.

"Greece?" said the President. "Why Greece?"

"Because," Lardner said, "my wife doesn't like Great Neck."

Granny and several friends were leaving Toots Shor's a few weeks before his death. There was some confusion just inside the revolving door and one of the group was aware, without looking back, of strangers hesitating behind him, uncertain whether to push through or wait for the way to clear. He heard a woman say, "A lovely man. Let them go."

The small sounds of departure covered the question that must have been asked, but the woman's reply came clearly to the sidewalk.

"Mr. Grantland Rice."

She spoke quietly, but her tone was like banners.

BOYCOTT THE MOSCOW OLYMPICS

New York Times News Service *February 1980*

Neville Trotter is as right as two martinis at lunch. He is the Conservative member of the British Parliament who has asked the Prime Minister, Margaret Thatcher, to lead a worldwide boycott of the Olympics in Moscow to protest the Soviet invasion of Afghanistan.

"Another venue should be found," Mr. Trotter says, "and if necessary the games should be postponed for a year. This is the one lever we have to show our outrage at this naked aggression by Russia. We should do all we can to reduce the Moscow Olympics to a shambles."

The boycott movement hasn't gained much momentum as yet. It was discussed as a possibility at a meeting of the North Atlantic Treaty Organization nations in Brussels. On the *MacNeil/Lehrer Report* on television, Senator Carl Levin of Michigan said a boycott should be considered and Senator Richard G. Lugar of Indiana said it would be "small potatoes."

At the International Olympic Committee headquarters in Lausanne, Switzerland, Lord Killanin, president of the I.O.C., declared the games would go on and pleaded for politicians to stay out of Olympic affairs. If horses ran as true to form as the Olympic oligarchy, the favorite would never lose. Ever since they learned to speak with heads buried in sand, the badgers have been saying that politics has no place in the Olympic movement, and as long as any of them can remember, the games have been a stage for political discord and social protest.

The official—and inflexible—position of the Olympic brass on these matters was enunciated almost half a century ago by the noblest badger of them all, the late Avery Brundage. In 1935 there was strong sentiment in this country against participation in the 1936 Olympics in Berlin, on the grounds that sending a team to that carnival of Nazism would be tantamount to endorsing Hitler.

"Frankly," said Brundage, then president of the United States Olympic Committee, "I don't think we have any business to meddle in this question. We are a sports group, organized and pledged to promote clear competition and sportsmanship. When we let politics, racial questions or social disputes creep into our actions, we're in for trouble."

The boycott movement was defeated, and Avery in victory was even franker than before. "Certain Jews must now understand," he wrote, "that they cannot use these games as a weapon in their boycott against the Nazis."

Hitler's anti-Semitism eventually led to the unspeakable Holocaust, but in 1935 the only known fatality was the suicide of Fritz Rosenfelder after his expulsion from an athletic club in Württemberg.

When Americans look back to the 1936 Olympics, they take pleasure only in the memory of Jesse Owens' four gold medals, in the discomfiture of Joseph Goebbels at the success of America's "black auxiliaries." Except for that, we are ashamed at having been guests at Adolf Hitler's big party.

We should have known better. As early as 1933, Julius Streicher's *Der Stürmer* had carried this comment on Rosenfelder's suicide: "We need waste no words here. Jews are Jews and there is no place for them in German sports. Germany is the Fatherland of Germans and not Jews, and the Germans have the right to do what they want in their own country."

We didn't know better, and we were painfully slow to learn. General Charles E. Sherrill, an American member of the I.O.C., asked that Helene Mayer be invited to compete for Germany to prove that Jews would not be discriminated against. Daughter of a Christian mother and a Jewish father, she was a champion fencer who had represented Germany in the 1928 and 1932 Olympics. On his return to America, General Sherrill said: "I went to Germany for the purpose of getting at least one Jew on the German Olympic team, and I feel that my job is finished. As for obstacles placed in the way of Jewish athletes or any others in trying to reach Olympic ability, I would have no more business discussing that in Germany than if the Germans attempted to discuss the Negro situation in the American South or the treatment of the Japanese in California."

Jews were barred from swimming facilities in Germany, from the ski resort of Garmisch-Partenkirchen and from all private and public practice fields, and of course they were not permitted to compete in Olympic tryouts. Yet Frederick W. Rubein, secretary of the United States Olympic Committee, said: "Germans are not discriminating against Jews in their Olympic tryouts. The Jews are eliminated because they are not good enough as athletes. Why, there are not a dozen Jews in the world of Olympic caliber."

Said General Sherrill: "There was never a prominent Jewish athlete in history."

The Olympic brass won that time. We did not meddle in the internal affairs of Germany.

The games went on in Australia almost immediately after Soviet tanks crushed a revolt in Hungary, though blood flowed when Hungarians met Russians in water polo. The games went on in Mexico City two weeks after Army machine guns massacred more than thirty students in the Plaza of the Three Cultures. The games went on in Munich while Arab terrorists were murdering eleven members of the Israeli delegation. On that occasion, though, they took time out for a memorial service that Avery Brundage turned into a pep rally.

"We have only the strength of a great ideal," Avery said. "I am sure the public will agree that we cannot allow a handful of terrorists to destroy this nucleus of international cooperation and good will we have in the Olympic movement. The games must go on."

That day it was written here: "The men who run the Olympics are not evil men. Their shocking lack of awareness can't be due to callousness. It has to be stupidity."

WRITING LESS—AND BETTER?

New York Times *1982*

[This was one of Smith's last columns before his death.]

Up to now, the pieces under my byline have run on Sunday, Monday, Wednesday, and Friday. Starting this week, it will be Sunday, Monday, and Thursday—three columns instead of four. We shall have to wait and see whether the quality improves.

Visiting our freshman daughter (freshwoman or freshperson would be preferred by feminists, though heaven knows she was fresh), we sat chatting with perhaps a dozen of her classmates. Somehow my job got into the discussion. A lovely blond was appalled.

"A theme a day!" she murmured.

The figure was not altogether accurate. At the time it was six themes a week. It has been seven and when it dropped to six that looked like roller coaster's end. However, it finally went to five, to three and back to four, where it has remained for years.

First time I ever encountered John S. Knight, the publisher, we were bellying up to Marje Everett's bar at Arlington Park. He did not acknowledge the introduction. Instead, he said: "Nobody can write six good columns a week. Why don't you write three? Want me to fix it up?"

"Look, Mr. Knight," I said. "Suppose I wrote three stinkers, I wouldn't have the rest of the week to recover. One of the beauties of this job is that there's always tomorrow. Tomorrow things will be better."

Now that the quota is back to three, will things be better day after tomorrow?

The comely college freshman wasn't told of the years when a daily column meant seven a week. Between those jousts with the mother tongue, there was always a fight or football match or ball game or horse race that had to be covered after the column was done. I loved it.

The seven-a-week routine was in Philadelphia, which reminds me of the late heavy-weight champion, Sonny Liston. Before his second bout with Muhammad Ali was run out of Boston, Liston trained in a hotel in Dedham.

I was chatting about old Philadelphia days with the trainer, Willie Reddish, remembered from his time as a heavyweight boxer in Philadelphia.

"Oh," Willie said apropos of some event in the past, "were you there then?"

"Willie," I said, "I did ten years hard in Philadelphia."

There had been no sign that Liston was listening, but at this he swung around. "Hard?" he said. "No good time?"

From that moment on, Sonny and I were buddies, though it wasn't easy accepting him as a sterling citizen of lofty moral standards.

On this job two questions are inevitably asked: "Of all those you have met, who was the best athlete?" and "Which one did you like best?"

Both questions are unanswerable, but on either count Bill Shoemaker, the jockey, would have to stand high.

This little guy weighed 96 pounds as an apprentice rider thirty-two years ago. He still weighs 96 pounds and he will beat your pants off at golf, tennis, and any other game where you're foolish enough to challenge him.

There were, of course, many others, not necessarily great. Indeed, there was a longish period when my rapport with some who were less than great made me nervous. Maybe I was stuck on bad ballplayers. I told myself not to worry.

Some day there would be another Joe DiMaggio.

Langston Hughes

Arna Bontemps has written of Langston Hughes (1902–67): "He has been a minstrel and troubadour in the classic sense. He has had no other vocation . . . Hughes has worked competently in all the literary forms." He became a columnist in 1943 for the Chicago Defender, *which, echoing the white-edited* Tribune, *dubbed itself "The World's Greatest Weekly." Here was published on February 13 a dialogue with "My Simple Minded Friend," Jesse B. Simple, a Virginia-bred Harlem black. The Simple dialogues caught on, and Hughes's friend appeared in about a quarter of the weekly columns the poet wrote over twenty-three years. Hughes mined the column for five books, beginning with* Simple Speaks His Mind *(New York, 1950). In due course, Simple was endowed with a wife, Joyce; an after-hours girlfriend, Zarita; and a hard-luck cousin, Minnie. In Simple's eyes, all whites (with the unimpeachable exception of Eleanor Roosevelt) were suspect, while his middle-class locutor tried vainly to show things weren't as bad as they seemed. As with Dooley, Simple addressed the anomalies of an unjust world without sanctimony. The Simple dialogues are appraised in* Langston Hughes: Black Genius *(New York, 1971), edited by Therman O'Daniel, and in the second volume of Arnold Rampersad's biography,* I Dream a World *(New York, 1988).*

THERE OUGHT TO BE A LAW

Chicago Defender *circa 1943–44*

"I have been up North a long time, but it looks like I just cannot learn to like white folks."

"I don't care to hear you say that," I said, "because there are a lot of good white people in this world."

"Not enough of them," said Simple, waving his evening paper. "If there was, they would make this American country good. But just look at what this paper is full of."

"You cannot dislike *all* white people for what the bad ones do," I said. "And I'm certain you don't dislike them all because once you told me yourself that you wouldn't wish any harm to befall Mrs. Roosevelt."

"Mrs. Roosevelt is different," said Simple.

"There now! You see, you are talking just as some white people talk about the Negroes they *happen* to like. They are always 'different.' That is a provincial way to think. You need to get around more."

"You mean among white folks?" asked Simple. "How can I make friends with white folks when they got Jim Crow all over the place?"

"Then you need to open your mind."

"I have near about *lost* my mind worrying with them," said Simple. "In fact, they have hurt my soul."

"You certainly feel bad tonight," I said. "Maybe you need a drink."

"Nothing in a bottle will help my soul," said Simple, "but I will take a drink."

"Maybe it will help your mind," I said. "Beer?"

"Yes."

"Glass or bottle?"

"A bottle because it contains two glasses," said Simple, spreading his paper out on the bar. "Look here at these headlines, man, where Congress is busy passing laws. While they're making all these laws, it looks like to me they ought to make one setting up a few Game Preserves for Negroes."

"What ever gave you that fantastic idea?" I asked.

"A movie short I saw the other night," said Simple, "about how the government is protecting wild life, preserving fish and game, and setting aside big tracts of land where nobody can fish, shoot, hunt, nor harm a single living creature with furs, fins, or feather. But it did not show a thing about Negroes."

"I thought you said the picture was about 'wild life.' Negroes are not wild."

"No," said Simple, "but we need protection. This film showed how they put aside a thousand acres out West where the buffaloes roam and nobody can shoot a single one of them. If they do, they get in jail. It also showed some big National Park with government airplanes dropping food down to the deers when they got snowed under and had nothing to eat. The government protects and takes care of buffaloes and deers—which is more than the government does for me or my kinfolks down South. Last month they lynched a man in Georgia and just today I see where the Klan has whipped a Negro within a inch of his life in Alabama. And right up North here in New York a actor is suing a apartment house that won't even let a Negro go up on the elevator to see his producer. That is what

I mean by Game Preserves for Negroes—Congress ought to set aside some place where we can go and nobody can jump on us and beat us, neither lynch us nor Jim Crow us every day. Colored folks rate as much protection as a buffalo, or a deer."

"You have a point there," I said.

"This here movie showed great big beautiful lakes with signs up all around:

NO FISHING—STATE GAME PRESERVE

But it did not show a single place with a sign up:

NO LYNCHING

It also showed flocks of wild ducks settling down in a nice green meadow behind a government sign that said:

NO HUNTING

It were nice and peaceful for them fish and ducks. There ought to be some place where it is nice and peaceful for me, too, even if I am not a fish or a duck.

"They showed one scene with two great big old longhorn elks locking horns on a Game Preserve somewhere out in Wyoming, fighting like mad. Nobody bothered them elks or tried to stop them from fighting. But just let me get in a little old fist fight here in this bar, they will lock me up and the Desk Sergeant will say, "What are you colored boys doing, disturbing the peace?" Then they will give me thirty days and fine me twice as much as they would a white man for doing the same thing. There ought to be some place where I can fight in peace and not get fined them high fines."

"You disgust me," I said, "I thought you were talking about a place where you could be quiet and compose your mind. Instead, you are talking about fighting."

"I would like a place where I could do both,." said Simple. "If the government can set aside some spot for a elk *to be a elk* without being bothered, or a fish *to be a fish* without getting hooked, or a buffalo *to be a buffalo* without being shot down, there ought to be some place in this American country where a Negro can be a Negro without being Jim Crowed. There ought to be a law. The next time I see my congressman, I am going to tell him to introduce a bill for Game Preserves for Negroes."

"The Southerners would filibuster it to death," I said.

"If we are such a problem to them Southerners," said Simple, "I should think they would want some place to preserve us out of their sight. But then, of course, you have to take into consideration that if the Negroes was taken out of the South, who would they lynch? What would they do for sport? A Game Preserve is for to keep people from bothering anything that is living.

"When that movie finished, it were sunset in Virginia and it showed a little deer and its mama laying down to sleep. Didn't nobody say, 'Get up, deer, you can't sleep here,' like they would to me if I was to go to the White Sulphur Springs Hotel."

" 'The foxes have holes, and the birds of the air have nests; but the Son of man hath not where to lay his head.' "

"That is why I want a Game Preserve for Negroes," said Simple.

A TOAST TO HARLEM

Chicago Defender *circa 1943–44*

Quiet can seem unduly loud at times. Since nobody at the bar was saying a word during a lull in the bright blues-blare of the Wishing Well's usually overworked juke box, I addressed my friend Simple.

"Since you told me last night you are an Indian, explain to me how it is you find yourself living in a furnished room in Harlem, my brave buck, instead of on a reservation?"

"I am a colored Indian," said Simple.

"In other words, a Negro."

"A Black Foot Indian, daddy-o, not a red one. Anyhow, Harlem is the place I always did want to be. And if it wasn't for landladies, I would be happy. That's a fact! I love Harlem."

"What is it you love about Harlem?"

"It's so full of Negroes," said Simple. "I feel like I got protection."

"From what?"

"From white folks," said Simple. "Furthermore, I like Harlem because it belongs to me."

"Harlem does not belong to you. You don't own the houses in Harlem. They belong to white folks."

"I might not own 'em," said Simple, "but I live in 'em. It would take an atom bomb to get me out."

"Or a depression," I said.

"I would not move for no depression. No, I would not go back down South, not even to Baltimore. I am in Harlem to stay! You say the houses ain't mine. Well, the sidewalk is—and don't nobody push me off. The cops don't even say, 'Move on,' hardly no more. They learned something from them Harlem riots. They used to beat your head right in public, but now they only beat it after they get you down to the stationhouse. And they don't beat it then if they think you know a colored congressman."

"Harlem has a few Negro leaders," I said.

"Elected by my *own* vote," said Simple. "Here I ain't scared to vote —that's another thing I like about Harlem. I also like it because we've got subways and it does not take all day to get downtown, neither are you Jim Crowed all the way. Why, Negroes is running some of these subway trains. This morning I rode the A Train down to 34th Street. There were a Negro driving it, making ninety miles a hour. That cat *were really driving* that train! Every time he flew by one of them local stations looks like he was saying, 'Look at me! This train is mine!' That cat were gone, ole man. Which is another reason why I like Harlem! Sometimes I run into Duke Ellington on 125th Street and I say, 'What you know there, Duke?' Duke says, 'Solid, ole man.' He does not know me from Adam, but he speaks. One day I saw Lena Horne coming out of the Hotel Theresa and I said, 'Huba! Huba!' Lena smiled. Folks is friendly in Harlem. I feel like I got the world in a jug and the stopper in my hand! So drink a toast to Harlem!"

Simple lifted his glass of beer:

> "Here's to Harlem!
> They say Heaven is Paradise.
> If Harlem ain't Heaven,
> Then a mouse ain't mice!"

"Heaven is a state of mind," I commented.

"It sure is *mine,*" said Simple, draining his glass, "From Central Park to 179th, from river to river, Harlem is mine! Lots of white folks is scared to come up here, too, after dark."

"That is nothing to be proud of," I said.

"I am sorry white folks is scared to come to Harlem, but I am scared to go around some of *them.* Why, for instant, in my home town once before I came North to live, I was walking down the street when a white woman jumped out of her door and said, 'Boy, get away from here because I am scared of you.'

"I said, 'Why?'

"She said, 'Because you are black.'

"I said, 'Lady, I am scared of you because you are white.' I went on down the street, but I kept wishing I was blacker—so I could of scared that lady to death. So help me, I did. Imagine somebody talking about they is scared of me because I am black! I got more reason to be scared of white folks than they have of me."

"Right," I said.

"The white race drug me over here from Africa, slaved me, freed me, lynched me, starved me during the depression, Jim Crowed me during the war—then they come talking about they is scared of me! Which is why I am glad I have got one spot to call my own where I hold sway—Harlem. Harlem, where I can thumb my nose at the world!"

"You talk just like a Negro nationalist," I said.

"What's that?"

"Someone who wants Negroes to be on top."

"When everybody else keeps me on the *bottom,* I don't see why I shouldn't want to be on top. I will, too, someday."

"That's the spirit that causes wars," I said.

"I would not mind a war if I could win it. White folks fight, lynch, and enjoy themselves."

"There you go," I said. "That old *race-against-race* jargon. There'll never be peace that way. The world tomorrow ought to be a world where everybody gets along together. The least we can do is extend a friendly hand."

"Every time I extend my hand I get put back in my place. You know them poetries about the black cat that tried to be friendly with the white one:

> The black cat said to the white cat,
> 'Let's sport around the town.'
> The white cat said to the black cat,
> 'You better set your black self down!' "

"Unfriendliness of that nature should not exist," I said. "Folks ought to live like neighbors."

"You're talking about what ought to be. But as long as what *is* is— and Georgia is Georgia—I will take Harlem for mine. At least, if trouble comes, I will have *my own window* to shoot from."

"I refuse to argue with you any more," I said. "What Harlem ought to hold out to the world from its windows is a friendly hand, not a belligerent attitude."

"It will not be my attitude I will have out my window," said Simple.

Harry Golden

Few outside North Carolina had heard of Harry Golden (1902–81) until 1956, when the national press picked up his Vertical Negro Plan for ending school segregation by using stand-up desks since Southerners seemed to object to mixed sitting, not mixed standing. Then it was discovered that Golden edited a paper in Charlotte called The Carolina Israelite, *that he was born on New York's Lower East Side, where his immigrant father was an editor of* The Jewish Daily Forward, *and that he had been a pamphleteer for Henry George's single-tax movement and worked for various New York papers before settling in North Carolina to found his monthly paper in 1941. According to Golden, the only two papers south of the Mason-Dixon line to welcome the 1954 desegregation ruling were his own and the* Louisville Courier-Journal. *After* Only in America *(New York, 1958) became a bestseller, it came to light that Golden had been in Federal prison in the 1930s after pleading guilty to mail fraud charges. ''This story ties me closer to him,'' said his neighbor and friend, Carl Sandburg. The book's success led to Golden's writing a column for the Bell-McClure Syndicate, which in turn produced material for four more books. His autobiography is entitled* The Right Time *(New York, 1969)*

A GOLDEN MISCELLANY

The Carolina Israelite *1950–58*

Introduction to *Only in America,* March 1958

I live in a high-porched house built before the Great Wars on Elizabeth Avenue in Charlotte, North Carolina. Here, I write and publish *The Carolina Israelite.*

I should explain that *The Carolina Israelite,* five columns, sixteen pages, is published monthly, is printed entirely in English, and that more than half its subscribers are non-Jews. I print no "news," personals, socials, or press releases. And my last "obituary" was on the assassination of Julius Caesar in 44 B.C.

With the exception of a few "letters to the editor," I write the entire paper myself, 15,000 to 25,000 words a month. I arrange my reading matter in the form of editorials, set in 8-point Century, 24½ picas, with 10-point bold-face heads in caps. I put a short rule between the items, which range in length from twenty words ("How Dr. Samuel Johnson Prepared Oysters") to a 3,000-word article ("Sweet Daddy Grace, the Southern Father Divine").

I draw heavily on history, literature, philosophy. I do not run the editorials in chronological or departmental sequence. I merely try to arrange the columns so that a long article is usually followed by two or three short pieces.

Each month I set the ads first, about one hundred "card" advertisements, with no displays and with as little copy as possible. Then I cram my editorials into every other inch of available space. I sort of slither them in and around the ads on each of the sixteen pages, and with only one beginning and one end. Many of my subscribers have tried to pick individual items of special interest to them; all in vain. The only chance they have is to begin at the top left-hand column on page one, and keep going to the end.

Some of my readers make "an evening" with *The Carolina Israelite.* When the paper arrives other events are put aside, and the high school son or daughter reads my paper from cover to cover to the assembled family. I tell them stories of the Lower East Side of New York where I was born. "It's like a personal letter to me," they write, or "It reminds me of my father when he came home from the shop and read the paper to us."

I have found that people are hungry for this "personal touch." They like to read of the great news stories of the past. I tell them about Dorothy Arnold's disappearance; the sinking of the *Titanic;* how "Hell's Kitchen" got it's name; about the old gang wars; about the time someone threw missiles on the funeral procession of a famous rabbi; about the Tong Wars of Chinatown; and about the Thalia Theatre where magician Thurston made a woman disappear from a cage suspended over the audience. I tell them about the first half-dollar I ever saw, given to me by "Big Tim" Sullivan, the Tammany Hall Sachem, and many other such tales, stories, anecdotes, and reminiscences.

For the first few years of its existence, my paper was just another one in that vast stack of periodicals which comes to the "exchange" desk of every daily newspaper. These fall into categories: the "labor" press, the "Negro" press, the "Anglo-Jewish" press, the official organs of lodges and fraternal orders, house organs of government departments and big corporations, organizational media, church periodicals, and magazines that have "home," "mother," "religion," "ladies," or "women" in the title.

I made up my mind to do something about it. I decided to turn my paper into a "personal journal," and put to use the results of almost forty-five years of uninterrupted reading. I also suspected that many Southerners were hungry for a word "above the battle."

There were difficulties. Because they are so far from the main concentrations of the Jewish population, the small Jewish communities of the South live in deadly fear that one Jew may say or do something that "will involve the whole Jewish Community." The Jews have a point, of course, since so many Gentiles believe that each of us is "spokesman" for the entire community. But I also knew that since this "mass responsibility" idea involves also the credit jeweler, the pawnbroker, and the textile manufacturer, the Jewish communities of the South would survive a personal journalist, too.

In one issue I asked the question, "Am I a Tar Heel?", and immediately some of the mighty daily papers in the State wrote editorials extending this honor to me. Indeed, they said, Harry Golden is a Tar Heel; to which I replied in their Open Forums that I never carried a designation with greater pride.

All of this is a lesson in sociology. I have found that the Southerner (and any other man, for that matter) arches his back at the fellow who throws a brick over the wall and runs away. But when that same fellow becomes *part of the community,* the attitude toward him changes. He may not particularly endear himself, but he does acquire status, and he is not only respected, but actually welcomed as part of the ebb and flow of daily life.

I believe too that *"Let us not stick our necks out"* or *"Let someone else talk on that subject"* are watchwords of the ghetto, and I do not believe in ghettos: white, black, or Puerto Rican.

Why Other Planets Have Not Contacted Us

In all our space literature we automatically picture the Martians or the other Visitors From Outer Space trying to wipe us out and grab our women. Big deal. We are always worried about someone carrying off our women. This is chutzpah (arrogance). I believe the reverse is closer to the truth. I think the Martians and other Visitors From Outer Space are afraid they'll get killed the minute they set foot on this nervous, inhibited, frustrated, and trigger-happy little Earth.

Another thing must worry them. The position with respect to the Sun is, of course, different with each planet; the climatic and atmospheric conditions are different. It is unlikely that any visitors from outer space can conform to the necessary physical requirements we have established —narrow hips, tall, clean-shaven, and no "frizzy hair," which Westbrook Pegler recently pointed out was utterly "foreign." And then what about color? That's important. Suppose the color of their skin is, for instance, navy blue, or even magenta, what then? Wouldn't that set us off in a frenzy the minute we saw one of them? They know it. After all we are comparative newcomers. Some of those planets are not four billion years old, but seventy billion years old.

Old hands. They keep watching and keep saying, "Not yet, Charlie." They have decided to wait. Or maybe George Bernard Shaw was right. May they all use Earth as a sort of interplanetary lunatic asylum.

Forget Victuals

The prettiest words in the English language begin with the letter *m*— murmuring, Monongahela, mackinaw, Madagascar, maiden, majesty, Majorca, and marinated (especially herring). The ugliest word in the language is—victuals. You can't say or write it. The best thing is to forget it.

Wink at Some Homely Girl

Some years before his death, H. L. Mencken asked his friends to "wink your eye at some homely girl!" in remembrance of him. What nonsense! Sheer nonsense! Mencken acknowledged, for many years, a reputation as a woman-hater, which is the external sign of adolescence. A woman-hater is no expert on these matters. The idea is presumptuous, this "winking at a homely girl." Some of the finest loving on this earth has been due to the initiative, ingenuity, and kindness of the "homely" girl. And how does one go about deciding who is a "homely" girl?

Is there really a "homely" girl anywhere in the world? By whose standards? Mencken's, the casting director's, or the girl's husband? And they have husbands, you know—by the million.

"Wink at some homely girl." Every newspaper and literary journal in the country fell for this Mencken kid stuff, while hundreds of thousands of wonderful schoolteachers, social workers, YWCA secretaries, telephone operators, and stenographers laughed themselves sick. Some of them laughed so hard at this nonsense that their eyeglasses fell off.

The Vertical Negro Plan

Those who love North Carolina will jump at the chance to share in the great responsibility confronting our Governor and the State Legislature. A special session of the Legislature (July 25-28, 1956) passed a series of amendments to the State Constitution. These proposals submitted by the Governor and his Advisory Education Committee included the following:

A. The elimination of the compulsory attendance law, "to prevent any child from being forced to attend a school with a child of another race."
B. The establishment of "Education Expense Grants" for education in a private school, "in the case of a child assigned to a public school attended by a child of another race."
C. A "uniform system of local option" whereby a majority of the folks in a school district may suspend or close a school if the situation becomes "intolerable."

But suppose a Negro child applies for this "Education Expense Grant" and says he wants to go to the private school too? There are fourteen Supreme Court decisions involving the use of public funds; there are only two "decisions" involving the elimination of social discrimination in the public schools.

The Governor has said that critics of these proposals have not offered any constructive advice or alternatives. Permit me, therefore, to offer an idea for the consideration of the members of the regular sessions. A careful study of the plan, I believe, will show that it will save millions of dollars in tax funds and eliminate forever the danger to our public education system. Before I outline any plan, I would like to give you a little background.

One of the factors involved in our tremendous industrial growth and economic prosperity is the fact that the South, voluntarily, has all but eliminated VERTICAL SEGREGATION. The tremendous buying power of the twelve million Negroes in the South has been based wholly on the absence of racial segregation. The white and Negro stand at the same grocery and supermarket counters; deposit money at the same bank teller's window; pay phone and light bills to the same clerk; walk through the same dime and department stores, and stand at the same drugstore counters.

It is only when the Negro "sets" that the fur begins to fly.

Now, since we are not even thinking about restoring VERTICAL SEGREGATION, I think my plan would not only comply with the Supreme Court decisions, but would maintain "sitting-down" segregation. Now here is the GOLDEN VERTICAL NEGRO PLAN. Instead of all those complicated proposals, all the next session needs to do is pass one small amendment which would provide *only* desks in all the public schools of our state—*no seats.*

The desks should be those standing-up jobs, like the old-fashioned bookkeeping desk. Since no one in the South pays the slightest attention to a VERTICAL NEGRO, this will completely solve our problem. And it is not such a terrible inconvenience for young people to stand up during their classroom studies. In fact, this may be a blessing in disguise. They are not learning to read sitting down, anyway; maybe standing up will help. This will save more millions of dollars in the cost of our remedial English course when the kids enter college. In whatever direction you look with the GOLDEN VERTICAL NEGRO PLAN, you save millions of dollars, to say nothing of eliminating forever any danger to our public education system upon which rests the destiny, hopes, and happiness of this society.

My WHITE BABY PLAN offers another possible solution to the segregation problem—this time in a field other than education.

Here is an actual case history of the "White Baby Plan To End Racial Segregation":

Some months ago there was a revival of the Laurence Olivier movie, *Hamlet,* and several Negro schoolteachers were eager to see it. One Saturday afternoon they asked some white friends to lend them two of their little children, a three-year-old girl and a six-year-old boy, and, holding these white children by the hands, they obtained tickets from the movie-house cashier without a moment's hesitation. They were in like Flynn.

This would also solve the baby-sitting problem for thousands and thousands of white working mothers. There can be a mutual exchange of references, then the people can sort of pool their children at a central point in each neighborhood, and every time a Negro wants to go to the movies all she need do is pick up a white child—and go.

Eventually the Negro community can set up a factory and manufacture white babies made of plastic, and when they want to go to the opera or to a concert, all they need do is carry that plastic doll in their arms. The dolls, of course, should all have blond curls and blue eyes, which would go even further; it would give the Negro woman and her husband priority over the whites for the very best seats in the house.

While I still have faith in the WHITE BABY PLAN, my final proposal may prove to be the most practical of all.

Only after a successful test was I ready to announce formally the GOLDEN "OUT-OF-ORDER" PLAN.

I tried my plan in a city of North Carolina, where the Negroes represent 39 percent of the population.

I prevailed upon the manager of a department stores to shut the water off in his "white" water fountain and put up a sign, "Out-of Order." For the first day or two the whites were hesitant, but little by little they began to drink out of the water fountain belonging to the "coloreds"—and by the end of the third week everybody was drinking "segregated" water; with not a single solitary complaint to date.

I believe the test is of such sociological significance that the Governor should appoint a special committee of two members of the House and two Senators to investigate the GOLDEN "OUT-OF-ORDER" PLAN. We kept daily reports on the use of the unsegregated water fountain which should be of great value to this committee. This may be the answer to the necessary uplifting of the white morale. It is possible that the whites may accept desegregation if they are assured that the facilities are still "separate," albeit "Out-of-Order."

As I see it now, the key to my Plan is to keep the "Out-of-Order" sign up for at least two years. We must do this thing gradually.

They Never Met a Payroll

1. Copernicus
2. Galileo
3. Newton
4. Einstein

My Mother and God

My father was an intellectual and our home was filled with talk. We are a vocal people to begin with, and it was not unusual for intellectuals to spend hours discussing the meaning of a single sentence in the Law, or for that matter, a single word. We are the greatest hairsplitters in the world. A pilpul is what they call these complicated discussions. But my father went far beyond the Biblical text. He was at home with Henry George, and Eugene V. Debs, and Benjamin Franklin, and the rationalists of the past. He conducted a sort of philosophy discussion group every Saturday afternoon. Five or six of the men would be gathered around the table and my mother was busy serving them. She walked on tiptoe not to disturb the great men, as she brought the platters of boiled potatoes or haiseh bubbles (chick-peas), and silently went her way. But there was always a trace of a smile on her face; if not cynical, certainly one of amusement.

My mother, I would say, was a primitive woman. She spoke only Yiddish. She could read the prayers out of the book but that was all. She spent all her time cooking, cleaning, sewing; sewing for the family as well as professionally for the neighbors. I think my intellectual father guessed at my mother's "amusement." I have had the feeling that he knew that she was not overly impressed. My mother, of course, thought all those discussions were nonsense. What does a person need but God? And she had God. Sometimes I smile at all the goings-on over the radio about God. Whose God are they talking about anyway—what do they know about God? My mother talked with God all the time, actual conversations. She should send you on an errand and as you were ready to dart off into the crowded, dangerous streets, she turned her face upward and said: "Now see that he's all right." She smiled at the boy, but was dead serious when she spoke to Him. She gave the impression that this was a matter-of-fact relationship, part of the covenant. "In the home that boy is my obligation but once he is out on the street, that is Your department and be sure to see to it." And she never permitted a single expression involving the future to be uttered without that covering clause, "With God's help." And this had to follow hard upon the original assertion. Thus if you ran down the hallway saying, "I'll go to the library tomorrow," she chased after you to make sure that there was no great lapse between your stated intention and the follow-up, "With God's help."

I do not know of any people *less* chauvinistic than the Jews. Just imagine if another race had produced the Ten Commandments, for instance. Think of the place that event would have held in history? But the Jews have always insisted that they had nothing to do with any of these wonderful things. God merely used them to establish His moral code among the peoples of the world. This idea influenced our entire history and every phase of our lives. If a dish happened to turn out well, do you think my mother would take credit for it? Not at all. She said it was an act of God. God helped her cook and sew and clean. And sometimes you have to wonder about it. I am thinking of Mother's potato latkes (pancakes) and holishkas (chopped beef and spices rolled in cabbage leaves and cooked in a sweet-and-sour raisin sauce) and kreplach (small portions of dough folded around chopped beef, boiled, and then dropped into a steaming hot platter of golden chicken soup), and I will say this, "If God did not really help her prepare those dishes (as she claimed), how is it that I haven't been able to find anything to equal them in all these years?" This is the kind of evidence that would even stand up in a court of law.

Art Buchwald

Who is Art Buchwald (1925—   )? *"The greatest satirist in English since Pope and Swift," according to a jacket blurb on one of his twenty or so books.* And who solicited the blurb? *Most likely a tall, dark, and handsome newspaperman who was raised in Long Island, served in the Marines during World War II, attended the University of Southern California, migrated to Paris in 1948 to become a nightclub reporter on the European edition of the* Herald Tribune, *and moved to Washington in 1962 to write what he calls a Washington-type column; in 1982 he received a Pulitzer Prize. He is happily married to his first wife and for recreation likes to eat in great restaurants with beautiful women on his expense account—all of which is culled from jacket copy.* What's the trick in writing a Buchwald column? *As he told the* New York Times Book Review: *"You can't make up anything anymore. The world itself is a satire. All you're doing is recording it."*

POLITICAL POLL IN 1776

New York Herald Tribune *November 1, 1962*

The political pollster has become such an important part of the American scene that it's hard to imagine how this country was ever able to function without him.

What would have happened, for example, if there were political pollsters in the early days of this country?

This is how the results might have turned out.

When asked if they thought the British were doing a good job in administrating the Colonies, this is how a cross section of the people responded:

British doing good job / 63%
Not doing good job / 22%
Don't know / 15%

The next question, "Do you think the dumping of tea in the Boston Harbor by militants helped or hurt the taxation laws in the New World?

Hurt the cause of taxation / 79%
Helped the cause / 12%
Didn't think it would make any difference / 9%

"What do you think our image is in England after the Minute Man attacked the British at Lexington?

Minute men hurt our image in England / 83%
Gave British new respect for Colonies / 10%
Undecided / 7%

"Which of these two Georges can do more for the Colonies—George III or George Washington?"

George III / 76%
George Washington / 14%
Others / 10%

It is interesting to note that 80 percent of the people questioned had never heard of George Washington before.

The next questions was, "Do you think the Declaration of Independence as it is written is a good document or a bad one?"

Bad document / 14%
Good document / 12%
No opinion / 84%

A group of those polled felt that the Declaration of Independence had been written by a bunch of radicals and the publishing of it at this time would only bring harsher measures from the British.

When asked whether the best way to bring about reforms was through terrorism or redress to the Crown an overwhelming proportion of Colonists felts appeals should be made to the King.

Reforms through petition / 24%
Reforms through act of terrorism / 8%
Don't know / 66%

The pollsters then asked what the public thought was the most crucial issue of the time.

Trade with foreign nations / 65%
War with Indians / 20%
The independence issue / 15%

The survey also went into the question of Patrick Henry.

"Do you think Patrick Henry did the right thing in demanding liberty or death?"

Did a foolhardy thing and was a trouble maker / 53%
Did a brave thing and made his point / 23%
Should have gone through the courts / 6%
Don't know / 8%

On the basis of the results of the poll the militant Colonials decided they did not have enough popular support to foment a revolution and gave up the idea of creating a United States of America.

NIXON GOES TO THE MOUNTAIN

Washington Post *April 26, 1973*

Last weekend President Richard Nixon went to Camp David alone, without family or aides. Press Secretary Ron Ziegler denies it, but it has been reliably reported that the President went up to the top of the mountain to speak with God.

"God, God, why are You doing this to me?"

"Doing what, Richard?"

"The Watergate, the coverup, the grand jury hearings, the Senate investigations. Why me, God?"

"Don't blame me, Richard. I gave you my blessing to win the election, but I didn't tell you to steal it."

"God, I've done everything You told me to do. I ended the war. I defeated poverty. I cleaned the air and the water. I defeated crime in the streets. Surely I deserve a break."

"Richard, I tried to warn you that you had sinful people working for you."

"When, God?"

"Just after the Committee to Re-Elect the President was formed. When I saw the people you had selected to head up the committee, I was shocked. We've got a long file on them up here."

"Why didn't You tell me, God?"

"I tried to, but Ehrlichman and Haldeman wouldn't let Me talk to you on the phone. They said they'd give you the message I called."

"They never told me, God."

"It figures. Then I sent you a telegram saying it was urgent that you contact me."

"The only telegrams I read during that period were those in support of my bombing North Vietnam."

"Finally, Richard, I made one last effort. I showed up at a prayer meeting one Sunday at the White House and after the sermon I came up to you and said there were men among you who would betray you. Do you know what you did, Richard? You introduced me to Pat and then you gave Me a ballpoint pen."

"I didn't know it was you, God. So many people show up at these prayer meetings. Is that why You're punishing me—because I snubbed You?"

"I'm not punishing you, Richard. But even I can do just so much. If it were merely a simple case of bugging at the Watergate, I could probably fix it. But your Administration is involved in the obstruction of justice, the bribing of witnesses, the forging of papers, wiretapping, perjury and using the mails to defraud."

"Good God, nobody's perfect!"

"I guess that's what the grand jury is saying."

"Look, I've got less than four years in which to go down as the greatest President in the history of the United States. Give me a break."

"You've got to clean house, Richard. Get rid of everyone who has any connection with the scandal. You must make it perfectly clear you were hoodwinked by everyone on your staff. You must show the American people that when it comes to the Presidency, no one is too big to be sacrificed on the altar of expediency."

"God are You asking for a human sacrifice?"

"It would show your good faith, Richard."

"All right. I'll do it. Will You take Jeb Magruder, Richard Kleindienst and John Dean III?"

"What kind of sacrifice is that?"

"John Mitchell?"

"Keep going."

"Haldeman and Ehrlichman?"

"That's more like it."

"And then, God, if I sacrifice them, will You keep me out of it?"

"Richard, I can't work miracles."

JOKES THAT BOMB

Washington Post *August 21, 1984*

"Mr. President, can I have a voice level please? We go on the air in a few minutes."

"My fellow Americans, I'm pleased to tell you today that I've signed legislation that would outlaw Russia forever. We begin bombing in five minutes."

"Mr. President, you're not coming through very well. What I'm getting on my earphones is that you said we were going to begin bombing the Russians in five minutes."

"You heard me correctly. There is nothing wrong with the sound."

"Mr. President, you're not serious, are you?"

"Of course not. It's just a joke, like 9-8-7-6-5-4-3-2-1."

"I don't know that joke."

"Well, there's two guys on this American nuclear submarine and they start wondering what would happen if they put both their keys into the missile computer at the same time."

"Mr. President, you're fading on me. Could you speak up just a little?"

"How's this? I'm sick and tired of the commies turning down all my disarmament plans. I say let's nuke 'em, and get it over with. How was that?"

"Your voice was loud and clear, but I'm not sure I heard the message correctly. Did you say something about nuking the Russians?"

"I'm just having a little fun with the mike test. We've not on the air, are we?"

"Not yet, sir. But we're awfully close. Could we try it once more? Why don't you recite 'Mary Had a Little Lamb?"

"I don't know that one. How about 'Give my regards to the Kremlin, say hello to the big Red Square. Tell all the folks on Gorky Street that we'll soon be there.' "

"That was good, Mr. President. Are you comfortable with the volume?"

"Why shouldn't I be comfortable with the volume? Are you taping all these tests?"

"Yes, sir."

"Good, you never can tell when I want to use one. How much time do we have?"

"Two minutes. Would you like to try another test?"

"How's this? I never saw a mushroom cloud. I never hope to see one. But if I did I know I'd rather see than be one."

"Mr. President, I hope I'm not out of line, but do you know something I don't know?"

"How's that?"

"Well, all these voice tests indicate there seems to be something on your mind. A lot of guys in the control room are calling their wives."

"Don't be ridiculous. I'm just trying to make the voice tests more interesting. They have no right to call their wives because anything I say before my radio broadcast is off the record."

"Yes, sir. But suppose the Russians pick up on the tests and think it's the real thing?"

"Just let them try it and see how far they get."

"That's not the point, sir. I think when we're going for a voice level we ought to stick to safer subjects such as Mondale and taxes."

"I'm the president and I can say anything I want to when I'm testing. It's my mike and I paid for it."

Russell Baker

For consistency of performance, Russell Baker (1925–   ) has an obvious rival, his friend Art Buchwald. But Buchwald works with a spatula, Baker with a nail-file. Buchwald burlesques what is ludicrous in the day's tidings, while Baker contemplates the same imbecility with an old Roman's melancholy. Baker's memoir of his Virginia childhood, Growing Up *(New York, 1982), evokes a simpler, harsher era when jobs were scarce and families close. He won a second Pulitzer Prize for the book, the first being in 1979 for his then thrice-weekly column in the* New York Times. *Part of his boyhood was spent in Baltimore, where (after a wartime service in the Navy) he was hired by* The Sun. *His work earned an assignment to London; there he showed a talent for unorthodox reportage in his weekly "Window on Fleet Street." Lured to the* Times *by James Reston, Baker first wrote about politics, then in 1962 started his "Observer" column. His pieces have been collected in ten volumes, and in 1989 he added another panel to his memoirs,* The Good Times.

A BETTER COLUMN

New York Times   *May 3, 1981*

The editor has imported a Japanese newspaper columnist. Apparently I am not cost-efficient. The editor has not said it in so many words, but the calligraphy is on the wall. Suddenly there is ominous talk about getting rid of outmoded old word guzzlers who consume high-priced newsprint at insupportable rates.

It is said that when the Japanese columnist comes into production he will be able to deliver an 800-word column idea in 150 words that can be easily digested in stop-and-go city reading. For highway reading, the same idea can be delivered in 129.6 words.

This is nothing compared to what is yet to come. I am told that in Japan they are already developing a miniaturized newspaper columnist capable of reducing an 800-word column to 25 words.

Thanks to the spreading popularity of speed-reading this means that Americans will soon be able to read as many as 128 newspaper columns in 60 seconds. Exposure to 128 newspaper columns, even over the span of a year, can do devastating damage to human brain cells; consuming 128 in 60 seconds, you might think, would be catastrophic.

Japanese newspaper-column miniaturizers, aware of the danger, have already perfected numerous safety devices. These include the newspaper-column bumper, which automatically stops the reader from colliding with a fourth consecutive column and bounces him off into the comic strips. An air bag is also being developed. Activated by the reader's physical collapse, this will spring out of the classified ads and surround the stunned column reader with life-restoring oxygen.

Costs of the air bag are still exorbitant, but Japanese industrialists are confident that they can be brought within the financial capability of the average American newspaper once Japanese technology produces a silicon chip to miniaturize the average American column reader.

The Japanese columnist on our payroll, I hear, will go into production as soon as he passes the entrance examination of the American Newspaper Columnists Association. This requires applicants only to demonstrate that they are more qualified to be President of the United States than the President is.

I have moved that the association adopt more difficult rules. Specifically, these would forbid membership to any columnist who arrives at the job before noon, leaves later than 12:45 P.M. and refuses to use at least 800 words when writing a column about a 50-word idea.

My Japanese competitor fails on each of these tests. He sleeps on a cot in the office, rises at 6:30 A.M., and does 50 push-ups to prepare his mind for the work of reducing 800-word ideas to 150-word columns.

Instead of leaving for lunch at 12:45 P.M. and discovering it necessary to rendezvous with highly reliable sources at a movie theater and keep cocktail-hour appointments with invaluable Government insiders, he eats at his desk at 4 P.M.

By this time, he has already computerized 145 columns—an entire year's output for old, outmoded word guzzlers. Does he relax over lunch with the crossword puzzle? No. He asks other columnists to join him for an exchange of ideas about how their work can be improved and columns made even shorter.

My last good card is the union. As usual, the union is threatening to go on strike. "Do you have strikes in Japan?" I asked him.

"Certainly," he said. "We are very fond of strikes in Japan."

That made me like him a bit more. "We like strikes here too," I said. "Of course, I'm not crazy about walking round and round the building for weeks at a time, especially in rainy weeks, but yelling at people who cross the picket line is one of the great ways to let off steam."

"Americans quit working when they go on strike?" he asked.

"Why do you think we call it a 'job action'? Because there's no action on the job. Don't you have job actions in Japan?"

"In Japan," he said, "we have what I suppose you Americans would call 'job inactions.' When we strike, we put on armbands to show we are unhappy and we go into the plant and work twice as hard as usual to prove to the boss how valuable we are."

I anticipate he will have some agreeable union problems which may save my skin. If he doesn't, Congress can just call me Chrysler and give me a pass to the Treasury. The economic significance of this Japanese invasion is obvious to ——.

Sorry, but it's 12:45. I must meet a highly placed source over a bottle of Bordeaux and a duckling.

THE ONLY GENTLEMAN IN HUCK FINN

New York Times *April 14, 1982*

The question of what books are fit for young eyes arose in unusually absurd form in the Washington suburbs recently when authorities fell to arguing whether the Mark Twain Intermediate School of Fairfax County should drop Mark Twain's *Huckleberry Finn* from the curriculum. My immediate question was, what's it doing in the curriculum in the first place?

It's a dreadful disservice to Mark Twain for teachers to push *Huckleberry Finn* on seventh, eighth- and ninth-graders. I had it forced on me in the 11th grade and, after the hair-raising opening passages about Huck's whisky-besotted "Pap," found it tedious in the extreme. Thereafter I avoided it for years. It had been poisoned for me by schoolteachers who drove me to it before I was equipped to enjoy it.

I had similar experiences with Shakespeare *(As You Like It* and *Macbeth),* George Eliot *(Silas Marner),* Charles Dickens *(A Tale of Two Cities)* and Herman Melville *(Moby Dick).* Schoolteachers seemed determined to persuade me that "classic" was a synonym for "narcotic."

Ever since, it's been my aim to place severe restrictions on teachers' power to assign great books. Under my system any teacher caught assigning Dickens to a person under the age of 25 would be sentenced to teach summer school at half pay.

Punishment would be harsher for assigning *Moby Dick,* a book accessible only to people old enough to know what it is to rail at God about the inevitability of death.

Huckleberry Finn can be partly enjoyed after the age of 25, but for the fullest benefit it probably shouldn't be read before age 35, and even then only if the reader has had a broad experience of American society.

Unfortunately, this sensible reason for pruning the school curriculum was not advanced in Fairfax County's case for dropping *Huckleberry Finn.* Instead of pointing out that assigning the book to adolescents damages Mark Twain, the authorities argued that Mark Twain damages the students.

John H. Wallace, one of the school's administrators, made the case in the *Washington Post.* The book "uses the pejorative term 'nigger' profusely." (It does.) "It speaks of black Americans with implications that they are not honest, they are not as intelligent as whites and they are not human."

While this is meant to be satirical, and is, Mr. Wallace concedes, it also "ridicules blacks," is "extremely difficult for black youngsters to handle," and therefore subjects them to "mental cruelty, harassment and outright racial intimidation."

I suppose a black youngster of 12, 13 or 14 might very well suffer the anguish Mr. Wallace describes, and even white youngsters of that age might misread Twain as outrageously as Mr. Wallace has in thinking the book is about the dishonesty, dumbness and inhumanity of blacks. This is the kind of risk you invite when you assign books of some subtlety to youngsters mentally unprepared to enjoy them.

Mr. Wallace thinks Mark Twain aimed only to be "satirical," but only in the loosest sense can *Huckleberry Finn* be called satire. It is the darkest visions of American society, and it isn't satire that makes it a triumph, but an irony full of pessimism about the human race and particularly its white American members.

Irony is the subtlest of artistic devices, and one of the hardest for youngsters to grasp. It requires enough experience of life to enable you to perceive the difference between the world as it is and the world as it is supposed to be. Many adults have trouble seeing that the world Huck and Jim traverse along the Mississippi is not a boyhood adventure land out of Disney, but a real American landscape swarming with native monsters.

The people they encounter are drunkards, murderers, bullies, swindlers, lynchers, thieves, liars, frauds, child abusers, numbskulls, hypocrites, windbags and traders in human flesh. All are white. The one man of honor in this phantasmagoria is black Jim, the runaway slave. "The nigger," as Twain has his white trash call Jim to emphasize the irony of a society in which the only true gentleman was held beneath contempt.

You can see why a black child nowadays, when "nigger" is such a taboo word that even full-blooded racists are too delicate to use it, might cringe and hurt too much to understand what Twain was really up to. It takes a lot of education and a lot of living to grasp these ironies and smile, which is why adolescents shouldn't be subjected to *Huckleberry Finn.*

The sensible thing for the Mark Twain Intermediate School of Fairfax County to do, I thought, would have been to meet the race issue head-on, put aside some other things and conduct a schoolwide teach-in to help its students understand what Huck and Jim are really saying about their world.

When the great teach-in was over, a few might even have understood why Mark Twain, if he'd surprised himself by landing in Paradise, would be watching them and laughing and laughing and laughing.

MY CUNNING NEW YORK EYES

New York Times *April 18, 1984*

My eyes are New York eyes. They are wary, wise, cunning eyes. They are eyes trained in swarming streets and fetid subways to identify menace and maniacs at 50 paces and signal the brain to order the feet to change to a safer course.

They are eyes that know how to see everybody while looking at nobody, to notice everything without glancing at anything. They are veteran eyes, hardened by years of New York survival.

Last week I brought my New York eyes to Leesburg, Va., where they instantly spotted a parking place at the curb on the main thoroughfare. They were astounded. They hadn't seen an unoccupied curb parking space in years.

Sweeping the battlefield, they reported even more amazing information: There wasn't a single enemy anywhere in traffic seizing position to do battle for this parking space. I moved into it without being threatened by sidearms or even cursed.

"This is weird," the eyes said. "This town gives us the creeps. Let's vamoose out of here."

I am not one to be terrified by a suspicious parking situation. What's more, I was hungry and determined upon lunch, but after leaving the car I reminded the eyes not to make contact with anyone on the sidewalk. They began scanning with their New York cunning and immediately swept across an approaching couple—a man of about 60 and a woman, possibly his wife, about the same age.

The eyes looked through them to a point 500 miles away, noticed no suspicious bulges and reported, "They may be traveling without weapons."

Judging them safe, I let the distance close between us when the eyes reported that both woman and man were not only struggling to make eye contact, but also smiling. "Don't let them make eye contact!" I shouted to the eyes.

Too late. My cunning New York eyes, fatigued perhaps by too much driving, were maybe lulled to a false sense of security by the absence of Mayor Koch on the corner crying, "Have you heard how well my book's doing?"—my marvelous New York eyes succumbed and locked in eye-to-eye embrace, first with the man, then with the woman.

The contact lasted only a thousandth of a second, but that, I knew, was long enough for disaster.

Then an eerie sequence of events failed to take place:

1. The man failed to ask me for a cigarette while the woman failed to maneuver behind me and produce a knife to cut off my retreat route. **2.** Simultaneously, the woman failed to seize my lapel, shriek that her babies had all been kidnapped by pygmy tribesmen and implore me to go their rescue. **3.** The man, holding his smile firmly in place, failed to pinion my hands, stuff my pockets with pamphlets and threaten to pursue me through the streets until I came to Jesus, subscribed to the Manichaean heresy or gave him money for the propagation of Zoroastrianism. **4.** A millionth of a second later, despite having me locked in eye contact, the woman failed to insist I sign a petition demanding unilateral disarmament, bilingual education or a Federal program for the spaying of stray cats. **5.** This failure by the woman was followed almost instantly by the man's failure to lean into my ear and chant either, "Acid and grass, acid and grass—What'll it be? Acid or grass?" or, "I've got the butyl, I've got the amyl; I've got the amyl, I've got the butyl."

As I say, all these things failed to occur in the fraction of a second that my New York eyes failed in their duty. Brief though that instant was, though, I experienced an eternity of anxiety, and although amazed at what had not happened, I realized that something worse might be coming. This, after all, was not New York, but Leesburg, Va. Leesburg has probably developed tricks still undreamed of in Manhattan. If there had been time, I would have broken out in a cold sweat, for they were now abreast of me, close enough to reach out and do their worst. The man struck first. "Good morning," he said. Then the woman: "Nice day, isn't it?" she said. Before I could recoil they were beyond me and walking away.

"Check 'em out," I told my New York eyes. The eyes reported both man and woman proceeding unmenacingly away from us. "I told you this town was weird," said the eyes. "Weird, schmeird," I said. "This place is crazy." We got out of town fast.

Jimmy Breslin

When Jimmy Breslin (1929–   ) was a columnist on the New York Daily News, *it was a matter of professional honor to disagree with the paper's editorials on every possible issue. So complained his editor, Michael J. O'Neill, who pleaded: "Can't you agree with us just once?" Breslin said no, it would ruin his reputation and anyway his dissent meant the paper covered all sides of an issue, thus pleasing a maximum of readers. The story is told by O'Neill in his foreword to* The World According to Breslin *(New York, 1984). Since breaking in as a columnist with the* Herald Tribune *in 1963, Breslin has steadfastly remained an urban populist, moving shop in 1987 to* Newsday. *His commentary won a Pulitzer in 1986.*

DIES THE VICTIM, DIES THE CITY

New York Daily News   *November 1976*

They were walking along in the empty gray afternoon, three of them, Allen Burnett, Aaron Freeman, and Billy Mabry, Burnett the eldest at seventeen, walking up Bedford Avenue in Brooklyn and singing out Muhammad Ali rhymes into the chill air. As they reached the corner of Kosciusko Street, it was Allen Burnett's turn to give his Ali rhyme: "AJB is the latest. And he is the greatest."

"Who AJB?" one of them said.

"Allen J. Burnett."

They were laughing at this as they turned the corner onto Kosciusko Street. The three wore coats against the cold. Burnett was in a brown trench coat; Freeman, a three-quarter burgundy leather; and Mabry, a three-quarter beige corduroy with a fox collar. A white paint stain was on the bottom at the back of Mabry's coat. Mabry, walking on the outside suddenly was shoved forward.

"Keep on walking straight," somebody behind him said.

Billy Mabry turned his head. Behind him was this little guy of maybe eighteen, wearing a red sweater, dark pants, and black gun. Aaron Freeman, walking next to Mabry, says he saw two others besides the gunman. The three boys kept walking, although Mabry thought the guy in the red sweater had a play gun.

"Give me the money."

"I don't have any money," Allen Burnett said.

The guy with the gun shot Allen Burnett in the back of the head. Burnett pitched into the wall of an apartment house and went down on his back, dead.

The gunman stood with Allen Burnett's body at his feet and said that now he wanted coats. Billy Mabry handed back the corduroy with the paint stain. Freeman took off his burgundy leather. The gunman told the two boys to start running. "You don't look back!" Billy Mabry and Aaron Freeman ran up Kosciusko Street, past charred buildings with tin nailed over the windows, expecting to be shot in the back. People came onto the street and the guy in the red sweater waved his gun at them. The people dived into doorways. He stuffed the gun into his belt and ran up Bedford Avenue, ran away with his new coats. Some saw one other young guy with him. Others saw two.

It was another of last week's murders that went almost unnoticed. Allen Burnett was young. People in the city were concentrating all week on the murders of elderly people. Next week you can dwell on murders of the young, and then the killing of the old won't seem as important.

Allen Burnett's murder went into the hands of the Thirteenth Homicide Squad, situated on the second floor of a new police building on Utica Avenue. The outdoor pay phone in front of the precinct house has been ripped out. The luncheonette across the street is empty and fire-blackened. At first, a detective upstairs thought the interest was in a man who had just beaten his twenty-two-month-old child to death with a riding crop. That was unusual. Allen Burnett was just another homicide. Assured that Burnett was the subject, the detective pointed to Harold Ruger, who sat at a desk going through a new manila folder with Burnett's name on it. Ruger is a blue-eyed man with wavy dark-brown hair that is white at the temples. The twenty-four years he has spent on the job have left him with a melancholy face and a soft voice underlined with pleasant sarcasm: "They got two coats. Helluva way to go shopping. Looks like there was three of them. That leaves one guy out there without a coat. I'll look now for somebody who gets taken off for a coat tonight, tomorrow night, the next few days."

In a city that seems virtually ungoverned, Harold Ruger forms the only municipal presence with any relationship to what is happening on the streets where the people live. Politicians attend dinners at hotels with contractors. Bankers discuss interest rates at lunch. Harold Ruger goes into a manila folder on his desk and takes out a picture of Allen Burnett, a young face covered with blood staring from a morgue table. In Allen Burnett's hand there is a piece of the veins of the city of New York.

Dies the victim, dies the city. Nobody flees New York because of accounting malpractice. People run from murder and fire. Those who remain express their fear in words of anger.

"Kill him for nothing, that's life—that's what it is today," his sister Sadie was saying. The large, impressive family had gathered in the neat frame house at 30 Van Buren Street. "He was going into the army in January and they kill him for nothing. That's the leniency of the law. Without capital punishment they do what they want. There's no respect for human life."

Horace Jones, an uncle said, "The bleeding hearts years ago cut out the electric chair. When the only way to stop all this is by havin' the electric chair."

"We look at mug shots all last night," Sadie said. "None of them was under sixteen. If the boy who shot Allen is under sixteen, there won't be any picture of him. How do you find him if he's under sixteen? Minors should get treated the same as everybody else. Equal treatment."

"Electric chair for anybody who kills, don't talk to me about ages," Horace Jones said.

The dead boy's mother, Lillian Burnett, sat with her head down and her hands folded in her lap.

"Do you think there should be an electric chair?" she was asked.

"I sure do," she said, eyes closed, head nodding. "Won't bring back my son, but I sure do want it. They tied up three old women and killed them. If they had the electric chair I believe they would rob the three women, but I don't believe they'd kill them."

The funeral was held two days later, at the Brown Memorial Baptist Church, on Washington Avenue. A crowd of three hundred of Allen Burnett's family and friends walked two by two into church. Walked erectly, solemnly, with the special dignity of those to whom suffering is a bitter familiarity. Seeing them, workmen in the street shut off pneumatic drills. Inside the church, the light coming through the doorway gleamed on the dark, polished wood of the benches. The doors were closed, an organ sounded, and the people faced the brutality of a funeral service; a baby cried, a woman rocked and screamed, a boy sobbed, a woman fainted, heads were cradled in arms. The mother screamed through a black veil, "My baby's gone!"

An aunt, Mabel Mabry, walked out of the church with lips trembling and arms hugging her shaking body. "My little nephew's dead," she said loudly. "They find the ones who killed him. I'm tellin' you, they got to kill them too, for my nephew."

The city government, Harold Ruger, just wants to find the killer. Ruger was not at the funeral. "I got stuck in an eighty-floor elevator," he said when he came to work yesterday. "I was going around seeing people. We leave the number, maybe they'll call us. That's how it happens a lot. They call." He nodded toward a younger detective at the next desk. "He had one, an old man killed by a kid. Information came on a phone call, isn't that right, Al?"

"Stabbed eight times, skull fractured," the younger detective said.

Harold Ruger said, "What does it look like you have? Nothing. And he gets a phone call, see what I mean? The answering is out there and it will come." His finger tapped the file he was keeping on the murder of Allen Burnett.

TAKE MY KIDS, PLEASE

New York Daily News *February 1982*

There was a television interview in which Koch the Mayor, a bachelor, was quoted as saying that he would like to have children. Somebody put his head out of the mayor's office and said, yes, that was right, the mayor had said that he wanted children.

"Children?" somebody asked.

"You bet. He hears they're terrific."

The word went about City Hall. Everybody began saying, "Yeah, children!"

Immediately, there came the sound of feet running out of the offices of City Council President Carol Bellamy, who also has never married. Here was a young woman assistant to Bellamy calling out to all in City Hall:

"Put down that Carol wants children too."

"Carol wants children?" somebody said.

"Sure Carol wants children, Carol says children are fantastic. Make sure you put it in the story that Carol Bellamy wants children. Don't just put down that Koch is the only one in City Hall who wants children. Remember Carol Bellamy wants children too." When these stories reached my house at night, I read them with the eye of a criminal. Certainly, there was great political maneuvering on the part of Koch and Bellamy, for it is becoming clear that the long period of admiration by newspeople for those living isolated, insulated lives is about over. How, people are wondering, can anybody be so effective in high office in a place like New York when the person appears to be a product of an estranged, alienated culture? Apparently, Koch and Bellamy have decided, after reading a poll perhaps, that it would look great with the electorate if they had children. But in order for them to have children at this point, right now, today, in time for elections, somebody would have to give them the children.

I was in the kitchen of the house I now run and one corner of the room was blocked by garbage cans.

"Why aren't they outside?" I asked.

"Because there is no room in the big garbage cans," a son, P. Breslin, said.

"Why isn't there room?"

"Because I didn't put them out this morning. The truck came too early."

"What's too early?"

"The truck came before I woke up."

"Bellamy," I said.

"What?" P. Breslin asked.

"I said you go to Bellamy," I said.

I went to the stairs and called up, "How are you doing up there?" For the last hour, son C. Breslin had been in his room, supposedly doing a history report for school.

There was no answer. I called up again. Still no answer. I went up the stairs and found that the reason he didn't answer was that the music was on so loud in his room that he would have been unable to hear a dynamite explosion. C. Breslin was jiggling to the music, and on the desk in front of him was a sheet of paper as blank as his stare.

"Two for Bellamy," I said.

"What?" C. Bellamy shouted.

I screamed, "I said Bellamy now has two kids!"

I asked downstairs for the whereabouts of K. Breslin. She did her homework and went out, I was told. "She said, don't worry, she'd call."

"What time did she go out?"

"About eight."

At seven-thirty, on my way home, I had called daughter K. Breslin and she said she had two hours of schoolwork to do and was just getting to it. Daughter K. Breslin is the one who does so well in school that the teachers no longer send me home failure forms; the teachers now take out large sheets of blank paper and fill them with the accounts of her scholastic career.

As promised, she did call me that night. I had been sitting up waiting for her, and at 1:00 A.M. she called me from a phone booth on Queens Boulevard and said she had been trying for hours to get into the house, but nobody heard her ringing the bell. I had been sitting approximately four and one half feet inside the front door for the entire night. She arrived home on roller skates, and I immediately took a deep breath and tried to scream in order to alleviate chest pains. She held up her hands and said, "You're making a nervous wreck out of me." She skated through the living room and went upstairs to bed.

"Koch!" I shouted.

"What?" she called.

"You're going to Koch."

And now yesterday I sat in the house with three children of mine, each of whom I can well do without, and with a daughter from another family, L. Oliver, a daughter whose family wishes she would run away for good. Oh, yes, there are people who wish their children would disappear on them. Many people. The children are generally over the age of twelve. For of course anybody can take care of children up to the age of twelve. In those years before twelve, children are dazzling and humorous and can be cared for with a large amount of work, but at the same time the work is uncomplicated. At age twelve, however, all children reach into the air and from somewhere grasp a lance and from then on they hold the lance at the chest of the parent. If the parent tries to back away from the lance, the child simply extends the arm so the lance remains, still digging into the chest of the parent. There is no escaping the lance, and of course one day, the child over twelve simply pushes the lance and it pierces the parent and kills him.

So here I was with these four kids over the age of twelve, four kids who are unwanted for a good reason, and on the table in front of me was this great fresh news that Koch the Mayor and Carol Bellamy need sudden children to save their political careers.

"They have them," I announced.

"Have what?" one of the kids said.

"They have two children. You two go to Bellamy on Monday morning," I said to C. and P. Breslin. "She lives in Brooklyn." I turned to K. Breslin and L. Oliver. "You two go down to City Hall on Monday morning and sit on a bench and if anybody asks you who you are just say, 'We're the Koch children.' When he comes out to go home, you go with him. Good-bye and good luck."

I felt tremendous. I not only had removed a problem from my life, one that was about to kill me, but I had done the city a tremendous favor —I had helped the political careers of our great leaders who suddenly found themselves in need of children. In one day, I had done a service for the city in which I live and in which I once attempted to raise children, which I announce today is impossible for me to do, so I am giving them away to a couple of needy people.

Pete Hamill

Pete Hamill (1935–) used to rush out at lunchtime to buy the early editions of the New York Post *to read Kempton and Cannon. In 1960, he wrote an impassioned letter to the* Post's *editor, James Wechsler, who invited him to try out as a reporter. Five years later, he was writing a column so readably radical that he was assailed in 1970 for his "irrational ravings" by Vice President Spiro Agnew. Hamill borrowed the phrase for a collection of his pieces published in 1971. The Brooklyn-born son of Belfast Catholics, Hamill came to journalism via the Korean War, night school, commercial art, and the varied influence of Bill Mauldin, Walt Kelly, Hemingway, and Orwell.* The Invisible City *(New York, 1980) collects his fictional vignettes from the* Daily News *and the* Post, *where the peregrinating Hamill is again a columnist.*

SAIGON: A TREACHEROUS CITY

New York Post *January 2, 1966*

In the morning we ate breakfast in the My Cahn riverboat restaurant. There, sitting at a wooden table, eating an omelet, sipping coffee, you felt a part of that hoked-up Orient from the old books and the half-remembered movies. Did not that sea captain with the gold earring step from some Michael Curtiz sound stage? That slender girl in the flowing *ao dai* dress—did she not come out of Conrad?

As we ate, junks moved through the muck-colored Saigon River, their tenders dressed in shorts and conical hats, paddling with even, patient strokes. Across the river, a long, jumbled row of huts was squashed against the waterline, while billboards rose above them, advertising transistor radios and lubricating oil.

The city itself was almost mournfully silent. Occasionally, the toneless boom of a foghorn could be heard from downriver, and then the steady chukkachukkachukkachukkachukk of a motorboat. Before us, the great merchant ships of the world were moored to the teeming docks, being emptied of the materials of war by scurrying platoons of Vietnamese longshoremen. The ships bent away from us in a diminishing arc, vanishing behind the bend of the river, their spiky wilderness of masts and spars forming an almost Oriental calligraphy against the foggy morning. Halfway through breakfast, we heard artillery.

Puh-phwoom! Puh-phwoom! It was coming from perhaps 15 or 20 miles away, and John Harris, an old friend from the Hearst papers, said it was probably 105-mm artillery. "I'm glad," Harris said, "they're firing that way."

This restaurant itself had been bombed by the Vietcong some months before, because it was favored by many Americans. A bomb filled with steel slivers had gone off in a corner, scattering human beings before it, and then, when the survivors rushed across the small gangway that connects the boat to the shore, a second bomb went off and killed them, too. This was, of course, an act of murder, and no veneer of political necessity could alter its basic indecency.

In Saigon today, no one ever looks directly at the person he eats dinner with; he is too busy watching the other customers. If a man walks into the men's room carrying a briefcase and emerges without it, one is advised to hit the floor. This is known as civilization.

One of the terrible things about all of this is that Saigon is a city of great physical charm, the sort of place one might visit on a vacation. It reminds me of certain sections of Mexico City and Barcelona, with broad, open avenues and squares, trees ruffled by a slight breeze from the river, flower markets, restaurants opening onto the streets, newsboys walking barefoot. The architecture is a kind of 1920's Modern gone through a sea change, with rounded "streamlined" corners, streaking concrete façades and hotel lobbies filled with bad paintings.

I have seen old photographs of the city, in the days when the main streets were covered with outdoor cafés, tables strewn across sidewalks, people sitting at them reading the papers fresh in from Paris. The French invented modern Saigon, and it is a city built for pleasure. Looking out across it from the roof of the Hotel Caravelle, with the red-tile roofs reaching through the trees for the river, you could be looking at a slice of Provence.

But today it is a city under siege, its very heart laced with treachery. No one can say precisely how many Vietcong sympathizers there are among the city's 2,000,000 inhabitants. But simply because of that, there is a sinister atmosphere about the place. Americans are the target, and you can never be sure if the shoeshine boys or the hall porter might be your killer. The Vietnamese are a people of great manners, dependent on nuance and ritual; they make most Americans feel clumsy and boorish. But walking the city, you feel more than that. In addition, you feel vulnerable, probably as vulnerable as you ever will in your life. In combat conditions, you expect to be shot at. You don't expect it while reading Malraux over your morning omelet.

This afternoon I stopped for a beer in the Imperial Bar on Tu Do Street, which under the French was the Rue Catinat, one of the most exclusive shopping streets in the Orient. The outdoor tables are gone now, and the sidewalks are filthy. Most of the great restaurants have their windows covered with wire mesh and elaborate grilles (the Saigon *Post* even contains advertisements for barbed wire, cut to your specifications). Sitting in the Imperial, sipping a 33 beer, I suddenly realized that I was drinking in a cage. No one even sat near the rim of the cage, because if a hurled grenade bounced off and exploded, fragments could still blind or maim you. The best table in the house was behind a two-foot-thick pillar near the bar. Sitting at it was a large-breasted Frenchwoman in her forties. She was the owner.

The streets come alive about ten in the morning here, though occasionally a government jeep will drive around earlier, blaring exhortatory speeches from its loudspeakers and then playing atonal music for the alleged benefit of the listeners. Some of the correspondents call this character Nguyen the K. The first time I heard it I couldn't tell whether he was running for office, shilling for a department store, or announcing a coup.

The city's tempo is swift but erratic. Each day, war or no war, everything shutters from twelve to three for siesta, and if you are looking for a high government official at those hours, you will probably locate him in a state of unconsciousness on his couch. Bicycles and a kind of open, motorized rickshaw called a cyclo are everywhere. The cyclos seem to burn river mud instead of gasoline, and between them and the hundreds of tiny blue-and-cream Renault taxis, the city is usually covered with a pale-blue exhaust fog.

There are also a lot of Vietnamese cops around, usually in pairs, dressed in white (the Americans here call them "white mice"). They watch everything, and as I write this, one of them is across the street from the hotel, standing in front of the building which the French intended as an opera house, which served for a while as the National Assembly, and which is now a kind of ghostly town hall. The cop is down there, watching me. I wave. He waves. I smile. He smiles. I suppose if I took off my shoes, he would take off his shoes.

But more than anything else, you are constantly reminded of the war. Jeeps, trucks, Navy vehicles (Saigon is 50 miles up a river from the sea) filled with soldiers and Marines whiz back and forth with great haste. The important buildings, like the American Embassy, the USIS, and some of the hotels serving as military billets, are surrounded by cops, MP's, Marines, all of them behind a fence of round concrete blocks, three feet in diameter, strung together with lines of bared wire. These are to prevent suicide squads from driving truckloads of dynamite to the doorstep, machine-gunning the guards, and letting the fronts of the buildings vanish in an explosion. The technique was successfully used at the Metropole Hotel a few months ago and at the U.S. Embassy last year. The Metropole is still a gutted half ruin, although the bars across the street, also damaged, have reopened.

Still, it is a pleasant enough city, and war adds its own peculiar exhilaration. This afternoon I sat in the large open-front restaurant of the Continental Palace Hotel, eating lunch. I was reading the Saigon *Post,* a newspaper of consummate style, when I saw the following ad: "Mr. Simon will offer you his Italian specialties and his famous pizza in a rustic cadre and vertical atmosphere." How can you completely dislike a city whose advertising men write in the style of the late Colonel John R. Stingo?

The Continental Palace, by the way, is not paying off the Vietcong to keep the wire mesh off its windows. No, this hotel has the great fortune to house the Polish members of the International Control Commission. The Vietcong would certainly not want to violate the Geneva Agreements by blasting some of their fellows off the map.

THERE ARE NO YOUNG GUYS HERE

New York Post *January 10, 1966*

Perhaps we, who come from the fortunate places of the earth, shall never understand about places like Cam Ne. Where I come from, a place with shattered windows and no steam heat in winter is thought of as a slum. We think it criminal if rats scurry between the walls, or if children are forced to work at sixteen, or if a man loses one shot at decency and comfort because his education was incomplete or the color of his skin was unacceptable to others.

But in the Cam Nes of the world, to live past three is a success, and to make it to thirty is a triumph. I wish I could bring you here somehow; I wish you could see the faces of the old women, the light in their eyes extinguished, their small, shrinking heads looking dumbly from under conical hats, their skin eroded, clay-dry, pitted with the half-healed gashes of the swamp leech.

When they smile, which is seldom, their teeth show tar black from chewing betel nuts. If I could make that clear, make clear that these women have ceased being women at all, that their bodies have gone fallow and bone-hard like some strange new vertical beast of burden, make clear how disease has sapped them, and the filth of the rice paddies has flaked their skins, and ruined their blood, and shortened their very lives—if I could make that clear to you, you would begin to understand something about Cam Ne and perhaps about Vietnam.

You would begin to understand about Vietnam and the wretchedness of the land, if you could see the roads in the morning, clogged by people on the move, all of them old men and old women and young children. They carry on their backs all that they own: bamboo struts that make up their houses, small sacks of clothing, chickens, and an occasional pig. It is all they have. No books, no paintings, no radios, none of the soft ornaments of the twentieth century. Last year alone, 750,000 people in this country moved their place of residence, trying to keep a few hundred feet ahead of the violence. The whole country has been doing this for a quarter of a century.

But perhaps you should see Cam Ne on a trip with a couple of Marines. On this morning, I took a walk with two sergeants, Chuck Burzamato and Harold R. Hoerning. Burzamato is a short, red-faced guy, who came from Mott Street, lived in Brooklyn, and has been in the Marines for seventeen years. His wife and children live in San Clemente, California, and his parents live at 1402 East Third Street in Brooklyn. Hoerning has been a Marine for twenty-one years. His wife and four kids live in Oceanside, California, and his father, a retired New York cop, lives in Bayside, Long Island. They are professional Marines. They are also men.

We walked together down the dirt road which leads from the Marine camp to Cam Ne. The thick, gluey mud of Asia stuck to our boots and made a sucking sound as we walked. Hoerning was carrying a carbine, and Burzamato held a shotgun.

"The funny thing in Cam Ne," Burzamato said, "is that we got through to the children, and we even are getting through to the old people. But there's no young guys here. None."

"That's right," Hoerning said. "The young guys are all off with the VC."

The town is a collection of scattered huts and houses, laced with thick, crawling jungle. The Marines have been urging the people who live there to cut down the undergrowth, to clear the area of jungle. We saw an old man, with a Ho Chi Minh beard, slashing at the tangle with a machete.

"Hey, Pop!" Burzamato shouted. The old man stopped and smiled. "He's got swell-lookin' gums, don't he?" Burzamato said. We walked into the bush to talk to him. "Numbah One!" Burzamato said, using the local phrase which means something is very good. "You do number one job, Pop. You come today one o'clock, see boxie [doctor], get food! Numbah one."

The old man bowed, and shook his head yes, and said, "Numbah one, Numbah one." Burzamato gave him a cigarette and we moved into the jungle.

Everywhere in the jungle we saw trenches and long, narrow slots dug under the roots of trees. When the battle was fought here, the Vietcong were dug into the holes, covered with foliage, firing machine guns at the Marines, as they moved in. "If Charlie's dug into one of those holes," Hoerning said, "you'd need a direct hit with artillery to get him out."

The jungle itself had a sinister quality. I suppose if you come from cities, there is always something treacherous about uncontrolled nature. If it is in Vietnam, the possibility of violence around each turn makes it even more so. It must have been terrible to fight here; we literally could not see 20 feet on any side of us. "You could have twenty VC in there," Burzamato said, gesturing toward a dripping dark area to the right, "and never see them."

Suddenly we came to a small clearing. On a knoll, up above a small untended private rice paddy, stood a brick house. There had once been a walk leading to the door, but it was cracked and smashed now, with scrub growing in the broken places. Bougainvillaea ran up the sides of the house, and on its porch stood a small young girl, maybe six or seven, a boy about four, and no one else. All of it—the children, the house, the small 10-foot-by-20-foot rice paddy—all seemed about to be swallowed or suffocated by the jungle.

"Hal-looo," Burzamato shouted. We walked up to the porch. The house was bare and empty. Not a single piece of furniture, no food, nothing. In one corner stood a neat Buddhist altar. The girl looked terrified. "Don't be afraid, beautiful. Me numbah one." He reached in his pocket and pulled out some candy. The little girl was afraid to take it. The boy reached out, and Burzamato gave it to him, explaining with gestures that he should share it with his sister. We started to leave when I saw something move in the corner. It was an infant, huddled in a kind of thatch nest, covered with flies. The child's skin was gray, its eyes clamped shut, its stomach swollen. It was obviously dying.

"Jesus," Burzamato said. "Jesus, Jesus, Jesus."

I thought he was going to cry.

We walked through the village for two hours. Everywhere the old people were clearing away the tangle of undergrowth, chopping away the 35-foot bamboo, taking cigarettes from the two Marines. One old man had cut away about 10 square yards, and Burzamato gave him a whole pack of smokes.

We came across one little girl whose eye had been split by a piece of flying bamboo, and Burzamato called a Marine corpsman, David Luck, from St. Paul, Minnesota, and had the eye cleaned and treated. He gave out eighteen packs of candy, and when we started back later, the children followed him all the way to the base camp. He looked like a squat, gun-toting pied piper.

At one o'clock, the people of Cam Ne had lined up, and the Marines had spread the donated clothing across sheets of cardboard. The children came into the compound two at a time, and the Marines sorted out the clothing, trying to find things of the correct size. "Lookit this," Burzamato said, holding up a sheepskin-lined jacket. "That's for when it goes under a hundred."

He took out a nightgown, made of a diaphanous material. "Just what the mama-san needs for a big Saturday night in Cam Ne." There was a lot of joking and laughter, but the children walked away from that place, past the ruined shell of the house which served as headquarters, and they were smiling.

Perhaps we do not have either a legal or moral right to be in this country, and certainly the war itself is a disgusting and an abstract thing. But believe me, the Americans who are here are as decent as anyone I've ever met. It should be unnecessary to say so, but the Marine Corps is not the Wehrmacht, and if I had my choice of dinner companions between Staughton Lynd and Chuck Burzamato, I would not be long in the choosing.

THE ROOTS OF WAR

New York Post *January 11, 1966*

We saw the cemetery at half past six in the morning, perched on a small knoll, its stone crosses jutting through the thick morning fog about 300 yards from the road. The driver pulled the jeep over, and we walked through the damp, wheezing earth to the front gate. Only a two-foot iron fence separated the dead from the living. We walked in.

The tombstones were in disarray, leaning at angles like monuments to cripples, and all of them were sinking slowly into the damp, black soil of Asia. Lichen spattered their faces, and snaky tendrils of wild swamp weeds pulled at their shafts. The jungle was taking them back.

There were the names—Lataud, Broussard, Michaux—and there were the years of their deaths: 1931, 1927, 1940, 1919. What had brought these men to this terrible rendezvous in a sinkhole of a town on the South China Sea? They had crossed continents and oceans to get here. Perhaps there were dreams of fortunes to be taken like plunder from the jungles and paddies. Perhaps they were in flight from some tamer place.

But they had been dead a long time, and jets roared over the graves now, and in another twenty years, the crosses that marked their bones would be gone, too, swallowed by the very earth.

"Let's get the hell out of here," the driver said. "This place gives me the creeps."

We walked back to the jeep. A bent, aged Vietnamese had stopped at the side of the road, staring at us. He had seen the French die in his country, and there was something in his face that said he would see the Americans die here, too.

The French presence has not, of course, completely vanished. You feel it everywhere, but especially in and around Saigon, which, more than anything else, is a European city clamped astride an Asian base. The French built it upon the backs of the Vietnamese, and even today it feels more like a rather large town in the south of France than a capital in the tropics.

The French, it has been said, colonized with boulevards and brothels, and in this country, at least, the saying was true. There was a time when Saigon housed the largest brothel in all of Asia, a fantastic structure called the House of Mirrors, because of the decorations in the cubicles, stocked with good wines, a good bar, and 1,200 girls. Today it is gone, but more modest brothels exist in profusion. So do the boulevards. Both, I am told, by people who remember the old days, are somewhat seamier now.

The French are most visible here in the restaurants. They travel in groups of five or six, their language utilized as a shield against the British, American, or Vietnamese around them. They sit at tables, conversation purring, talking about De Gaulle and the films of Truffaut and the currency exchange on the black market. Most of them sneer at the American effort here; after all, if France could not conquer, what possible chance could the Americans have?

There are still about 12,000 French nationals here, with a financial investment of about $500,000,000. Most of this is in rubber, and one of the minor scandals of this war is the way we have been forced to fight in the rubber plantations. Some soldiers have told me that in some rubber plantation battles, they had been given orders not to shoot the rubber trees because the United States would have to pay for them. I've asked some people in authority about this, and they have confirmed the story, but never, of course, for attribution.

The reasoning seems to be that rubber is one of the few natural resources South Vietnam can use for export purposes. If the war ever ends and the rubber plantations are destroyed, the country will be in even more miserable economic shape than it is now. So the Vietcong hide behind rubber trees, and American soldiers get their legs shot off mindful of the future economic security of the businessmen of Saigon.

Perhaps the long-range reasoning is correct, rational, and perceptive. But it is also abstract and rather sleazy. The fact is that every French rubber plantation owner here pays off the Vietcong to stay in business. The Vietcong takes that money and buys guns which are used to kill Americans and Vietnamese. The plantations also provide shelter. A good number of the plantation employees work the rubber by day and don the black pajamas of the Vietcong irregulars at night. When large troop movements are necessary the Vietcong can move across country almost entirely through French rubber plantations.

One military man here told me of one patrol into a rubber plantation where he came across almost 600 Vietcong, marching merrily, four abreast, through the main road of the plantation, with all the abandon of the South Hartford Marching and Chowder Association on its annual picnic and beer party. The French never said a thing. If this smells of collaboration with the enemy, well, the French have a talent for it.

The government of Vietnam, of course, should put a stop to all of this. They should either nationalize the rubber plantations or have the French shot as collaborators. They will do neither. For one thing, there are few Vietnamese in the country, I'm told, who are capable of running the plantations, so nationalization would be disastrous. Second, the government desperately needs the revenues it can extract from the sale of rubber to France. Start shooting Frenchmen, and you lose your market.

The French intellectual tradition has also held on with an iron tenacity. In the University of Saigon, the old system of the dictatorial professor reading notes and expecting to have them parroted back, jokes and all, still holds. Most of the professors hold America and its culture in contempt and assume that all literature ended with Proust. The local painters work in the tradition of the French Impressionists, rather than in any original or even Oriental tradition. The dream of most students in the universities here is to take a degree and leave for Paris. One nice thing about that, of course, is that they can beat the draft. Paris is filled with thirty-nine-year-old medieval scholars on the lam from their own country.

One should probably not be so upset over this situation. But this war did not begin in 1960, when the National Liberation Front was founded; or in 1954, with the fall of Dienbienphu. This war began more than a hundred years ago when the French shot their way into this country. Once again in this century, the United States has come in to try to clean up the filthy moral refuse left behind by Europeans. But this time it is an even more terrible situation than ever, because this time we might be wrong.

I just can't believe that the Vietnamese peasant, over whose loyalty this war is being waged, can really tell the difference between the French legionnaire and the American soldier. Could a Red Hook longshoreman tell the difference between a Vietnamese and a Chinaman? I doubt it. That's why we're in trouble.

William F. Buckley, Jr.

Critics of William F. Buckley, Jr. (1925–   ) spoof his intimidating words on his long-running talk show, "Firing Line," e.g.: maieutic, eristic, velleity, *and* energumen. *This is the look-at-me Buckley who writes thrillers about superspy Blackford Oakes; who ran for Mayor of New York in 1965, saying his first move if elected would be to demand a recount; who won his early renown in 1951 by arguing that Yale was godless and leftist. But a second Buckley was also evident from 1962, when he began his syndicated column. He appeared not just in rightwing newspapers but in the then-liberal* New York Post. *In addressing a wider audience, Buckley seemed to move from the hard right and closer to the once despised center. With the victory of Ronald Reagan, a charter subscriber to Buckley's fortnightly* National Review, *both columnist and country had changed. His early strictures on free markets had become a commonplace, and he had come to agree that government did have a role in expanding civil rights. Unchanged is Buckley's scorn for the blandness of political discourse. Buckley has published seven collections of his columns and magazine pieces, and is subject of a biography,* William F. Buckley, Jr. *(New York, 1988), by John B. Judis.*

THE THEORY OF JACK ANDERSON

Washington Star Syndicate   *April 21, 1972*

I said to Jack Anderson, "Mr. Anderson, I'd like to know whether you believe that I have the right to go through your files and to disclose their contents in my newspaper column?" And Jack Anderson said, "No. I don't think you have that right because I am not a public official." And I said, with that succinctness for which I am famous: "(a) The Supreme Court, in its rulings on libel, has pretty much dismissed the distinction between a public official and a public figure; (b) there is no question about it that you, Mr. Anderson, are a public figure; indeed (c) you are more influential than most public officials—so if you are entitled to see the files of Presidents and Senators and Cabinet Ministers, why am I not entitled to see your files?" To which Mr. Anderson replies—lamely, I think—that okay, he'll show me his files, if I'll show him mine. To which I reply: No, I won't let you see mine, but my position is consistent, because I don't assert the right to see the private files of the President. But yours is inconsistent because you assert the right to see theirs, while denying them the right to see yours.

So it went—so it goes—and it is very difficult indeed to wrest from Mr. Anderson the theory by which he exercises the right gleefully to disclose and to dwell upon the working papers of government officials. I tried another tack. . . .

Look, I said, I think you are right when you say that there is a conflict of interest implicit in the arrangement whereby the same man who classifies a document as confidential has the sole authority to declassify it, and I grant that authority is usually exercised in a self-serving way. That is, public officials tend to release documents that make them look good, and suppress documents that make them look bad. Now: wouldn't you agree that by the same token there is a conflict of interest as regards your publication of secret documents? I mean, here you are telling us that you would not in fact give out secret documents that come to you if they imperil the national interest. But as a newspaperman and a sensationalist, aren't you naturally inclined to further your interests rather than your country's interests, even as you accuse the politicians of doing?

Well, said Mr. Anderson, he would like it if a perfectly impartial tribunal (by the way, there is no such thing) were in charge of decisions about what documents should be kept secret and what documents should be declassified.

Okay, I said, but why shouldn't there then be a tribunal that passes on which of the documents that come into your possession should be publicized by you and which should be kept secret? Surely if a tribunal is appropriate to guard against self-serving tendencies of public officials, a tribunal is equally appropriate to guard against self-serving tendencies of newspapermen?

Well, said Mr. Anderson, if the government agrees to set up such a tribunal, I'd agree to go along. So, said I: What is the reason for waiting for the government? Isn't it an approach toward what is desirable to set up a tribunal to pass on your own disclosures?

Dead end.

Mr. Anderson's difficulty, as a theorist, is that he cannot accost the question of public privacy except in terms of evildoing. Now it is absolutely and obviously and unmistakably clear that public officials are very frequently engaged in such evil activity as hypocrisy, cynicism, dissimulation, the whole bit. Everybody who is running for President at this very moment is engaging in the kind of rhetoric that an undereducated mule would not take seriously. But it does not follow from this that a government official is required to send a copy of all his private papers to Jack Anderson, to do with as Anderson sees fit.

When he disclosed the minutes of the special White House group that faced the problem of the India-Pakistan war, Anderson justified himself by saying that there was a great discrepancy between what Henry Kissinger had said was official U.S. policy (namely, neutrality), and what the minutes actually disclosed *was* U.S. policy. The White House denied the discrepancy, whereupon Mr. Anderson gave out the whole of the minutes. Now these included—as an example—the statement by one U.S. official talking at the round table with a dozen assistants of the President: "The Department of Agriculture says the price of vegetable oil is weakening and it would help us domestically . . . to ship oil to India." And, from the Chief of Naval Operations, "The Soviet military ambition in this exercise is to obtain permanent usage of the port of Visakhapatnam." Both of these expressions are, to put it formally, intimate: and their disclosure has nothing whatever to do with the hypocrisy imputed to Henry Kissinger.

What is the theory of our right to hear such spontaneously expressed opinions?—which would simply not have been expressed in the first place if it were known that they would end on Jack Anderson's desk. The gentleman, in fact, has no theory of his right to the information. He has, merely, a squatter's right, and is better off forgetting the theory and confining himself to saying: I'll do it as long as I can get away with it. That is a theory of sorts.

THE ETHICS OF JUNKETEERING

Washington Star Syndicate *December 22, 1973*

My friend Mr. Mike Wallace, a gentleman who is given to protracted concern with scruple, called recently in his diligent way to inquire what are *my* rules concerning "junkets," by which he means trips paid for by someone else. He proposes to do a television program on the subject, and I am grateful to him for his maieutic inquiry about *my* own views, which had not crystallized.

(1) When a columnist is invited on a trip, he should begin by asking whether the host expects that his guest will write about the trip. Obviously if you are invited, say, to look in at the opening of Disney World, your hosts expect that you will write about Disney World. If you are invited (as I along with 1,000 others were) to travel in great luxury to witness the dedication of a new refining plant in a distant Arctic archipelago, your host clearly does not expect that you will write about the plant.

(2) In the second case, then, there is obviously no inhibition in accepting the invitation. In the first, you need to say to yourself: If I find that Disney World is a great bust, will I feel altogether free to say so notwithstanding that the trip down was paid for, as also the hotel bill? That question is only coped with by the injunction: to thine own self be true. But here a qualification is appropriate. Going to Florida and back is not a very big deal, in this peripatetic age. So that the indebtedness of the visiting journalist is not really as heavy as, say, a trip to, well, Mozambique, to examine the policies of the Portuguese government there.

(3) Here the situation becomes more complex, and more interesting. On the one hand there are newspapers that pridefully insist that no journalist should travel to a foreign country at that country's expense because implicit obligations are incurred. That point of view is defensible.

But there is another point of view. It is part of a journalist's duty to move about, and to report to his readers on what he sees. Most often he will use his own money to pay the fare. But—and here are more subqualifications—sometimes (a) the trip is too expensive to justify the amount of time the journalist reasons he can devote to the subject (how many columns can one write about Mozambique?); and (b) sometimes the journalist is simply not certain whether the trip will produce anything interesting enough to justify the trip.

It is my opinion that in such circumstances the journalist should feel free to accept the round-trip fare, cutting his potential losses to his own time. But—once again—he must know that he will feel free to write as he sees the situation, without any inhibition deriving from the auspices. This is especially difficult because one often tends to lean over backward to establish one's analytical independence, and that is as unjust as to shill. To say the problem is easily solved by simply avoiding the temptation is to take the easy way out.

Some examples, from my personal experience.

I traveled, at the expense of South Africa and the Portuguese government, eleven years ago, to South Africa and Mozambique. I wrote three columns, and a long essay piece. Where I came out on the general subject of South African domestic policies is, I suppose, best situated by saying that a pro-South African committee in the United States made reprints of my essay, while the government of South Africa refused to distribute it.

I traveled over one hectic weekend, at the expense of the government of Northern Ireland, to view the situation there just before Orangeman's Day. It was an excruciatingly uncomfortable trip of seventy-two hours. I wrote three columns, in which I find not a hint of servility to my hosts.

I traveled, at the expense of the United States government, from New Zealand to the Antarctic, and stayed five days, visiting the South Pole, and writing about U.S. operations there. This is one for the naturalists, and though I treasure the experience, I would not undergo it again, even to bring peace with honor to the nations that contend there for scientific advancement.

The general impression is that such jaunts are offered daily to columnists. That has not been my experience. Of course, it is possible that from afar they smell in me that incorruptibility that causes the angels and the saints to chant my name.

ON LEVELING WITH THE READER

Washington Star Syndicate *March 8, 1975*

In recent weeks several correspondents, thoughtfully sending me copies, have triumphantly advised editors of newspapers in which this feature appears, that "Mr. Buckley was himself a member of the CIA," and that under the circumstances, that fact should be noted every time a newspaper publishes a comment by Mr. Buckley on the CIA.

Now the Boston *Phoenix,* which is that area's left-complement to the John Birch Society magazine, publishes an editorial on the subject that begins with the ominous sentence, "*William F. Buckley, Jr.'s past is catching up with him. In the 50's he served as E. Howard Hunt's assistant in the Mexico City CIA station. . . .*" Accordingly, the *Phoenix* has protested to the editor of the Boston *Globe,* and reports to its readers, "Ann Wyman, the new editor of the *Globe*'s editorial pages, is now considering whether to append Buckley's past CIA affiliation to his column, which appears regularly in the *Globe.* Wyman intends to consult with other *Globe* editors. . . . The *Globe* may finally be on to him."

If so, it would indeed have taken the *Globe* a very long time, since it began publishing me in 1962, and my CIA involvement (a twenty-five-year-old friendship with Howard Hunt) is, among newspaper readers, as well known nowadays as that Coca-Cola is the pause that refreshes. But one pauses to wonder what is the planted axiom in the position taken by the Boston *Phoenix*?

It is true that I was in the CIA. I joined in July 1951 and left in April 1952. Now the assumption, not always stated, is that obviously anybody who was ever a member of an organization, defends that organization. But one wonders: Why should this be held to be true?—the most prolific critics of the CIA are in fact former members of it.

I attended Yale University for four years. Is it the position of the Boston *Phoenix* that, therefore, everything I write about Yale is presumptively suspect, because as a Yale graduate I am obviously pro-Yale? But it happens that shortly before entering the CIA I wrote a book very critical of Yale. And, as a matter of fact, I have in recent years written critically about Yale on a dozen occasions. So consistently, indeed, that Miss Wyman may feel impelled to identify me, at the end of every column I write about Yale, in some such way as: "Mr. Buckley, a graduate of Yale, is, as one would expect, a critic of that university."

I am a Roman Catholic, and have written, oh, twenty columns in the last ten years critical of a development within the Catholic Church. Should I be identified as a Roman Catholic?

I like, roughly, in the order described, (1) God, (2) my family and friends, (3) my country, (4) J. S. Bach, (5) peanut butter, and (6) good English prose. Should these biases be identified when I write about, say, Satan, divorce, Czechoslovakia, Chopin, marmalade, and New York *Times* editorials?

I wonder if Miss Wyman is being asked, implicitly, to label the religious or ethnic backgrounds of all her columnists? *"Mr. Joseph Kraft, who writes today on Israel, is a Jew."* That would presumably please the editors of the Boston *Phoenix*.

Or, *"Mr. William Raspberry, who writes today about civil rights in the South, is black."*

Or how about: *"Mr. John Roche, who writes today in favor of federal aid to education, receives a salary from Tufts whose income depends substantially on federal grants."*

Pete Hamill, who laughed his head off a few years ago at the hallucinations of Robert Welch, asks in the *Village Voice:* "Is Bill Buckley still a member of the CIA? Have any of Buckley's many foreign travels been paid for by the CIA?" One columnist recently wrote that *National Review*'s defense of the CIA, and my own friendship for Mr. Howard Hunt, might suggest that the CIA had indeed put up money for *National Review* over the years, though he conceded that if that were the case, the CIA was indeed a stingy organization—Mr. Garry Wills knows, at first hand, something of the indigence of that journal. Unfortunately Mr. Wills is the exact complement of Mr. Revilo Oliver, who was booted out of the John Birch Society for excessive kookiness sometime after he discreetly revealed that JFK's funeral had been carefully rehearsed. Both are classics professors by background.

Perhaps one should identify anyone who writes about politics and is also a classics professor as being that? The Boston *Phoenix* and Miss Wyman should ponder that one.

Herb Caen

Herb Caen (1916–   ) started as a columnist in Sacramento High School, then worked as a police reporter for the Sacramento Union *before taking a job in 1936 as a radio columnist for the* San Francisco Chronicle. *He persuaded the paper to let him try a local interest column, which made its debut on July 5, 1938. Except for its absence during Caen's World War II service, it has appeared six days a week since. From 1950 to 1958 Caen worked for the* Examiner, *then returned to his old paper. Most days the column consists of shorts items separated by three dots and laced with puns and neologisms (''beatnik'' was his coinage). The gossip comes from informants, and is tossed on slips into an ''item smasher.'' On Sundays he writes on a single theme, local lore being a staple. His dozen-odd books include* Baghdad-by-the Bay *(New York, 1949),* Don't Call It Frisco *(New York, 1953), and* Only in San Francisco *(New York, 1960), from which the following* Examiner *columns are taken. Barring a contender unknown to your editor, Caen probably holds the longevity record for a daily column.*

TWENTY YEARS AS A COLUMNIST

San Francisco Examiner *July 5, 1958*

On July 5, 1938, this column first saw the dark of print in the cold light of day. It was the signal for a series of strange happenings. Shortly after its inception, Hitler decided everybody should have Anschluss without Weltschmerz, Churchill assumed control in Britain, a disastrous World War rocked the world, and the Atomic Age began. As thrones tumbled and crowned heads rolled, the column marched relentlessly on, like the brainless creature it is, and the conductor himself grew steadily in stature from postpubescence to preadolescence, where he remains to this day, head lolling as he attempts to sit upright in his high chair. Next stop, according to plan: the ol' rockin' chair.

Twenty years. A long time to stand in the corner of a newspaper, scrawling inanities, illiteracies, and even obscenities on a paper wall. Twenty years of unflagging devotion to items, tritems, sightems, slightems, and even frightems; to the highly forgettable fact, the reminiscence nobody remembers, the flash that didn't pan out, the fallen arch remark; to the flopsam and jetsam, the abjectrivial and the three-dotty ephemera of a city's day-by-daze. A long time to be coining words, turning a golden language into pure caenterfeit.

Twenty years. Almost six thousand columns and six million words. Put them all together and they smell, Mother. Who was dancing with whom and where, but why? Marriages recorded, births noted, divorces granted—sometimes all in the same family, for this is a family town. Tycoons observed at work, drunks observed at play—sometimes the same people, for this is the city that never sleeps (and sometimes, as Frank Norris observed bitterly, "the city that never thinks"). Two decades of decadence and destiny, of beauty and beastliness in a fog-misted dream world I like to think of as Baghdad-by-the-Bay—its pennants sometimes brave, sometimes drooping.

Twenty years—long enough to watch a city grow away from you even as you stand in the middle of it. Long enough to have memories that flash through your mind with the jerky speed of an old newsreel: The Bay Bridge reaching out across the water, its shadow lengthening gradually over the ferries it was about to doom. The last ferry to Sausalito, the crowd singing "Auld Lang Syne," 'whiskey bottles bobbing in the moonlit wake, the skipper crying silently. Debutantes who are now grandmothers dancing with the boys they didn't marry in the Mark's Peacock Court. Little Joe Strauss, who built the "impossible" Golden Gate Bridge, shrugging—"The redwoods will last longer." The rickety roller coaster at the Beach, the half trolley, half cable on the Fillmore hill, the handsome mounted cops on the downtown streets (especially Jack Allen, who broke a thousand hearts), Jake Ehrlich and his toadies lunching daily in old Fred Solari's in Maiden Lane, Joe Vanessi, Bill Saroyan and Artie Shaw playing pool till 4 A.M. every night in Mike's on Broadway, Harry Bridges winning the rumba contest at La Fiesta on Bay, Sally Rand presiding daintily over her all-night Blue Room in the back of the Music Box, the lavish gambling casino at 111 O'Farrell, Sunday-night hobby-horse races at Roberts'-at-the-Beach, doorman Joe (Shreve) Foreman reigning like a king at Post and Grant, Anita Zabala Howard slumming at Izzy Gomez's (her boy friends, we said, were "running the gamut from Anita to Zabala"), the last night of the fair on Treasure Island, when the moon was full and the enchanting lights went down one by one, and, as we headed home with one last glance at the sudden darkness, we knew we would never be so young again.

Twenty years at soft labor, chipping at a rock with a feather. Answering the phone: "I'm jumping off the Bridge in twenty minutes. Be there if you want the story" (no jump, no story, ever); "If I don't get a plug for my client, I lose my job" (life is rough); "Man, I've got the scoop of the year for you" (it isn't); "Now, promise me you won't print this" (I won't, somebody else does); "How DARE you print that?" (I didn't, somebody else did); "I lost my dog (cat, parakeet, car, brief case, bottle, wife) and you've gotta help me find it" (I'm lost myself, friend); "I've got a gun and I'm on my way to your office to kill you" (wait till the next edition, we just went to press).

And reading the mail: "I loved your column about——," "I hated your column about——," "Why don't you write more about——," "Why do you write so much about——," "What do you have to do to get in your column——," "What do you have to do to stay out——," "I've stopped reading your column, all that scandal——," "I've stopped reading your column, not enough dirt——," "I've stopped reading your column." I still read the mail.

Twenty years of trying to keep in step with the passing parade. Of reporting that Hilton will build a hotel here, of saying the city has grown too big for its bridges, of keeping a snoreboard at opera openings, of calling Bayshore Skyway the Bayshore Dieway, of making Berkeleyans mad by calling it Berserkeley, Oaklanders mad by calling it Brookland, Sausalitans mad by calling it Souselito. I apologize, twenty years' worth.

But best of all, twenty years of holding a mirror to the city and never tiring of the sight, twenty years of being in love with the view around the corner, streets that climb the stars, hills losing their heads in the fog, cables running from here to day before yesterday, and history walking at your elbow down a dark alley at midnight. On any birthday—20th or 120th—the perfect gift.

I WONDER WHAT BECAME OF . . .

San Francisco Examiner *circa 1950s*

Men who remove their hats when there are ladies in the elevator . . . Opium smokers in Chinatown . . . The proposal for a horse-drawn carriage concession in Golden Gate Park . . . Saturday-afternoon tea dances in hotels (and the kids who used to "lobby dance" to avoid the cover charge) . . . After-hours joints with peepholes and passwords . . . Erma Leach, the blonde who rose to fame as a flagpole sitter for Horsetrader

Ed . . . The Government's hate affair with Harry Bridges . . . Parking spaces.

Whatever happened to hair stylists who used to feature "marcels" . . . The rage for Boston bulldogs . . . The man who announced he had perfected a self-lighting cigarette . . . Yellow slickers with bright sayings inked all over 'em . . . The plans for a playground and park on Angel Island . . . All the coffee you can drink for a nickel . . . And what happened to "Wha' hoppen?"

Long-gone but far from forgotten: Wooden hobbyhorse races in night clubs . . . Those "ruptured-duck" emblems you wore in your lapel to show you'd served in the armed services . . . The Bay Area's tennis supremacy . . . Scales that return your penny if you guess your own weight correctly . . . Fortune tellers who actually use crystal balls . . . Good hot jazz in the Fillmore District . . . Five-cent jukeboxes . . . Matching "Sweetheart Suits" for guys and dolls.

Where oh where has my little dog gone, and also the brave girls who wore Empress Eugenie hats . . . Those cardboard figures that fitted atop the spindle on your phonograph and danced as the record revolved . . . "Elimination dance contests" in night spots (with a bottle of cheap champagne for the winners) . . . The "call board" for limousines at the Opera House . . . Free samples in candy stores (and free hunks of bologna in butcher shops) . . . Orange Julius stands—and hot summer nights to enjoy 'em during.

I wonder what happened to the overalls that somebody threw into Mrs. Murphy's chowder . . . "My name is Yon Yonson, I come from Wisconsin, I worked in the lumber mill there—" . . . The days when it was hip to say "hep," and the jivey set was bigger than the Ivy set . . . Platform shoes with ankle straps, and little fat furs called "chubbies" . . . Boys who wore long curls till they were five and thought they had it made when their mothers took 'em out of Lord Fauntleroy suits and put 'em into Buster Brown collars . . . "Walk Upstairs and Save $10."

Among my souvenirs are songs like "Among My Souvenirs" . . . Live broadcasts by name bands from hotel supper rooms . . . Wide hand-painted ties (featuring nudes) that would never tie . . . Model T Fords with Ruckstel axles, gas tanks in the rear, and rare witticisms painted all over the sides, i.e., "Abandon hope, all ye who enter here" . . . Milk-shakes so thick you had to spoon 'em . . . Clever chaps who knew all the table-top tricks, especially how to flip a spoon into a glass by using a fork for a lever . . . The nights when a certain cab company's "celebrity tour" of the city included a visit to a madam in a Russian Hill mansion.

I wonder what's become of Sally, also Mabel, Irene, and Mary . . . "Little Audrey" stories (When the cannibals put her in the pot, Little Audrey just laughed and laughed; she knew there wasn't enough to go around) . . . Couples who practiced at home so they could dance the

Lambeth Walk, the Big Apple and the Balboa Hop in public . . . The game known as "Handies" . . . Cords with bell-bottom trousers—and stores that gave free baseball bats with boys' suits . . . Collar-button dispensers . . . Old ladies gliding around in electric cars that had silk curtains and cut-glass vases for flowers.

Put me out on a limbo with the excellent lads who bought Ruf-Neks and Home Run Kisses because they contained pictures of baseball players that, properly waxed, could be used for "laggers" . . . People who ended every argument with "So's your old man" . . . Big white lapel buttons with "Damfino" on them (to plug the candy bar of the same name) . . . The kids who'd holler, "Whadda ya feed yer mudder for breakfast?" just before the "Rags, bottles, and sacks" man would call out his miserable message . . . "Knock-knock." "Who's there" "Albie." "Albie who?" "Albie down to getcha in a taxi, honey."

I wonder what's become of merry-go-round restaurants, with the food slowly moving past on a conveyor belt . . . Golden Gate Park squirrels that would dive into your coat pocket after a peanut . . . Ice that tastes as good as the ice you used to steal off the back of an ice wagon . . . Kids who were merely "crazy mixed up" instead of juvenile delinquents . . . Candy dispensers on the backs of theater seats . . . Drummers whose bass drums lighted up with each kick to show a couple drifting across a moon-swept lake in a canoe.

I wonder what you'll do today that'll help fill a chapter like the above—twenty years from now. I wonder if there'll be a twenty years from now.

AH, SAN FRANCISCO

San Francisco Examiner *circa 1950s*

Twin Peaks blooming in green freshness at the end of drab Market Street, the sun reflecting fiercely on the windshields of a thousand cars filing across the Bay Bridge, the East Bay hills softly golden at the end of a perfect spring day, Alcatraz looking suddenly like an enchanted isle as it sits and stares dreamily at its own perfect reflection in the still waters . . . Ah, San Francisco.

Harold Zellerbach riding democratically in the front seat with his chauffeur as his black limousine drones through the Broadway Tunnel, the Oriental sharpies with patent-leather shoes and hair to match enjoying the warm sun on the fringe of Portsmouth Square, the plain-clothes coppers cruising around in the overly plain sedans that shout, "Police!" louder than any sign, the good jazz of Cal Tjader filtering out of the Blackhawk and getting lost under the stars that shine down on Turk Street, too . . . Oh, San Francisco.

The red-and-blue stacks of an American President Liner shimmering in the sun as the great liner floats like a mirage past the Marina, the silhouette of the clipper ship *Balclutha* adding a touch of square-rigged glamour to the waterfront, an ancient trolley rattling past Lotta's Fountain exactly as it has done for forty long years, the red beacons atop the Clay-Jones and Bellaire apartments blinking back and forth from Nob Hill to Russian, like pulse beats in the night sky . . . Ah, Baghdad-by-the-Bay.

The big signs that invite you to the little joints along Broadway, the newsboy who was once a stockbroker hollering his headlines off in the ticker-taped heart of Montgomery Street, the motorists who drive the wrong way in one-way streets and answer your warning honk with a pickle-faced look, the dozens of mailboxes clustered at the entrances to the smelly tenements of Chinatown, jammed together like the people who live upstairs . . . Oh, cool gray city of love, hate, filth, beauty and incense and garlic in the air.

The people who take a Sunday drive through Golden Gate Park and gnash their ulcers in irritation because they can't drive through its beauties fast enough, the bobby-soxers and the grandmothers eying each other curiously as they wait at the Curran stage entrance for an autograph from Jack Benny, the withered little man who peddles his sacks of lavender on Grant Avenue near the entrance to flower-bowered Podesta Baldocchi, unemployed B-girls wandering along Mason Street in search of a guy who'll buy them a real drink for a change . . . Ah, Queen City of the West, with white, pennant-topped towers that reach to the sky and gutters that need cleaning and mansions with marble halls and streets that need sweeping.

A Powell Street cable waddling slowly across the Broadway Tunnel's concrete bridge as though it's not quite safe for cables to cross bridges, longshoremen playing catch like kids in the shadows of the great piers that have known so much strife, the thin-faced artists of Little Bohemia showing their wares on the sidewalks and trying not to look disappointed when passers-by keep passing by, the polo players and the soccer players in the Park and the yachtsmen in their self-conscious blues in Yacht Harbor and the near-nudes baking to a turn and turning to bake on the Marina Green . . . Oh, Pearl of the Pacific, treasure of the trade winds, mecca of the mariner, sanctuary of the screaming sea gull, port of call for half the world and beloved landmark for the other half.

The Post Street clerks who pause to stare appraisingly at the $10,000 trifles in the windows of Shreve's and then move on with a silent shrug to indicate that they weren't quite satisfied with the quality, girls shrieking on the rides at the beach and sounding exactly like every girl who has ever shrieked on any ride anywhere, the old men who sit alone in side-street hotel lobbies and then walk alone to eat in cafeterias where they can be alone together and share their misery in unspoken understanding . . . Ah, city of sophistication and culture and hammer murders and shakedowns, and people who are overcrowded together and never speak to each other.

The Negro children of Sutter Street staring out of their cracked windows at the tennis players in their white-white shorts chasing a white-white-white ball across the courts of the California Tennis Club, a jet plane leaving a vapor trail smudged across the blue like a sky writer who started a message and then forgot it, springtime's young lovers parked at Land's End to enjoy the view of each other, the grown-up kids who hang around Earle Swenson's ice cream parlor at Union and Hyde and say, "Gee, thanks" (just like their children) when Earle rewards them with a free sample, the unemployed guys in their dirty non-working clothes clustered on Howard Street to discuss their last meal and wonder where the next one is coming from . . . Oh, big-little town of wide views from dark alleys.

An extravagant sunset fading so fast you can't fully enjoy all the work that must have gone into it, a long white sail fluttering home at dusk past the amber lights of the bridge that only a dreamer could have built, the ceaseless nighttime hum of life and tires and lights and horns in this worldly town that never quite finds time to go to sleep, and then—the moon rising fast out of the far-off east to beam whitely down on the hills and valleys and restless waters of the tiny city that has no boundaries . . . San Francisco. Ah, San Francisco.

Mike Royko

Mike Royko (1932–   ) is so potent a circulation lure in Chicago that the Sun-Times, *sued to prevent his quitting in 1984. He resigned the day after the paper was sold to Rupert Murdoch, the Australian-born press lord, who had provoked Royko's jibe that no self-respecting fish would be wrapped in a Murdoch paper. A Cook County judge dismissed the* Sun-Times's *contention that the paper would be irreparably harmed if Royko were allowed immediately to write for the* Tribune; *for four days in January 1984, the column appeared in both papers. Royko is a native Chicagoan, whose column first appeared in the* Daily News *from 1963 until the paper folded in 1978 and he moved to the* Sun-Times. *Royko won a Pulitzer for commentary in 1972, and is the author of* Boss *(New York, 1971), a portrait of the late Mayor Richard J. Daley, and six column collections.*

IN ALIEN'S TONGUE, "QUIT' " IS "VACATION"

Chicago Tribune   *January 12, 1984*

A Chicago politician called today and chortled: "Congrats, you're one of us now, you sly devil, you."

One of you? What are you talking about? I've never been indicted, convicted or even nominated.

He chuckled knowingly and said: "C'mon, you turned out to be a real double-dipper."

A double-dipper? Me?

"Sure. And you remember how many times you've rapped us for double-dipping, don't you?"

You mean for somehow managing to be on two payrolls at the same time?

"Right, you slicker, you. But now you've done it yourself. When are you going to run for alderman? Believe me, you've got all the instincts.'

Despite my protests, he was still chuckling when he hung up.

A moment later my Uncle Chester called and said: "I want to apologize. I just told your aunt that you're not as dumb as I always thought you were."

I appreciate that. But what changed your mind?

"Because I see that you managed to get two papers to print your stuff at the same time. How'd you swing that? I was always amazed that even one would do it."

Me, too, but this isn't my idea. I'm against it.

"Then I'm wrong. You really are dumb."

Let me explain.

"Don't bother. You probably don't understand it yourself. G'by."

He might be right, but I'd like to try to explain this bizarre situation anyway.

As people who read both Chicago newspapers might have noticed, my columns have appeared in both of them the last couple of days.

The columns in this paper are new. The ones in the other paper are reprints of columns that were written and published in past years.

The reason there are new columns in this paper is that I now work here.

The reason old columns are appearing in the other paper is that I don't work there anymore. But The Alien who now owns it doesn't seem to understand that. So he keeps printing my old columns and saying that I'm on vacation.

I don't know why The Alien is doing that. Maybe it's a custom in his native land, which is about 6,000 miles from Chicago.

If so, it is a very strange custom.

I mean, in this country, most employers know when somebody does or doesn't work for them.

Around here, if somebody walks into the boss' office and says something like, "You're kind of a disreputable character and I don't want to work for you, so I quit, and here is my resignation," the boss would surely understand.

And the boss would say something like: "Good riddance. Turn in your key to the underlings' washroom."

But apparently it doesn't work that way in The Alien's native land. There, I suspect, when a person quits and walks out, the boss smiles brightly and says: "Ah, he has gone on vacation."

If so, they must have some really confused payroll departments.

Or maybe there's another explanation. It could be that The Alien, in trying to learn about our customs, has been studying City Hall.

If that's the case, then I can understand why The Alien is acting so strangely.

In our city hall, it's always been difficult to tell if people are working, on vacation, retired, or even dead or alive. And it's made little difference. The work level has been about the same.

There have been documented cases of aldermen's young nephews being hired as city inspectors and immediately vanishing, not to be seen again until they showed up for their retirement party.

It is said that a City Hall supervisor once showed up at the wake of a foreman from streets and sanitation. As he stood over the coffin, somebody said: "Did you know him well?" The supervisor said: "He worked for me for 30 years, so I came here to see what he looked like."

But if that's what The Alien believes, somebody should straighten him out. That's the way it's done in City Hall, but not in the private sector. The custom is for the rest of us to work in order to support our ancient political tradition.

I suppose this is the kind of confusing problem that we're going to have to get used to in this modern world, with rich foreigners running in and out of each other's countries to buy up each other's businesses.

And it could be worse. As an anthropologist friend said:

"It's a good thing for you the other paper wasn't bought by somebody from the wealthy but distant and remote nation of Manumbaland."

Why?

"It is the custom there that when somebody resigns from his job, he is beheaded."

I guess I was lucky.

But there's still time.

I'VE FULLY PAID MY VICTIM TAX

Chicago Tribune *June 28, 1984*

The detective put two photo albums on my desk and said: "If he's in there, you might recognize him."

The albums were huge. They were filled with front and side pictures of thugs, stick-up men, muggers, pimps, purse snatchers and all-around thieves.

Halfway through, I said: "You must have every bum in Chicago in here."

The cop shook his head. "No. Those are just the ones from your area."

It was an awesome thought: all those mean, moronic mugs plying their trade in my neighborhood alone. And if you add together all the mugs in the other neighborhoods, they're the size of an army. In fact, there are probably many nations whose armies aren't as big as Chicago's street mug population. Or as mean. Or as well armed.

In a way it made me feel better about having been robbed. If there were that many of them out there, being robbed seemed almost natural. Not being robbed was unnatural.

As a friend of mine put it: "It's sort of like a tax we pay for living in the city. A victim tax."

If so, I'm a solid taxpayer. Looking back, my victim tax adds up like this:

Robbery: This is my second one. The first time was about 15 years ago, when I was covering a story on the South Side and a group of large punks took my money and watch. They didn't have guns, though. They relied on muscle.

I also had a near-miss a few years ago. They came out of a dark gangway, but because they were wearing those goofy high-heeled shoes, I outran them.

Burglaries: Three. The last time was when they battered down the door of my old house. Eventually they were caught and turned out to be kids from prosperous suburban families. When he learned that his son was a burglar, the father of one of them pitched over with a massive attack. The judge sentenced them to see a shrink.

Car theft: Three. Once, they went joy riding, then dumped the car in front of an alderman's house. I don't know if that was a coincidence or they had a sense of humor. The next time they stripped it and left it in a forest preserve. The last time, it vanished forever, probably in a chop shop.

Actually, I'm getting a little tired of it. My neck is stiff from looking over my shoulder. While city life has its charms, I don't like having to guess whether the guy walking toward me on the dark side street has a shiv up his sleeve. He probably doesn't like the feeling either.

It's not the money. The two punks in my building's outer lobby got about $95. My stolen cars were insured. So was the stuff taken in my house burglaries.

But I don't like looking down the barrel of a gun. Especially when the guy holding it is nervously hopping from one foot to another, twitching, breathing hard, and I'm wondering if the stupid bastard is going to blow me away.

It's also the indignity of it all. I'm a grown man. I've fought in a war, married, raised my kids, buried my dead, paid my bills, worked hard at my job and never mooched a dollar from anyone.

And there I was, wondering if it was all going to end on an impulse by an illiterate punk with the IQ of a turnip. It lasted less than a minute, but it isn't a pleasant feeling to know that your life is in some punk's grubby hands.

But what do you do? An hour after I was robbed, I was depressed because I realized I wasn't my father's son.

It happened to the old man many years ago. He was a milkman. One morning, before dawn, a guy with a knife started to climb into his truck. The old man kicked him in the face. The guy got up and ran. The old man slammed his truck into gear, drove on the sidewalk, floored the gas pedal, and—bump, bump—the world had one less stick-up man.

I could have done the same. My car was at the curb, only a few feet away. I could see them running down my street. In a few seconds, I could have caught them. I'm sure the insurance would have covered any damage to my bumper.

But it didn't even cross my mind. In my father's day, people fought back with ferocity. In my day, we pay the victim tax and wonder what sociological forces brought the poor lad to a life of crime.

All things considered, running them over is a much better idea.

Of course, his mother would probably sue me. And collect. So I guess I got off cheap.

WORDS FLY, BUT I WON'T BUDGE

Chicago Tribune *April 23, 1984*

Letters, calls, complaints and great thoughts from readers:

Clara Watson, Chicago: I don't understand how a person such as you, who seems to have a logical and analytical mind, can have such a foolish fear of flight.

I might understand it if you were totally ignorant of the facts. But since you are aware of the remarkable safety record of commercial airlines, and the astronomical odds against anything happening to you, I can't believe that you can't reason through this fear and put it behind you.

A fear that is contrary to reality is childish. It's like being afraid of the dark. You don't sleep with your lights on, do you?

Comment: Of course I don't sleep with my lights on. That would be silly because it would make it easier for the bogyman to see me.

R. E. Varga, Des Plaines: While reading your columns, I enjoy some and some not. But sometimes, besides, large things bothering me, it is also the little things, such as grammar, that bother me. In your field, grammar, language and writing should be proper. The enclosed paragraph boxes in red, is not correct. Should not the word "if" be in place of whether, or, if you wanted to use whether, should you not have included "or not", after "whether"? Proper grammar in the newspapers, even, if "Johnny Can't Read," persons, will get some help in the paper if not in school.

Comment: You know what makes me crazy, Mr. Varga? People who go through an entire column, spot one tiny error in grammar or punctuation, and send me a nit-picking letter about it. So I've turned *your* letter over to Pete Maiken, an ace grammarian on this paper's copy desk. He found five commas—FIVE, Mr. Varga!—that were used improperly. He found a "which" that should have been a "that." And he said: "The main problems in this letter are clumsiness, wordiness, redundancy and befoggery of meaning." Take that, Mr. Varga! And let this be a warning to the rest of you nit-picking, frustrated English teachers out there: Send in your nits at your own peril. You, too, will be exposed. This is war. I aren't kidding.

Paul Jackson, Chicago: I've just finished your column about Jesse Jackson and Louis Farrakhan, in which you said Black Muslims killed Malcolm X, and I must say that I am not surprised at your absolute bigotry. I am writing this letter to let you know that your intent does not go unnoticed.

You well know that the Nation of Islam had nothing to do with Malcolm X's killing. Yet you, who are a party to the real murderers and liars in this country, have the unmitigated gall to print that type of smut.

Comment: I don't know who you think killed Malcolm X, but I'll tell you who's in prison, serving life terms. There's Mujahid Halim, the former Thomas Hagen; Muhammad Abd al-Aziz, the former Norman Butler, and Khalil Islam, the former Thomas Johnson. All were in Elijah Muhammad's Black Muslim Nation, and members of the Fruit of Islam, the hard-eyed army of bodyguards. Hagen has admitted that they killed Malcolm X for bad-mouthing Elijah. So don't blame me for the murder. I wasn't even in New York that day.

Rensley Gron, Hawaii: Stop wasting your time on trash. Trying to stir up trouble between the sexes is trite and stupid. Your column about foreign wives and wimps, because of its inherent nastiness and lack of any real purpose, left me with a bad taste in my mouth.

Comment: Silly girl, it's supposed to be read, not eaten.

Jason Birch, Los Angeles: I resent your frequent jabs at the city that is my home. And I'm fed up with the stereotype of Californians as crazies and weirdos. I'm not and my friends aren't. I've been to Chicago several times, and I assure you that there are far more demented people in your city than in mine. In fact, anybody who would live there must be crazy.

Maybe what confuses you is that people out here are more sensitive, more concerned with the quality of life and their environment, more aware of basic values, than in a jungle like Chicago. Cultural matters are of importance here. I'm an artist, and I know.

Comment: Ah yes, let us look at the cultural, the sensitive, the artistic side of Californians. Recently, a radio station asked its listeners to vote on what it thought of the idea of making "California, Here I Come," the official state song. The results: The top vote-getter was the song "Thriller."

Next came the theme from "The Beverly Hillbillies." Other big votes went to "California Dreamin'," and "California Girls." I'm waiting for them to have a vote on the state bird. Put your money on the coocoo.

Henry Johnson, Chicago: I find shocking that you would come out in favor of legalizing drugs such as marijuana and cocaine. That's the most insane thing I have ever read in a newspaper. You should be fired for even hinting at such a thing. God help us from the ravings of people like you. How can anybody in his right mind want to unleash such a plague on our great country.

Comment: I don't know where you've been, but its been unleashed for years, and there is nothing that the government can do about it. So I simply propose that the less harmful stuff, such as grass and coke, be legalized and taxed. The only thing that would change would be that the government, instead of criminal dope dealers, would get revenue. And what's so insane about legalizing the smoking of a little piece of weed? We've already legalized little metal devices that you can hold in your hand, point, squeeze with one finger, and send a hunk of lead ripping through someone's bones, flesh, lungs, heart or brain, killing or maiming them. That's what I call insane.

William A. Caldwell

(Simeon Stylites)

William A. Caldwell (1906–86) personifies the plight of the columnist on a good suburban daily, in his case The Bergen *(New Jersey)* Record, *in the shadow of New York. From 1930 until his retirement in 1972, Caldwell wrote both the editorials and a column signed "Simeon Stylites," but their quality was a local secret even after the column won a Pulitzer Prize in 1971. Born in Butler, Pennsylvania, Caldwell grew up in Titusville, learning his trade from his father, managing editor of* The Herald. *He started writing sports for space rates in Bergen, moved on to reporting, then took over the editorial page. Over the decades he turned out 12,000 columns for some 350,000 readers. Mark A. Stuart has edited a selection,* In The Record *(New Brunswick, N.J., 1972) which opens with this fine quatrain, unknown to me, by Don Marquis:*

> *I pray Thee make my column read,*
> *And give me thus my daily bread.*
> *Endow me, if Thou grant me wit,*
> *Likewise with sense to mellow it.*

ACCOUNTING FOR SIMEON

Bergen Record *June 10, 1970*

She had said tactfully that the "Simeon Stylites" column of 2nd inst. had enchanted her, and when she added that she didn't remember what it was about, scatterbrain that she was, I knew she had something else on her mind.

"By the way," she said with the sudden brisk clarity that serves notice the by-the-ways are over and we're getting down to the nitty-gritty

—"by the way, that title of the column, does it have anything to do with apes, you know, gorillas?"

No, I said; that homonym is spelled "simian."

There was muffled shouting on the line. She had cupped her hand over the mouthpiece.

"I was just telling Mr. Smarty Pants here he was wrong, as usual," she said. "Now, about 'Simeon Stylites.'"

I told her about Simeon Stylites. It takes 36.9 seconds.

"See here, mister," she said; "there's an awful lot of people in the world that never heard of that cat." An edge of asperity had come into her voice. Evidently it does not please everyone to find he has missed the point of a small joke. "Maybe you better explain it, eh?"

I had just finished explaining it, I said.

"In print, eh?" she said. "So cats like Mr. Smarty Pants here can see for themselves, eh?"

Back in the fifties witty, tender Phyllis McGinley undertook to explain Simeon Stylites. He's the title of a poem in the collected *Times Three* (Viking, New York) that begins:

> On top of a pillar Simeon sat.
> He wore no mantle,
> He had no hat,
> But bare as a bird
> Sat night and day.
> And hardly a word
> Did Simeon say.

She explained Simeon Stylites. She didn't explain "Simeon Stylites."

To go back to the beginning, then, we shall have to go back to the fourth century, when the church was become an establishment governing the lives of millions of men. She had grown tolerant of human frailty. She may indeed have shared in frailties and enjoyed them. In Egypt and the Middle East there was angry dissent.

Anchorites and hermits, wedded to the good old austerities, formed communes, refused to wash, lived by weaving mats and baskets. One of these bizarre evangelists was Simeon Stylites (390?–459). It would be vanity to try to improve on Will Durant's synopsis. This is from *The Age of Faith,* the fourth volume of his titanic The Story of Civilization:

> At Kalat Seman, in northern Syria, about 422, Simeon built himself a column six feet high and lived on it. Ashamed of his moderation, he built and lived on ever taller columns, until he made his permanent abode on a pillar sixty feet high. Its circumference at the top was little more than three feet; a railing kept the saint from falling to the ground in his sleep.

> On the perch Simeon lived uninterruptedly for thirty years, exposed to rain and sun and cold. A ladder enabled disciples to take him food and remove his waste . . . .
>
> From his high pulpit on the column he preached sermons to the crowds that came to see him, converted barbarians, performed marvelous cures, played ecclesiastical politics, and shamed the money lenders into reducing their interest charges from 12 to 6 per cent.

"His exalted piety," says Will Durant, "created a fashion of pillar hermits which lasted for twelve centuries and, in a thoroughly secularized form, persists today."

Everything clear now? No?

Well, along about January 1930 the present Editor and Publisher of *The Record* thought it might be constructive to have on the editorial page a column of local origin rather than syndicated copy or serialized fiction, which was the usage. It would need a standing headline or title, some symbol of continuity, and people were invited to submit suggestions, with the usual embarrassing results, such as "Bon Mots Politique," which wasn't even grammatical, and "The Last Word," which was copyrighted. Oh, hell, said the Editor, let's call it "Simeon Stylites" until something more useful turns up, and for a few years we ran a legend under the title explaining things adequately, we thought:

"From the Top of a Column He Philosophizes About the Interesting Things of Life."

Simeon was a commentator on a column. "Simeon" would be a commentator in a column. The meaning seemed obvious. People began asking why we didn't eliminate the explanatory legend. We did. People began asking why we didn't explain our meaning more clearly.

It occurred to me last Jan. 11 that I'd been mumbling away on this perch ten years longer than the original columnist lasted, but I can't say I've acquired any other of his bad habits: no asceticism, no putrefaction, no worms that I know of. I am not sure that quite everything is clear, nor is Phyllis McGinley:

> And why did Simeon sit like that,
> Without a garment,
> Without a hat,
> In a holy rage
> For the world to see?
> It puzzles the age,
> It puzzles me.
> It puzzled many
> A Desert Father.
> And I think it puzzled the Good Lord, rather.

But at any rate affidavit is taken herewith that the facts hereinabove are correct.

NOTES AT QUARTER PAST FIVE

Bergen Record *December 1956*

In certain clandestine ways the most interesting item we shall never see under "It Was Today" is as follows:

"William Anthony Caldwell, uh, was born, 1906."

Almost everybody must have a sort of ex-officio concern, even enthusiasm, over the fact that he was born, and a man should be excused getting almost fanatically sentimental over his fiftieth birthday. We are aware, now that our second half-century is upon us, that a decent regard for the opinion of mankind dictates a slight rhapsody or a cry of bitterness. At his fiftieth birthday a man should produce an ode, a tear, at least a creak—he might at the very least say something astringent, carefully avoiding self pity, about the fact that the people who make "It Was Today" get there on the record of the years he'll never see again.

Can't.

Ten is the year of discovery, twenty is the year of high-hearted adventure, thirty is the year of fulfillment, forty is the year of desperation; but at fifty a man finds he has made his settlements with himself, and he finds that himself scarcely gives a damn.

You must not misunderstand this. Settlement needn't mean adjustment, and a part of settlement may be thanking whatever gods there be for his unconquerable soul—his contract with himself may stipulate that he is a rebel, angry, not a bit enchanted with the world he helped to botch.

But even the angry rebel is not sure at fifty that his anger matters much, even to himself. This is the way things are, and this is the way things will remain as long as the human race is a swine, and, while a man is content that there are truths and that they are good enough to die and live for, he knows his dying for them wouldn't even for him be a thing which would validate them.

He is not sure of anything, including this statement. From the grotesquely limited evidence available to him he gathers that time—in this gulf of the universe, at any rate—flows in one direction. It appears to him that growth, being conditioned by time, goes on. Evidently whatever now is, better or worse, will be past tomorrow. From time to time he regrets that he is involved in this process and is probably among the things which too must pass, but he is able to acknowledge that time and growth are wise to make no exceptions. If there were dispensations for him, so might there be for cancer and overpopulation and bishops who bless their nation's artillery in the name of the Almighty who is the father of us all.

He finds tears in his eyes more often—at the sight of a flag snapping, at the look of young people marching, at the sound of music, at being confronted with anything in the earth or sea or sky that is innocent. He finds it more difficult to harden himself against the tears—and yet more difficult than this to understand them.

"Swine," he says, but he knows men aren't swine on purpose: sometimes they know not what they do, but they do it for a decent purpose, reverently even when what they do is the ultimate irreverence which is death—even the killer is sick, and the name of his sickness is humanity, and he is not to be hated for it any more than is the cripple to be hated for his deformities.

He wishes he were young, but he no longer entertains the hallucination that matters would have turned out much differently, say, better; he knows that he is the product of such potentialities as wouldn't be altered the width of a chromosome by a swig of the elixir; and, in the next breath, he utters his thanks that he need not go again through the childbirth horror of any adolescence.

He has learned, he thinks, to value his own friendship; he needs his own good opinion, and sometime recently he thinks he found that of all the people he knows the hardest to deceive is himself. To his chagrin at first, later reassuringly, he found himself an indulgent friend, willing to accept his shortcomings—provided he does not lie about them.

There are nights when he's tired, but he's glad to be tired, because he's glad to be alive. There are times when he gets as far as knowing what happiness must be like. There are whole days when his faith—in time, in growth—gets the better of his doubt. He knows the cosmos, which will still be here a billion years from now, doesn't notice which are the faith days and which are the doubt ones. He's getting old, and all this reminds him is that he'd better—the day he stops, he'll really be in a jam.

HE BET HIS LIFE

Bergen Record *April 1968*

Martin Luther King, Jr., bet his life that the Negro could make his way into the mainstream of life in the United States by means of nonviolence, and he lost.

So a just cause has its martyr duly crucified, and it remains to be seen whether his truth goes marching on—or whether the doctrine of love and hope and gentleness which he preached and died for was ever the truth at all.

We had better not lapse into any such orgy of self-laceration as distorted the reality and trivialized the event when John F. Kennedy was murdered in Dallas.

Mea culpa, then cried tens of millions who had nothing on their conscience except the comprehension that not enough had been done to make the world conform to the bright young dream Kennedy had enunciated. It was not the truth. Kennedy had died the victim of our hate, we said, luxuriating in our confession, and everything we did from then on made liars of us.

The ultimate proof came forward instantly. Kennedy had been loved. We had not repudiated his dream. The stutter of the funeral drums had scarcely stopped echoing when the new President and the Congress proceeded to write into law all—but all—of the social legislation Kennedy had planned for the rest of his eight years in the White House.

In his death he had triumphed.

The analogy should not be pressed But it should be borne in mind. The difference between the contexts of the two assassinations is wide, and the difference has been made clear by the report from the President's Commission on Civil Disorder.

At the time of Dallas the crisis was years away. Now it is at hand. At the time of Dallas the thing most to be feared was callous repudiation of the white's contract with the Negro, signed 100 years before in the blood of the Civil War. The thing most to be feared now is the cleavage of the country into two societies, one white and one black, one having unlimited access to the good life and one herded into its dismal warrens and kept in repression under the guns of the superrace's police, both irreconcilably hostile. The danger is civil war. In a nation that affords any man with the price of a gun access to his private arsenal, the danger is sudden death. Nobody is immune.

Against such a hardening of hostilities Dr. King had bet his life.

There will be a time when, collecting and absorbing the totality of his work and works, we shall be able to appreciate more passionately the immense grandeur of this simple man. He was a prophet, and he knew it, and alone among the statesmen of his time he dared to utter his vision in the unabashed language of poetry. He had become a legend in his time. In the years to come he will come to tower over his time. It is a way martyrs have when their cause is just.

Unless!

Dr. King will be immortal, unless the justice of his cause required that memory of him be suppressed in a society which cannot afford to acknowledge that he was right.

Legally, the black is the equal of the white; socially and economically, he is inferior, not by reason of personal worth or genetics or traceable reason but by reason of an idiotic habit.

Idiotic habit pulled the trigger in Memphis. Let's not generalize the responsibility for any murder. Whenever blood is shed, in a kitchen brawl or on a battlefield, our hands are stained to some extent, but we cannot accept individualized blame for all human villainy.

Yet the death of Dr. King cannot be extricated from involvement in the cause he led. And the cause—the development of the struggle from reform to peaceful revolution—is intertwined in the life of each of us, white and black.

In the name of all the martyrs let no man yet say the assassination has proved that nonviolence will not work. Give Dr. King a chance to prove his case.

The black power he sought was a simple thing, so naive and innocent, so naked that it embarrasses most men and women to say its name. Its name is, for God's sake, love.

He bet his life.

Before we give way to rage and bitter tears let us beseech ourselves to consider the possibility that we won his bet.

Dorothy Day

Dorothy Day (1897–1980) wrote ''On Pilgrimage'' for thirty-eight years in The Catholic Worker, *a monthly published in New York. Dwight Macdonald described her column as ''an odd composite of Pascal's* Pensées *and Eleanor Roosevelt's 'My Day.' '' The paper, a penny a copy, was dedicated to the poor and to pacifism, but also approved private property and opposed the class war. ''Let us be fools for Christ,'' Dorothy Day said. ''Let us recklessly act out our vision, even if we shall almost surely fail, for what the world calls failure is, often, from a Christian viewpoint, success.'' Born in Brooklyn Heights of Republican and Protestant parents, she was raised in Chicago and found her vocation by way of the Socialist movement in Greenwich Village, where she worked with Max Eastman and John Reed on the irreverent left-wing* Masses. *She had already converted to Catholicism when she met the French-born Peter Maurin, an itinerant preacher who lived in a flophouse on the Bowery, and the pair founded* The Catholic Worker. *On her death,* Newsweek *called Dorothy Day ''the radical heart and conscience of American Catholicism.'' Her story is told in* Dorothy Day *(New York, 1982) by William D. Miller.*

MEDITATION ON THE DEATH OF THE ROSENBERGS

Catholic Worker *July 1952*

At eight o'clock on Friday, June 19, the Rosenbergs began to go to death. That June evening the air was fragrant with the smell of honeysuckle. Out under the hedge at Peter Maurin Farm, the black cat played with a grass snake, and the newly cut grass was fragrant in the evening air. At eight o'clock I put Nickie in the tub at my daughter's home, just as Lucille Smith was bathing her children at Peter Maurin Farm. My heart was heavy, as I soaped Nickie's dirty little legs, knowing that Ethel Rosenberg must have been thinking with all the yearning of her heart, of her own soon-to-be-orphaned children.

How does one pray, when praying for "convicted spies," about to be electrocuted? One prays always of course for mercy. "My Jesus, mercy." "Oh Lord Jesus Christ, son of the living God, have mercy on them." But somehow, feeling close to their humanity, I prayed for fortitude for them both. "Oh God, let them be strong, take away all fear from them, let them be spared this suffering, at least, this suffering of fear and trembling."

I could not help but think of the story in Dostoevsky's *Idiot,* how Prince Myshkin described in detail the misery of the man about to be executed, whose sentence was commuted at the last moment. This had been the experience of Dostoevsky himself, and he had suffered those same fears, and had seen one of his comrades, convicted with him, led to the firing line, go mad with fear. Ethel and Julius Rosenberg, as their time approached and many appeals were made, must in hoping against hope, holding fast to hope up to the last, have compared their lot to that of Dostoevsky and those who had been convicted with him. What greater punishment can be inflicted on anyone than those two long years in a death house, watched without ceasing so that there is no chance of taking one's life, and so thwarting the vengeance of the State. They had already suffered the supreme penalty. What they were doing, in their own minds no doubt, was offering the supreme sacrifice, offering their lives for their brothers Both Harold Urey and Albert Einstein, and many other eminent thinkers at home and abroad avowed their belief in the innocence of these two. They wrote that they did not believe their guilt had been proved.

Leaving all that out of account, accepting the verdict of the court that they were guilty, accepting the verdict of millions of Americans who believed them guilty, accepting the verdict of President Eisenhower and Cardinal Spellman who thought them guilty—even so, what should be the attitude of the Christian but one of love and great yearning for their salvation.

"Keep the two great commandments, love God and love your neighbor. Do this and thou shalt live." This is in the Gospel, these are the words of Jesus.

Whether or not they believed in Jesus, did the Rosenberg's love God? A rabbi who attended them to the last said that they had been his parishioners for two years. He followed them to the execution chamber reading from the Psalms, the 23rd, the 15th, the 31st. Those same psalms Cardinal Spellman reads every week as he reads his breviary, among those 150 psalms which make up not only the official prayer of the Church, but also the prayer which the Jews say. We used to see our Jewish grocer on the east side, vested for prayer, reciting the psalms every morning behind his counter when we went for our morning supplies. I have seen rabbis on all night coaches, praying thus in the morning. Who can hear the Word of God without loving the Word? Who can work for what they conceive of as justice, as brotherhood, without loving God and brother, If they were spies for Russia they were doing what we also do in other countries, playing a part in international politics and diplomacy, but they were indeed serving a philosophy, a religion, and how mixed up religion can become? What a confusion we have gotten into when Christian prelates sprinkle holy water on scrap metal, to be used for obliteration bombing, and name bombers for the Holy Innocents, for Our Lady of Mercy; who bless a man about to press a button which releases death on fifty thousand human beings, including little babies, children, the sick, the aged, the innocent as well as the guilty. "You know not of what spirit you are," Jesus said to his apostles when they wished to call down fire from heaven on the inhospitable Samaritans.

I finished bathing the children who were so completely free from preoccupation with suffering. They laughed and frolicked in the tub when the switch was being pulled which electrocuted first Julius and then his wife. Their deaths were announced over the radio half an hour later, jazz music being interrupted to give the bulletin, and the program continuing immediately after.

The next day the *New York Times* gave details of the last hours, and the story was that both went to their deaths firmly, quietly, with no comment. At the last Ethel turned to one of the two police matrons who accompanied her and clasping her by the hand, pulled her toward her and kissed her warmly. Her last gesture was a gesture of love.

They were children of that race to which Mary and Jesus and Joseph, the Holy Family, belonged. In their humanity they more closely resembled Jesus than we do who are not Jews. For that, too, we must love them. "Spiritually we are Semites," Pope Pius XI said. For that we must love them. "Father forgive them for they know not what they do." For that saying we must love them.

Let us have no part with the vindictive State and let us pray for Ethel and Julius Rosenberg. There is no time with God, and prayer is retroactive. By virtue of the prayers we may say in the future, at the moment of death which so appallingly met them, they will have been given the grace to choose light rather than darkness. Love rather than Hate. May their souls, as well as the souls of the faithful departed, rest in peace.

Nicholas von Hoffman

Prodded by what was then called the underground press, adventurous 1960s editors like Benjamin Bradlee of the Washington Post *started Style sections and began hiring interlocutors like Nicholas von Hoffman (1929–   ). Born in New York City, Von Hoffman worked for the* Chicago Daily News *before joining the* Post *in 1966. He wrote about antiwar demonstrations, civil-rights marches, the drug culture, and the feminist revolt in a Style column called "Poster." No writer for the* Post *generated more mail, according to the paper's historian, Chalmers Roberts. Since 1976, he has freelanced. His books include* Left at the Post, *with a Foreword by Bradlee (Chicago, 1970); and* Citizen Cohn *(New York, 1988).*

WAR PROTEST COMES TO WASHINGTON

Washington Post   *circa 1970*

Welcome, visitors, to Fat City.

There is a large, pro-American fifth-column population in Washington which will welcome you, feed you, and give you a place to sleep. You are not in totally hostile territory. But be careful. If you haven't got it completely together, stay indoors or go back home.

The masters of this city are old and crafty in the ways of power. They've ruled longer than you visitors have been alive. They've had visitors before—Martin Luther King, the Bonus Marchers, Coxey's Army—and they've outlasted them all, unmoved, unchanged, and unchanging. They know how to turn back and dissipate invasions from the United States of America.

Washington's first line of defense is its ability to inspire awe. The domed, colonnaded, and scrolled buildings, the marble boulevards, the parks and fountains humble a person and make him feel small and narrowly limited in life and power against the magic of the temples of government, against the pomp of imperial Disneyland. Every year massive invading armies come armed with their Instamatics and their flash cubes, shoot off their harmless weapons, plunder the souvenir shops, go on the rides, see real live senators snoozing at their desks, and are sent home happy and ignorant.

Washington's second line of defense is indifference. Washington will outwait you; Washington will vanish until you go away. It won't answer the phone or the door; everybody will be out of town; you will always be told you're in the wrong office and to go down the corridor. The Rev. Ralph Abernathy built Resurrection City and sat here for months, but it did no good. What is a year for a soldier in Vietnam or a mother waiting for a welfare check is a minute here—Washington can wait forever.

Guile, evasion, delay, tricks, entertainment, and *trompe l'oeil* all failing, Washington will use force. Believe that, and don't provoke them. Don't make the mistake of saying, "They can't be so stupid. They don't want *Pravda* printing pictures of them smashing in the heads of their own young people with the spire of the Washington Monument in the background."

The masters of this city aren't stupid. They deal in realities, not public relations. If that picture runs in *Pravda,* they don't mind. In fact, if it serves as a warning to the young and the peaceful in Russia not to start acting up, so much the better. The political and corporate bureaucrats of this world have instinctual ties of loyalty that supersede ideology. Marshal Grechko, the Russian minister of defense, will come to the aid of Melvin Laird before he helps you. On the scales of balance of power, Czechoslovakia is a counterweight to Vietnam.

The city itself is misleading. If you browse outside the tourist belt you'll run into the black slum, Washington's façade of poverty. The people who live in it are poor and powerless, but the city is neither. It is fat and rich judged by the standards of places you visitors come from. Washington's poor are a stagnant, ignored minority encircled by unbelievable per capita wealth.

Rich and splendid, Washington has nothing of a wartime capital about it. No sense of drabness or feeling that it might be in bad taste to display and enjoy nonchalant luxury when American soldiers are dying. The White House sets the tone of silver and velvet, brocade and satin, dinner jackets and footmen, ball gowns and wine. Call it the spirit of San Clemente, the decorum of mourning lasts only as long as the red dot on the TV camera glows.

The city is bereft of a sense of fitness of things, of a knowledge of what is meet and becoming conduct. It has confused pomposity with dignity and cannot remember that once these buildings, smaller and less opulent, were guarded not by soldiers but by citizens' affections and reverence.

The war doesn't exist here. Except when the people occasionally intrude, Washington's daily life is given over to fighting about boodle. The liberal, nonideological political scientist Harold Lasswell inadvertently summed up Washington in his little Eisenhower era book, *Politics: Who Gets What, When, How.*

"The study of politics is the study of influence and the influential," he wrote. "The influential are those who get the most of what there is to get. Available values may be classified as deference, income, safety. Those who get the most are elite; the rest are mass." So, welcome, visiting mass to the city of the elite.

People in town for a couple of days may not be able to find the elite and could mistakenly end up gawking at some little Congressman who merely keeps his nose clean and does what he's told. To find the elite, use the phone book and look up the National Highway Users Conference, the National Association of Real Estate Boards, and the Pharmaceutical Manufacturers Association. Don't forget the clubs where the white, Anglo-Saxon male eliters forgather in strictest seclusion to arrange who gets what, when, and how while grumping at a world that is beginning to figure them out.

The elite will ignore you as much as it can. To its members you're cheeky brats. Your refusal to wait out the seniority system of decision-making irritates them. When they think about you, they get angry and shout that there'll be none of this damn permissiveness around here.

Obey the laws and don't take chances. No holding, please, or, if you must, no more than you can swallow fast. Don't oink at the police. You can legitimately demand politeness, restraint, and lawful conduct from a cop, but if you oink at him, you're asking him to be a saint. We owe the cop the same politeness, restraint, and lawful conduct we demand of him.

The general situation here in Washington is better than Chicago but not so good as Woodstock. This estimate comes from Wes Pomeroy, the ex-Justice Department man who was at Chicago and ran the police operation at Woodstock.

He says, "The decision-makers here see their constituency as the silent majority and they think the silent majority want them to take a hard line. The Washington police department is sophisticated, loose, experienced, and professional at handling large crowds. Chief Wilson is one of the best chiefs in the United States. It's just unfortunate he's not making the decisions.

"Some of the things that have been happening here happened in Chicago. You see pictures in the paper of policemen practicing riot formation, statements about how many troops are in reserve. It's like preparing for a heavyweight fight. You have to have these preparations, but it's very foolish to play these games in the papers. It isn't going to scare anybody but it's going to be taken as a challenge.

"The entire fiasco in Chicago was almost solely the responsibility of a stubborn, unwise Mayor Daley who emasculated his police command. I went there twice before the convention as a messenger from the Attorney General asking Daley to let somebody from the government negotiate with somebody from the Mobilization. He didn't hear the message. The one thing Mayor Daley said was that if the Justice Department really wanted to help they could let him know when those out-of-town agitators were coming in to Chicago so he could take care of them."

It's better here, Pomeroy says: "Everybody's learned something. There isn't the runaway anger and hysteria of Chicago."

But it's no Woodstock either: "Woodstock was our turf and our issues, peace and music. For months ahead, we built in internal controls that would establish a cultural control. We had food, lawyers, ministers, and doctors for the people. Our policemen were peacekeepers. Nobody was armed. I designed a special uniform for them, bell-bottom trousers, T-shirts with a guitar and a dove on the front and the word peace on the pack. What you try and do is create an environment where people will do what you want them to do and they want to do."

So now you know, visitors.

Shalom.

George F. Will

George F. Will (1941–   ) writes well. Whether he reasons as well is oft debated. Ronald Steel, Lippmann's biographer, finds Will too partisan, intolerant, and predictable. Others fault the quotations dropped like raisins into his twice-weekly columns in the Washington Post *and twice-monthly in* Newsweek. *But if Will's stance is predictable—fiercely anti-Soviet, derisive of youth culture and the gods of the sixties—his arguments are not. Educated at Princeton and Trinity College, Oxford, Will reaches for often awkward moral issues and is no unctuous apologist for those in power. Speaking of his friend Ronald Reagan at a business banquet, Will recalled Cardinal Wolsey's words about Henry VIII: ''Be very careful what you put into his head. You will never get it out.'' Will appeared to be destined for academia (his father taught philosophy at the University of Illinois) when he wound up jobless in Washington and was soon writing for Buckley's* National Review, *where his work was spotted by Meg Greenfield, editorial page editor of the* Post. *His column started there in 1973, and Will's talents as a speaker became apparent on public television's ''Agronsky and Company'' and then on ABC News. He has published three collections of columns, the most recent being* The Morning After *(New York, 1986).*

TEN YEARS IN A COLUMN

Washington Post *December 18, 1983*

It is a decade since I, forever seeking madder music and stronger wine, found the perfect pleasure: writing a column. So today I take time out from vivisecting the rest of the world to say something about this vocation and to acknowledge a debt of gratitude.

My three children, watching me all day in my office at home, consider me unemployed, which in a sense I am. Nothing so pleasurable can be called a job. "Writing is not hard," wrote Stephen Leacock. "Just get paper and pencil, sit down, and write as it occurs to you. The writing is easy—it's the occurring that's hard." I disagree.

Today's world is endlessly provoking, at least to someone with a Tory sensibility and ordinary curiosity. I carry in my wallet a list of topics I am itching to get to, and I usually add to it about six topics a week. But I only write two columns a week, which is not enough to do justice to the train wreck of American manners, let alone public policy.

Even the abuse that comes a columnist's way is more entertaining than wounding, and it can be used to teach the young stoicism about slings and arrows. A chip off this old columnist came home from school upset because a playmate had called him a name. I opened for him the Los Angeles Times, which had recently run my column putting Napoleon in his place, and this day was full of letters from Napoleon's defenders, a sordid lot. My son especially enjoyed seeing his father called a "mad dog." I had not seen him so cheerful since I nearly dislocated my shoulder skiing.

I write in longhand, with a fountain pen, of course. I do so not as a political statement—although a Tory could hardly do otherwise—but because writing should be a tactile pleasure. You should feel sentences taking shape. People who use "word processors" should not be surprised if what they write is to prose as process cheese is to real cheese.

The columnist's craft has an alarmingly distinguished pedigree, beginning with Addison, Steele and Dr. Johnson. What Henry Adams said of the succession of presidents from Washington to Grant (that it refuted the theory of evolution) can fairly be said of the succession of columnists from Johnson to Will. But it cannot be said of the succession from Johnson to my favorite columnist, Murray Kempton.

Here is my expression of gratitude. I meandered into this craft, via university teaching and Senate staff work, but I know this: I am a columnist because 25 years ago, when I came East to college, I discovered the delights of Kempton, who then wrote for the *New York Post,* which then was a newspaper.

Every serious citizen must read the sports pages which are Heaven's gift to struggling mortals. But nothing is more optional than reading a column. Congress should make it compulsory, but will not. So a columnist needs three seductive skills: he must be pleasurable, concise and gifted at changing the subject frequently.

Changing the subject is easy: I write at least half my columns on subjects that are not on the front page, or often any page, of newspapers. It is an aim of my life to die without having written a column about who will win the New Hampshire primary. But, the, I may be the only journalist in Christendom who has never been to New Hampshire.

Most newspaper readers do not read columnists, and my guess is that 75 percent of my readers disagree with 75 percent of what I write. That is fine: it means the audience is opinionated, in need of instruction and capable of enjoying aggravation if it is inflicted with some felicity. Readers do not read a columnist because of his subject on a particular day. Rather, they read or do not read him because they like or dislike the way his mind ranges around the social landscape.

The amazing thing is that something this much fun is not illegal. Bobby Knight, Indiana University's basketball coach, who thinks of journalists the way Mussolini thought of Ethiopia, says: "All of us learn to write in the second grade. Most of us go on to greater things." But it is impossible to do anything—well, anything not done between the foul lines in a baseball park—more satisfying than writing.

JOURNALISM AND FRIENDSHIP

Newsweek *January 19, 1981*

Billy Loes, the patron saint of columnists, was a pitcher, mostly for the Brooklyn Dodgers. He was not, shall we say, all polish. He once lost a ground ball in the sun. He committed a memorable balk in a World Series game. (The ball squirted from his hand at the start of his windup. "Too much spit on it," he explained.) But he was a deep social thinker. ("The [1962] Mets is a very good thing. They give everybody a job. Just like the WPA.") And he was a major-league complainer. That quality inspired a poem, whose full text is "This is the trouble with Billy Loes,/ He don't like it wherever he goes." This week's column is about writing columns, which is in part, but only in part, the art of complaining usefully.

Fearful rumors are afoot that I may abandon the columnist's basic stance of thorough disapproval of all conduct but his own. Readers by the millions—well, okay, a couple of you, your eyes wet with unshed tears—are worried that on January 20 I shall succumb to conviviality, my captious spirit will vanish and I shall commence to carol like a lark. Caroling is fine from larks, but boring from columnists. Worriers note that last autumn I was not your basic undecided voter, and (believe it or not) they worry because the Reagans came to the Wills' for dinner. The latter was a small matter, but large enough to fill to overflowing the minds of some people.

The financial markets are gyrating, the budget is hemorrhaging, the currency is evaporating and the Russians are practically in Duluth, so there are questions more urgent than the following: Is journalistic duty compatible with feelings of friendship between journalists and those political people who do the work of democracy? And will this columnist be as critical of Reagan's Administration as he was of Carter's?

The answer are: yep, and I certainly hope not. I am moved to expand upon these answers only because journalism (like public service, with its "conflict of interest" phonetics) is now infested with persons who are little "moral thermometers," dashing about taking other persons' temperatures, spreading, as confused moralists will, a silly scrupulosity and other confusions.

We all have our peculiar tastes. Some people like Popsicles. Others like gothic novels. I like politicians. A journalist once said that the only way for a journalist to look at a politician is down. That is unpleasantly self-congratulatory. A journalist's duty is to see politicians steadily and see them whole. To have intelligent sympathy for them, it helps to know a few as friends. Most that I know are overworked and underpaid persons whose characters can stand comparison with the characters of the people they represent, and of journalists.

Friendship between journalists and politicians offends persons who consider a mean edge the only proof of "candor" in writing about politicians. ("To be candid, in Middlemarch phraseology, meant," George Eliot wrote, "to use an early opportunity of letting your friends know that you did not take a cheerful view of their capacity, their conduct, or their position . . .") But friendship, including relaxation in social settings, reduces the journalist's tendency to regard politicians as mere embodiments of ideas or causes, as simple abstractions rather than complicated human beings. The leavening of friendship may take some of the entertaining savagery from our politics, but then, politics has not recently suffered from an excess of civility.

The idea that only an "adversary relationship" with government is proper for journalists pleases some journalists because it seems hairy-chested, and because it spares them the tortures of thought. Continual thought about what to publish, and how to adapt to the nuances in a political city, is necessary for journalists who believe that they are citizens first. They have particular professional duties, but they and politicians are part of the same process, the quest for the public good.

John Kennedy showed a draft of his inaugural address to Walter Lippmann, who suggested a change (to refer to a hostile nation as an "adversary," not an "enemy"), which Kennedy made. On inauguration night Kennedy went to the home of Joseph Alsop, the columnist. He went for the best of reasons: friendship. Lippmann helped draft Woodrow Wilson's "fourteen points" (then became an astringent critic of Wilson and the Versailles Treaty). For five decades he preached detachment but practiced a decorous, public-spirited involvement.

Shortly after the Second World War, Senator Arthur Vandenberg asked Scotty Reston to read a speech about relations with Russia. Scotty did, considered it unfortunately negative and made a suggestion (that Vandenberg, formerly a leading isolationist, signal a turn toward internationalism), and the speech became a significant event. Lippmann was, Reston is, a citizen too: a citizen first.

Today good citizens ache for a chance to applaud the reasonably graceful exercise of power. I am hopeful; I am not *quite* of Evelyn Waugh's persuasion ("I have never voted in a general election as I have never found a Tory stern enough to command my respect"), and I do not covet the Billy Loes trophy for most incessant complaining. But . . .

The columnists I most admire, from Samuel Johnson to G. K. Chesterton to Murray Kempton today, have written about the "inside" of public matters: not what is secret, but what is latent, the kernel of principle and other significance that exists, recognized or not, "inside" events, policies and manners. Following, in my dim way, their luminous example, my columns are meditations on various principles. And in behalf of my own principles, which, I suspect, will not be fully and perfectly worshiped within the Reagan Administration. And unless there is unflagging worship at the altar of Will's creed, Will will emulate the great complainer, the sainted Billy Loes.

The most useful complaints are couched as arguments, and make a case for better ways, finer things. Often, the most useful arguments are the civil but spirited ones we have with friends. Invariably, it is this for which I write: the joy, than which there is nothing purer, of an argument firmly made, like a nail straightly driven, its head flush to the plank.

VILE LITTLE EMPEROR

Washington Post *May 30, 1982*

Was Napoleon poisoned? I certainly hope so.

Two authors of a heartening new book, *The Murder of Napoleon*, say arsenic, not cancer killed that dreadful Corsican. Not soon enough: by then he was in exile, fat as a pastry, long past careening around Europe making history and orphans. But today it is salutary, because instructive, to recall this vulgarian who prefigured the worst of modern politics.

He was the first modern tyrant, an absolutist without hereditary pretenses, an upstart compensating with brutality and cynicism for his lack of legitimacy, grounding his power in manipulation of the masses. He was (to borrow a Disraeli phrase) a self-made man who worshiped his creator. But he would refer to Louis XVI as "my uncle."

Like Hitler, who was not German, and Stalin, who was not Russian, Napoleon, who was not French, was a complete outsider, outside all restraints grounded in principles or affections. He had a megalomaniac's estimate of the importance of his undertakings. When planning to invade

England he said: "Eight hours of night in favorable weather would decide the fate of the universe."

He had what we now recognize as the totalitarian's thirst for revising as well as making history. One of the most famous of his melodramatic orders-of-the-day ("Soldiers, you are naked, ill-fed. . . .") was written 20 years later on Saint Helena. He combined a philistine's sense of culture and a martinet's reverence for the state: "People complain that we have no literature nowadays. That is the fault of the Minister for Home Affairs."

Believing that "the people must have a religion and that religion must be in the hands of the government," he pioneered a modern industry—the manufacture of ersatz religions for political purposes. His birthday (Aug. 15, a black crepe day in sensible households) became the Feast of St. Napoleon. But he felt he had been born too late, that Alexander the Great had had more fun:

". . . after he had conquered Asia and been proclaimed to the peoples as the son of Jupiter, the whole of the East believed it. . . . if I declared myself today the son of the eternal Father . . . there is no fishwife who would not hiss at me."

Correlli Barnett, a biographer of Napoleon, is what a biographer should be, "a conscientious enemy." He says Napoleon was "perhaps the earliest example of that phenomenon of the emerging mass society, the superstar." His career aggravated the tendency of 19th-century romanticism to celebrate "great men," the "geniuses" who are "artists of history-making." Romantic painters loved mountainscapes and despoiled many by painting Napoleon into the foreground, struggling through the Alps, like Hannibal.

Napoleon believed that in order to have good soldiers, a nation must always be at war. He gave France good soldiers—and good roads. Like Hitler with the autobahns, Napoleon adored roads for moving armies. He wanted 14 highways radiating from Paris toward any conquest his heart might desire. But highways can carry traffic two ways, and in May and June 1940, France's roads were enjoyed by troops from Germany—a nation that Napoleon helped make into a modern state by reforming its administration and inflaming its nationalism.

In the 18th century—"the age of reason"—better furnaces for iron-making resulted in better artillery, which conquered what until then had been the key to war—the fortress. Napoleon rose like a rocket as an artillery officer who understood how to use artillery when reasoning with a restive populace.

A century later, another Frenchman enamored of artillery (Marshall Foch) said: "Artillery conquers the ground, infantry occupies it." At dawn of the first day of the third battle of Ypres, British and French artillery fired 107,000 tons of shells and metal. 'Twould have been bliss that dawn to be alive for Napoleon, who boasted that he cared "little for the lives of a million men." He used conscription—"the nation in arms"—to make war bigger: 539,000 fought at Leipzig in 1813.

Barnett calls Napoleon "Byron-in-field-boots," the eternal fidget who turned Descartes' "I think, therefore I am," into "I order everybody about, therefore I am." He was the sociopath in politics, utterly without softening values, celebrating bureaucracy for others but willfulness for himself: "I am having three heads cut off here everyday and carried around Cairo. . . ."

The only lasting good he did was inadvertent: he excited Beethoven to compose "Eroica." At least Beethoven dedicated it to Napoleon temporarily. Beethoven then had better, second thoughts and changed the dedication. Thus did one of Europe's vilest lives intersect with one of Europe's noblest.

William Safire

When William Safire (1929–   ) was hired in 1973 to write a Washington column for the New York Times, *some ill-wishers hoped that he would fall on his kiester (regional slang for behind). The objection was that he had been a speechwriter in the Nixon White House, and was responsible for "nattering nabobs of negativism" and other alliterative assaults on the media by Vice President Agnew. Safire sometimes errs on the side of cuteness and is an unswerving conservative. But in the event, he foiled detractors with humor and a willingness to challenge leaders in his own camp. His columns on Bert Lance, President Carter's friend, won him a Pulitzer Prize in 1978. Safire's sideline is lexicography; having read the essays that follow, A. M. Rosenthal, then the* Times's *executive editor, proposed the "On Language" column in the Sunday Magazine that has resulted in five books. Safire has collected his political columns in* Safire's Washington *(New York, 1980).*

INTRODUCING "ESSAY"

New York Times   *April 16, 1973*

When word spread like cooling lava through the Nixon Administration that I was to become a columnist for *the New York Times,* speechwriters who stayed behind wanted to know: "Will you continue to stand up for the President, the work ethic, and the Nixon doctrine, or will you sell out to the élitist establishment and become a darling of the Georgetown cocktail-party set?"

Sipping a bourbon in one of those dreaded drawing rooms—more dens of inequity than iniquity—I was asked substantially the same question by a new colleague in the press: "Will you speak out impartially without fear or favor, or will you continue to be a slavish, craven parroter of the Nixon line, a flack planted in our midst?"

I have never ducked the tough questions; my answer, in both cases, was "yes and no," which when delivered with crisp authority inspires confidence. Truth to tell, the only way the reader or the writer of this column will find the answer is to watch this space for further developments.

But not on opening day; before pulling a long face to deal with public affairs as befits a serious columnist, let me trot around the bases to get the feel of this place.

On flackery: A young, nervous aide of Henry Kissinger called me one day a couple of years ago to ask a strange question: "What does the word 'flack' mean?"

I was gratified to be consulted on a matter of meaning and etymology, a lifelong field of interest, but I had learned the first rule of bureaucratic survival: Never give out information without first finding out why it is being sought. So I misinterpreted the question and replied: "The word 'flak' is an acronym coined in World War II to describe anti-aircraft fire, from the German words *Fleiger Abwehr, Kanonen.*"

Moments later, the aide called back to say: "Dr. Kissinger says he doesn't need you to teach him German, but a columnist just called him 'an Administration flack' and he wants to know whether he should take offense."

With that background tucked away for use, I passed along the current usage of "flack": an apologist, or paid proponent, with a usually pejorative but occasionally madcap connotation. To cheer Henry up, I added that the role, if not the word, could be an honorable one—a skilled advocate was needed to explicate policy—but when I saw him next, he gloomily informed me: "I decided to take offense." Perhaps I will, too, someday—but not for a while.

On vogue words: Readers of these essays will not be bombarded with any of the "dirty dozen": relevant, meaningful, knowledgeable, hopefully, viable, input, exacerbate, dichotomy, the use as verbs of program, implement, and structure, and ambivalent, though I am of two minds about ambivalent.

On choosing a title: Columnists for *the Times* tend toward using generic titles, thereby laying claim to great chunks of subject matter. "Washington" is spoken for by James Reston, leaving another Washington columnist only the word "column" to head his piece, which is apt but a little pretentious for a newcomer. Arthur Krock's "In The Nation" is carried on by Tom Wicker, "Foreign Affairs" belongs to C. L. Sulzberger, and amused detachment is the province of Russell Baker's "Observer."

Anthony Lewis, with whom I will appear on this page every Monday and Thursday (we are paired like a couple of Senators whose votes are fated to cancel each other out), calls his column "At Home Abroad" when he is overseas and "Abroad at Home" when he writes from the United States. This leaves available only "At Home at Home," which is self-satisfied, and "Abroad Abroad," which must be angrily rejected as sexist.

I have chosen "Essay," which sounds innocuous enough but flies in under everybody's radar. The word might not have the verve, sparkle, and rallying power of, say, "The New Federalism," but it offers room to ruminate and holds out no false promise of total topicality, since I may want to fiddle around with some way-out subjects.

I'm pleased to meet you. To essay means to make a beginning, to try, to put to a test, which I undertake with zest and determination, along with appropriate false humility. I hope you have the need for another point of view, for I hope to have something to say.

THE POLITICS OF LANGUAGE

New York Times *May 7, 1973*

Back when we had a half million troops in Vietnam, an irate citizen wrote a letter to the White House addressed not to the President but to his speechwriters. Why did the President keep warning about "precipitous withdrawals," the writer wanted to know, when the correct adjective was "precipitate"?

Heatedly, the citizen correcting the President pointed out the difference: while both words come from "precipice," or brink, precipi*tous* means steep (as in "a precipitous slope") and precipi*tate* means abrupt or headlong (as in "precipitate action").

Accordingly, a "usage alert" was distributed throughout the White House four years ago, and from that day on we all inveighed against "precipitate withdrawals."

When the President spoke on television recently about Watergate, and cautioned, "I was determined not to take precipitate action . . ." at least one viewer out of the 73 million who were watching said to himself: "By George, he's got it!"

Words, words, words. Attention to the use of language betokens a concern for orderliness and precision, and the people who write to Presidents or essayists to set them straight about the use of a work are a hardy bunch, good-humored about being stern, like traffic cops at school crossings.

Occasionally, appreciators of English blow their cool: After four years of listening to the metaphor of the computer favored by Press Secretary Ron Ziegler—input, programmed, time frame—his use of "inoperative" in withdrawing past statements set a great many teeth on edge. That was understandable, but why did the use of "compassion" by Henry Kissinger, about those who may be involved in acts of zealotry, stir such passionate response from Administration critics? Mario Pei's *Double-Speak in America* reports that the word compassion "has now become a shibboleth of the moderate left . . ." and as such, its use on Nixonian lips drove longstanding compassionaries up the wall.

Nor were they the only ones to take offense in the linguistic crossfire over Watergate: Every time the call went out for a special prosecutor of *unimpeachable* integrity, Nixon men took it as a not-too-subliminal affront, and silently seethed.

By and large, however, word watchers introduce a note of good sense and good grace, of interested disinterest, to the action and passion of their times. The word for a thing is not the thing itself—a pig is only a marking of letters on paper and not a flesh-and-blood animal, or a slang derogation and not a real police officer. The sense of dealing with a subject once removed from reality gives the lovers of language both their balance and their self-image of proud eccentricity, as they try to brush back a flood with a whisk broom.

In that spirit of respect, I must disagree with the gentleman who chastises me for using "proven" when Fowler's *Modern English Usage* says the proper word is "proved"; on that usage, the great book is outdated. I take issue with the lady who categorically rejects all euphemisms, because that would have denied F.D.R. and J.F.K. the use of "quarantine" to soften the diplomatic impact of a blockade. Nor will I dispense with "élitist" until convinced "racist" and "sexist" will sink with it.

What is more (it's easy to get worked up over this), nothing is wrong with the acceptance of colorful, if transitory, metaphors like "to blow your cool" or "to drive him up the wall," just as nothing is wrong with chewing gum in public, provided you do so quietly and dispose of it in a suitable wrapper. However, the advertising community's slavish following of the youth culture's "into"—as in "to be into Bach"—turns me off, conveying the image of a violated composer. ("Turns me off," by the way, turns me on—but it is a similarly foolish affectation, with drug-related overtones, and I'd better cut it out.)

There is a point to all this: The language of politics has much to learn from the politics of language.

In the language of politics, the apt alliteration of an Agnew competes with the colorful imagery of a Connally ("Ah wouldn't trust that feller further than Ah could throw a chimney by its smoke," he once expostulated), which is healthy; but there also exists a sinister tendency to deceive, to make the fluffy ponderous and to make the heavy light, and worst of all, to exaggerate and inflame, making a difference of opinion appear to be a difference of principle.

In the politics of language, the prescriptive school (don't say different *than,* say different *from*) is different from the descriptive school, which holds that general usage dictates "correctness." One is strict, the other permissive, one conservative and the other liberal, and they clash. But they go for each other's minds, not for each other's throats. They sally forth in civility and come home in compromise.

"English is my native tongue," writes a man who does not like my prose style at all, complaining rationally about a piece in this space. "I read the article twice. I still have no idea what Mr. Safire is trying to tell me." And then he concluded: "In fairness, I have a head cold."

* * *

Dear Mr. Safire:

A man who tacitly presumes under the heading of "Essay" to emulate in matter and manner his eighteenth-century predecessors should be careful not to dangle his modifiers so provocatively before "appreciators of English"—especially in an essay on usage.

In "The Politics of Language," which appeared yesterday in *The Times,* the modifier "After four years of listening to the metaphor," etc., modifies something incapable of listening, namely, Ron Ziegler's "use of 'inoperative.'" It is not just misplaced for "teeth"; even a "great many teeth" are similarly incapable of listening. No matter how many years pedants have listened to danglers, this one is going to set their teeth on edge, as a great many may already have told you.

WEBSTER'S BIRTHDAY

New York Times *October 15, 1973*

Tomorrow, on October 16, lexicographers the world over will celebrate the 215th anniversary of the birth of Noah Webster.

The writers of dictionaries will face toward West Hartford, Conn., his birthplace, and affectionately retell the apocryphal story of the time that wordy worthy was embracing the chambermaid when his wife unexpectedly burst into the bedroom.

"Noah, I'm surprised!" Mrs. Webster expostulated. Whereupon the great definer coolly responded, "No, my dear. You are amazed. It is we who are surprised."

But October 16 must not be only a day of anecdotage and merry puns: It is a day, too, for thoughtful second soberings. This has been quite a linguistic year.

Ron Ziegler has much to answer for in the coinage of "the President misspoke himself"; his computer program terminology, from "input" to "time frame," came a-creeper with "inoperative."

This essayist stubbed his pencil on "centered around," which several readers circled and sent in with "There's a neat trick," and on another occasion I threw down a gantlet instead of a gauntlet. (I miswrote myself.)

Watergate writers made the long leap from "caper" to "horror"; the "plumbers" gained a meaning beyond that of George Meany's profession; and the metaphysical cliché of the year became "at this point in time."

The most frequently misspelled new word at the Senate hearings was "misprision," a precise legal word rooted in the French *méprendre,* to err, and quickly replaced by the easier-to-spell "coverup," which soon lost its hyphen and thus its virginity. (Why am I impelled to point out here that "Senator" and "senile" come from the same root?)

"Hardball" is probably the most useful new coinage in politicalingo, to differentiate between the tough but legal activity that our side engages in, from the "dirty trick" that your side perpetrates.

The translation error most frequently made in the last year was that the name "Segretti" means "secret" in Italian: No less an authority than Mario Pei, America's leading vocabularian, writing in the quarterly *Modern Age,* held that "Segreti (single t) is the term for 'secrets,' and the single or double 't' gives the two words different pronunciations and derivations." (Mr. Pei is of Italian descent; he is sometimes thought to be Chinese when his name is confused with I. M. Pei, the architect, who is.)

The most furious inter-columnar battle over the meaning of an Americanism took place between David Broder and Mike Royko, on the word "clout." Washington's Broder used the word as defined in my dictionary—"power, or influence"—and Chicagoan Royko clobbered the "brazen misuse" of a word he takes to mean not power, but purely influence: "Clout is used to circumvent the law, not to enforce it," wrote Mr. Royko. When Mr. Broder sent this to me with "I relied on your goddam book and look what trouble I'm in," I turned to the ultimate arbiter, Arthur Krock, down the hall.

"Clout means impact," said Mr. Krock, disagreeing with all of us. "If I say something, it has an effect on public opinion, it has 'clout.' If you say it, it does not." That's it, gentlemen: Mr. Krock was probably a personal friend of Noah Webster.

The most brilliant pun of the year (there are a few of these awards envelopes still to open) was in a *Wall Street Journal* headline about a solvent developed to clean oil slick off sea gulls, harmless but temporarily befuddling: "Leaves no tern unstoned."

The most profound punster was Dan Rather of CBS, with his assessment of the energy crisis, which began: "You can fuel some of the people all of the time . . ."

The best new word that fills a gap in the language was minted by architectural writer Ada Louise Huxtable to describe a happy marriage of form and function: "beautility."

The best political derogation was Governor Ronald Reagan's characterization of the expensive medical program put forth by Senator Edward Kennedy: "Teddycare."

And the grammatical error most frequently made reflects a sad state of affairs: indictments are never "handed down," they are only "handed up."

Indictment grammar and etymology are worth studying because they metaphorically preserve truths too soon forgotten: An indictment is an accusation handed up to a high bench where a judge sits, for further adjudication. The Latin derivations in oldest dictionaries give us the most up-to-the-minute political guidance: An indictment means "the writing down of a charge," and only a verdict means "the speaking of the truth."

Happy birthday, Noah.

* * *

Oct. 15, 1973

Dear Bill Safire:

I enjoyed your Noah Webster/word/phrase column today (being so constituted that such matters interest me far more than the state of the Union).

The story you tell about Noah being surprised and his wife amazed I had always heard as being attributed to Dr. Johnson. With it, I always heard, went this one:

Dr. Johnson was, of course, notorious for not being overfond of clean linen, or bathing. At a dinner party once he sat next to a young lady who, after a while, couldn't help remarking: "Doctor, you smell."

"Ah, no, my dear," says Dr. Johnson. "*You* smell. *I* stink."

Totally apocryphal, no doubt, but it does—like your Noah Webster story—illustrate the nice distinctions between words which keep getting lost, to everyone's loss.

I was also interested in the business about indictments being hand up and not down. I've noticed that error repeatedly, and sometimes in *The Times.* Still—venturing to quibble—is that a *grammatical* error, as you say? More a mistake in usage, I should have thought.

Regards,
Dick Hanser
Mamaroneck, New York

Ellen Goodman

It has been the ill-luck of Ellen Goodman (1941–) to be promoted as the house feminist by editors and packagers of Op-Ed pages. Granted, she does deal with "motherhood, fatherhood, Walter Cronkite, fear of tunnels, falling asleep at late dinner parties, impotence, living alone," as the Boston Globe *noted in nominating her for a Pulitzer Prize (which she received in 1980). But doing so scarcely makes her a reflexive feminist, and in any case, as her editors noted, she also deals readably with national and international politics. A graduate of Radcliffe, she worked in the 1960s for* Newsweek, *the* Detroit Free Press, *and the* Globe, *beginning her column in 1974. Her pieces are collected in four books; these are taken from* At Large *(New York, 1981).*

RATING OUR KIDS

Boston Globe *February 1979*

The problem was that she had been visiting New York children.

Not that they looked different from other people's children. They didn't. They were even disguised in blue jeans so that they could pass in Des Moines or Tuscaloosa. Yet the New York children she met always reminded her of exotic plants encouraged to blossom early by an impatient owner.

She had spent one night with two thirteen-year-old girls who went to a school where they learned calculus in French. The next night she met a boy who had begged his parents for a Greek tutor so he could read the *Iliad*. He was nine.

This same boy came downstairs to announce that the Ayatollah Khomeini had just come to power. What, he wondered, would happen to Bakhtiar? Or oil imports?

She wondered for a moment whether this person wasn't simply short for his real age, say, thirty-two. But no, he was a New York child. She'd known many of them at camp and college. They wrote novels at twelve, had piano debuts at fourteen, impressed M.I.T. professors at sixteen. (She was, of course, exaggerating.)

The woman was thinking about all this on the plane. She was flying her flu home to bed. Sitting at 16,000 feet, enjoying ill health, it occurred to her that she was guilty of the sin of child rating.

It wasn't New York children; it was other people's children. It wasn't awe, she knew deep in her squirming soul; it was the spoilsport of parenting—comparing.

Her own daughter was reading *My Friend Flicka* while this boy was reading the *Iliad*. Her daughter regarded math as a creation of the marquis de Sade, and she was taught in English.

The girl could plot the social relations of every member of Class 5-C with precision, but she couldn't pronounce the Ayatollah to save her gas tank. At ten, she was every inch a ten year old.

The woman hunched down in her seat, wrapping her germs around her like a scarf. When her child was little, the woman had refused to participate in playground competitiveness, refused to note down which child said the first word, rode the first bike. She thought it inane when parents basked in reflected "A's" and arabesques and athletics.

Now she wondered whether it was possible for parents not to compare their children with The Others. Didn't we all do it, at least covertly?

In some corner of our souls, we all want the child we consider so extraordinary to be obviously extraordinary. While we can accept much less, at some point we have all wanted ours to be first, the best, the greatest.

So, we rank them here and there. But we are more likely to notice their relative competence than their relative kindness. More likely to worry about whether they are strong and smart enough, talented enough, to succeed in the world. We are more likely to compare their statistics than their pleasures.

And we are always more likely to worry. She had often laughed at the way parents all seek out new worry opportunities. If our kids are athletic, we compare them to the best students. If they are studious, we worry that they won't get chosen for a team. If they are musical, we worry about their science tests.

If they are diligent, we compare them to someone who is easygoing. And if they are happy, we worry about whether they are serious enough. Perhaps the boy's parents wish he could ski.

We compare our kids to the ace, the star. We always notice, even against our will, what is missing, what is the problem. We often think more about their flaws than their strengths. We worry about improving them, instead of just enjoying them.

She thought about how many parents are disappointed in some part of their children, when they should fault their own absurd expectations.

The woman got off the plane in the grip of the grippe. She got home and crawled into bed. The same daughter who cannot understand why anyone would even want to reduce a fraction brought her soup and sympathy. Sitting at the edge of the bed, the girl read her homework to her mother. She had written a long travel brochure extolling the vacation opportunities of a week in that winter wonderland the Arctic Circle.

The mother laughed, and felt only two degrees above normal. She remembered that Urie Bronfenbrenner, the family expert, had once written that every child needed somebody who was "just crazy" about him or her. He said, "I mean there has to be at least one person who has an irrational involvement with that child, someone who thinks that kid is more important than other people's kids."

Well, she thought, most of us are crazy about our kids. Certifiably.

A GIFT TO REMEMBER

Boston Globe *December 1980*

The man stood in the checkout line, holding onto the new bicycle as if it were a prize horse. From time to time, he caressed the blue machine gently, stroking the handlebars, patting the seat, running his fingers across the red reflectors on the pedals.

His pleasure, his delight, finally infected me. "It's a beautiful bike," I said to him, shifting my own bundles.

The man looked up sheepishly and explained. "It's for my son." Then he paused and, because I was a stranger, added, "I always wanted a bike like this when I was a kid."

"Yes," I smiled. "I'm sure he'll love it."

The man continued absentmindedly handling his bicycle, and I looked around me in the Christmas line.

There were carts and carts full of presents. I wondered what was really in them. How many others were buying gifts they always wished for. How many of us always give what we want, or wanted, to receive?

I've done it myself, I know that. Consciously or not. I've made up for the small longings, the silly disappointments of my own childhood, with my daughter's. The doll with long, long hair, the dog, the wooden dollhouse—these were all absent from the holidays past.

I never told my parents when they missed the mark. How many of us did? I remember, sheepishly, the tin dollhouse, the parakeet, the doll with the "wrong" kind of hair.

Like most children, I was guilty about selfishness, about disappointment. I didn't know what gap might exist between what my parents wanted to give and what they could give . . . but I thought about it.

I knew they cared and, so, even when it wasn't exactly right, I wanted to return something for my gift. I wanted to please my parents with my pleasure.

But standing in that line, I thought about what else is passed between people. Gifts that come from a warehouse of feelings rather than goods.

Maybe we assume other people want what we want, and try to deliver it. Maybe in every season, we project from our needs, we gift-wrap what was lacking in our own lives.

My parents, descendants of two volatile households, wanted to give us peace. They did. But I am conscious now of also giving my child the right to be angry. In the same way, I know parents who came from rigid households and busily provide now what they needed then: freedom. They don't always feel their children's ache for "structure."

I know others who grew up in poor households and now make money as a life-offering for their families. They don't understand when it isn't valued.

There are women so full of angry memories of childhood responsibilities that they can't comprehend their children's wish to help. There are men so busy making up for their fathers' disinterest that they can't recognize their son's plea: lay off.

Every generation finds it hard sometimes to hear what our children need, to feel what they are missing, because our own childhood is still ringing in our ears.

It isn't just parents and children who miss this connection between giving and receiving. Husbands and wives, men and women, may also give what they want to get—caretaking, security, attention—and remain unsatisfied. Our most highly prized sacrifices may lie unused under family trees.

Of course there are people who truly "exchange." The lucky ones are in fine tune. The careful ones listen to each other. They trade lists. They learn to separate the "me" from the "you." They stop rubbing balm on other people to relieve their own sore spots.

Perhaps the man in line with me is lucky or careful. I saw him wheel his gift through the front door humming, smiling. For a moment, I wondered if his son hinted for a basketball or a book. This time, I hope he wanted what his father wanted for him.

John Leonard

Not long ago, it would have been unthinkable in the New York Times *or other mainstream papers for columnists to dwell on such personal matters as divorce, pregnancy, infidelity, and professional insecurities. The way was opened in weeklies like* The Village Voice, *and the* Times *followed with "Private Lives" by John Leonard (1939–). For two years starting in 1976, Leonard dealt with such private matters as the reaction of friends to his second marriage ("Again? What for?") and the terrors of a power party. His pieces were collected in* Private Lives in the Imperial City *(New York, 1979).*

NEXT

New York Times *1976-77*

You are thirty-seven years old, which, according to Aristotle, is the ideal age for a man to marry. And so you have decided to marry, for the second time. It is odd how much morbid interest such a decision excites in a city otherwise prepared to tolerate almost any bizarre relationship among animals, vegetables and minerals. Marry? Again? How quaint! What for? Friends walk around your behavior, poke at it with a stick; take snapshots, as though they were anthropologists happening upon a Druid or the Ik.

If those friends are full of literary vapors, they will quote Dr. Johnson on second marriages: "The triumph of hope over experience." Or Sheridan: "Zounds! madam, you had no taste when you married me!" Or Ambrose Bierce:

> They stood before the altar and supplied
> The fire themselves in which their fat was fried

To which you may want to reply with a Dutch proverb: In marrying and taking pills it is best not to think about it too much.

But of course you have thought of little else as you've gone about your rounds of finding a ring, and finding a doctor who will bleed you and sign a bill of adequate health; and finding a clerk at City Hall who will sell you permission; and finding a judge who will come to your home and who will agree not to quote Kahlil Gibran in the service.

There are reasons of state for marrying a second time, instead of persisting in relationships (which are to marriage what gum is to nutrition): income tax returns, life insurance, health insurance, passports, joint checking accounts, custody of Bloomingdale's, all those slips of paper, cards of identity, by whose arithmetic one adds up to a personality in the modern world. Let's keep it simple for the computers.

There are moreover, two sets of children to consider. It would be better for them at school if, on explaining their domestic arrangements, they didn't sound like innocent bystanders in a particularly messy covert operation of the Cold War. Perhaps you think they would prefer to be elsewhere on the appointed Saturday. Don't let them be anywhere else. A son shouldn't go to the movies while his father is being married. If they stand around variously impersonating bookends, alarm clocks, ghosts, sentries, tourists or refugees, give them a poem to read or some flowers to hold. They are important witnesses. It isn't necessary that they altogether understand you; it is necessary that they take you seriously.

Whereas, bubonic plague wouldn't keep the two sets of parents, yours and hers, away from this occasion. They couldn't care less whether you are going through a mid-life "crisis" or a mid-life "passage"—so long as you're married. They will fly in from Bangkok or Reykjavik, to dust the furniture and hope for the best. They have been through these revolving doors before; you might even have attended one or two of *their* weddings. Let's keep doing it until we get it right. Along with the wheels of cheese and potted plants, wads of tissue and baby pictures, they will bring you a blank check on their loyalty.

As for your friends—those strangers, like your second wife, that you have taken into your house without the advice and consent of your parents—they will as usual finish off your Scotch, steal your books, burn holes in your patience, and make you laugh. They tiptoe up and down the stairs because, on the next landing or just coming out of a closet, there might be an ex-husband or an ex-wife or, for that matter, the Sullivanian analyst on the West Side to whom half of them had gone when they were sad. How unlike the first time, when we were all promising, confident, stupid, graduate students of ourselves, unblotched copybooks. Your friends now are halfway through the novels of their lives, and worried about the next couple of chapters. And yet, after all the jokes, they are genuinely happy for you, for the surprise twice in your plot. Weddings are nice; maybe it's part of the nostalgia craze.

That first time, everything was supposed to be grand. Someone had wanted to roast an ox on a spit in a pit of coals, and for there to be bagpipes in the apple trees and fox-trotting under the great elms and an acre of sunlight like a shield reflecting off the wineglass and the brass buttons of the blue blazers, as though we were advising Mars by semaphore of our golden youth.

Not so on a November Saturday in the city of New York in 1976. No bagpipes. Maybe a harpsichord or a guitar. Casual clothes. Scraps of Donne and Yeats:

> But let a gentle silence wrought with music flow
> Whither her footsteps go.

Is this a caution? Having domesticated the beasts of pride, does one these days walk one's luck warily around the block on a leash, instead of bragging about it? Is it safer to settle for what W. S. Gilbert called a "modified rapture"?

No. Having earned your modesty, this time be careful. You marry, not for reasons of state or children or parents or friends, but for yourself and the one you love, in gratitude. You are setting up a sanctuary, and know that it is fragile. You hope that you are finally a serious character in your own life, that everyone will believe this time you are and will behave like an adult, instead of one of those toys on the streets with the keys sticking out of their pineal glands, doing mindless damage. To quote Yeats again:

> How but in custom and ceremony
> Are innocence and beauty born?

You marry to be worthy of a gift, and want to say so out loud, but without shouting. One doesn't shout a prayer. Marriage is one of the few ceremonies left to us about which it is impossible—or at least self-demeaning—to be cynical. Then, in secret, rejoice.

THE POWER PARTY

New York Times *1976–77*

It will happen by accident. No one will have warned you. Why should they? They have their own erotic depths and nasal passages to worry about. You will leave the safety of your typewriter, the sanctuary of your home—and perhaps a provocative book, only half read, on genetic engineering or how to process cannabis in a Cuisinart. Going out is a way of not watching *Scenes from a Marriage* on public television. What you expect is an unlacing of the mind among loose friends. Off with the boots of duty! Let us talk small, gnaw chicken and risk a giggle.

But this is not to be. You know it is not to be at the door of your friend's apartment, which is opened by one of those servants rented for the evening from a bat cave in surprising Queens. He will hide your raincoat, sneer at your blue jeans and fetch you a drink. It is, then, a serious party. At serious parties, New Yorkers are not expected to fetch their own drinks; having to move from one side of a room to the other would cramp us in our seriousness.

On the other hand, batpeople tend to pin you into conversational corners. You can't, when some aggressive gnome starts in on psychoanalysis or astrology, excuse yourself to freshen your drink. The fresh drink materializes at your elbow, like a character defect.

Nor are you able to employ what is known in select circles as the Michael Arlen gambit. The Arlen gambit is to arm yourself at the bar with a couple of your own drinks. Two-fisted, then, you can make your escape from any ambush by maniacal bores: "I'm sorry, I really would like to hear more about the basic engram and operating thetanism, but Duchess Pittsburgh is waiting for her toddy of goat's bile." (It is, to be sure, a tricky gambit for smokers. Both hands occupied, what are you to do with the burning weed in the middle of your food hole? If only ears had thumbs.) A bat-person isn't going to fetch you two drinks at the same time, not unless he steals your ashtray.

You allow your eye to graze on the pasture of those present, and your bonhomie evaporates.

Can celebrityhood be said to glower? As if behind sandwich boards advertising their own famous names, they sit on sullen stools. It is wall-to-wall pout. See the famous TV anchorman, the famous magazine editor, the famous newspaper columnist, the famous courtesan. There are several novels, a Broadway play, half a Cabinet, two banks, one football team, a jazz musician and an Englishman. They are waiting for the latest edition of themselves, for the reviews, of an energizing principle.

There are too many celebrities and not enough sycophants. Such a discrepancy will be hard on the husbands and wives of the illuminati. Somebody has to grovel and sigh. Somebody has to be shouldered aside so that oaks may huddle before they're felled. This is more than serious. This is power, baby. This is the sort of party where you have to decide whether you are important enough to wait for the other fellow to say hello, or you have to say hello to the other fellow first.

Careers, like camels, hunker down to snooze with a wary eye on which way the fan is blowing.

You have been here before. It is always a mistake. It was a mistake three years ago, schlepping out to the Hamptons because you had never been there before, and how unworldly it was not to have seen the famous artists and writers behaving like debauched gazelles. And so you ended up in Sag Harbor as an uninvited guest at a stranger's birthday party, for which they imported Bobby Short.

Upon you they laid lobster and a grape of France and a What's-your-angel, sapajou? For yourself, you apologized. For the fifteen minutes they were famous, they famed and famed and famed. We end up, as Kurt Vonnegut has said, licking the boots of psychopaths.

The sad fact is that, taken singly, celebrities are interesting; in herds they low. It is a collective goiter; you want to give them a pill to reduce the swelling. Why are they driven to perform? In our art and science, the magic of our money-making, our sneaking politics, we adumbrate and counterfeit. A self is suggested. An image is projected.

Our children know better, but what a burden it is to pretend to be what we have made, to try to live up or down to an idea of us arrived at in some committee meeting or in desperation. It was, after all, just one idea among many.

What a trial to be Norman Mailer, Billy Carter, Farrah Fawcett-Majors; to have to grow a personality along the lines of the one you invented, the one that sold; to have to compete with other fabricated personalities, inflations of cunning, blimps of ego; to jostle at a power party. The more one distends, the more easily one is bruised. Friends would have forgiven you, instead of expecting a performance. You should have stood in bed, buttering your toast.

Suppose you are not clever, except at a typewriter. Suppose you are not sensitive, except in the middle of the night a month later. Suppose your generosity is theoretical, your courage wholly literary, your fast ball lacking jumping beans, your heartbreak is psoriasis. Suppose, deep down, you suspect that you are dull, and your public works are a form of vengeance. You talk a good poem, and think by numbers. Once upon a time, you were interesting; then Mother died and you had to give it up.

Friends ought to know who you really are, and invite you to dinner anyway. It needn't be served in a room where the walls are always white, the steel always stainless, the chicken always two hours late and the children stuffed into some hamper. Nor need it be oversubscribed with bylines who deep down in the Cuisinart feel just as fraudulent as you do, who do not at dawn presume to be wise. There is no wisdom; there are only punch lines.

Erma Bombeck

Writing well has been the best revenge for Erma Bombeck (1927–   ), laureate of suburbia, who started her column for $3 each in the Kettering-Oakwood Times *near Dayton, Ohio. This was in the early 1960s, when Bombeck was "too old for the paper route, too young for Social Security and too tired for an affair." She wrote on a plywood sheet slung across cinder blocks while cooking for, and tending to, several children and husband. In 1965, the* Dayton Journal-Herald *proposed a twice-weekly column; syndicated by Newhouse, it now appears in some 900 newspapers. Her books are bestsellers, their titles passing by osmosis into the language, e.g.:* The Grass is Always Greener Over the Septic Tank *(New York, 1976). The Bombecks have sold their contractor house and live in a hacienda near Phoenix, but the column has kept on the course fixed by Erma Fiste, the working-class girl who attended the University of Dayton, met Bill Bombeck, and found she had a talent.*

PRIMER OF GUILT: "BLESS ME, EVERYBODY, FOR I HAVE SINNED"

Motherhood, the Second Oldest Profession 1983

A

ABandoning children and responsibility, leaving them helpless and alone with a $200 babysitter, a $3,700 entertainment center, a freezer full of food, and $600 worth of toys while you and your husband have a fun time attending a funeral in Ames, Iowa.

B

BUying a store-bought cake for your son's first birthday.

C Cursing your only daughter with your kinky red hair and your only son with your shortness.

D Dumping cheap shampoo into a bottle of the children's Natural Herbal Experience, which costs $5 a throw.

E Explaining to "baby" of family why the only thing in his baby book are his footprint, a poem by Rod McKuen, and a recipe for carrot cake.

F Flushing a lizard down the toilet and telling child it got a phone call saying there was trouble at home.

G Going home from the hospital after hysterectomy and apologizing to kids for not bringing them anything.

H Hiding out in the bathroom when the kids are calling for you all over the house.

I Indulging yourself by napping and when caught with chenile marks on your face telling your children it's a rash.

J Jamming down the sewer three newspapers you promised your child with a broken arm you'd deliver for him.

K Keeping Godiva chocolates in TEA canister and telling yourself kids don't appreciate good chocolate.

L Laundering daughter's $40 wool sweater in hot water.

M Missing a day calling Mother.

N Never loaning your car to anyone you've given birth to.

O Overreacting to child who found your old report card stuck in a book by threatening to send him to a box number in Hutchinson, Kansas, if he talks.

P Pushing grocery cart out of store and forgetting baby in another cart inside until you have turned on the ignition.

Q Quarreling with son about homework only to do it for him and getting a C on it.

R Refusing to bail out daughter who lives by credit cards alone.

S Sewing a mouse on the shirt pocket of son who is farsighted and telling him it's an alligator.

T Taking down obscene poster from son's bulletin board just before party and substituting brochure for math camp.

U Unlocking bathroom door with an ice pick when a child just told you he'd not doing anything only to discover he's not doing anything.

V Visiting child's unstable teacher at school and telling her, "I don't understand. He never acts like that at home."

W Writing a postdated check to the tooth fairy for a buck and a half.

X X-raying for a swallowed nickel only after you heard it was a collector's coin worth $6.40.

Y Yawning during school play when your daughter has the lead—a dangling participle.

Z Zipping last year's boots on your son when you know they will never come off without surgery.

EVERYBODY ELSE'S MOTHER

Motherhood, the Second Oldest Profession 1983

She has no name. Her phone number is unlisted. But she exists in the mind of every child who has every tried to get his own way and used her as a last resort.

Everybody Else's Mother is right out of the pages of Greek mythology—mysterious, obscure, and surrounded by hearsay.

She is the answer to every child's prayer.

Traditional Mother: "Have the car home by eleven or you're grounded for a month."

Everybody Else's Mother: "Come home when you feel like it."

Traditional Mother: "The only way I'd let you wear that bikini is under a coat."

Everybody Else's Mother: "Wear it. You're only young once."

Traditional Mother: "You're going to summer school and that's that."

Everybody Else's Mother: "I'm letting Harold build a raft and go down the Ohio River for a learning experience."

A few mothers have tried to pin down where Everybody Else's Mother lives and what background she has for her expertise on raising children. They have struck out.

The best they can come up with is a composite that they put together by pooling their information.

As close as can be figured out, Everybody Else's Mother is a cross between Belle Watling and Peter Pan. She likes live-in snakes, ice cream before dinner, and unmade beds. She never wears gloves on a cold day and voted for Eugene McCarthy. She is never home.

She's never been to a dentist, hates housework, and never puts her groceries away. She sleeps late, smokes, and grinds the ashes into the carpet with the toe of her shoe.

She eats jelly beans for breakfast, drinks milk out of a carton, and wears gym shoes to church because they're comfortable. She never washes her car and doesn't own an umbrella.

Everybody Else's Mother moves a lot and seems to be everywhere at the same time. Just when you think she's moved out of the neighborhood, she reappears. She is quick to judge and has handed down more decisions in her time than the Supreme Court has in the last 200 years. She has only one child and a friend who "was a dear" carried it for her.

She has never used the word "No."

If Everybody Else's Mother showed up at a PTA meeting and identified herself, she would be lynched.

From time to time, the existence of Everybody Else's Mother is questioned. It is probably wishful thinking. Does she exist?

Oh yes, Virginia, she really does. She lives in the hearts of children everywhere who have to believe that somewhere there is an adult on their side. Someone who remembers the frustration of needing to belong to a peer group at some time of your life to do the forbidden . . . just because it's there.

Just because one has never seen her, does that prove she's not there? Would one question the existence of monsters that appear in bad dreams or tigers that crawl on the bed in the darkness and disappear when the lights go on?

Bill Vaughan

Bill Vaughan (1915–1977) was a local wine known and praised by travellers like Charles Kuralt of CBS and appreciated by distant colleagues like Russell Baker. Born in St. Louis and a graduate of Washington University, he joined the Kansas City Star *in 1939 and was a columnist (''Starbeams'') from 1945 until his death. His geniality was redeemed by a saving asperity in columns collected in* Bird Thou Never Wirt *(New York, 1962),* Half the Battle *(New York, 1963), and* Sorry I Stirred It *(New York, 1964).*

WITH A MUSE LIKE THIS, WHAT'S A WRITER TO DO?

Kansas City Star *circa 1960s*

I had a midnight visit recently from my Muse, a large and roseate lady who smokes cigars. I wish she wouldn't make these calls at unconventional hours. It causes talk among the neighbors, what with the chariot parked in front of the house and all. Still, she is one of the few Muses that make house calls any more. It sort of restores your faith in the profession.

As we know from reading the ancient poets, Muses were always rapping on their chamber doors with inspiration. Nowadays when you see a poet he is hurrying to the classroom or the library, which is where the Muses have their office hours. You have to go to them. Thursday is their day off.

I suppose this is one reason I stick with this particular Muse. Anyway, she perched on the side of the bed. Thank goodness nobody else woke up. You know how loosely wrapped Muses are. A perfectly respectable class of girl but always giving the impression that they are about to fall out of their clothes.

Explanations can be difficult: "Dear, I believe you know my Muse."

Chilly glances are exchanged.

This is apart from the subject. A writer can't operate without a Muse, and if there are misunderstandings, why it's all in the game.

The big trouble with this Muse, though, is that she has such punk ideas. Do you know what she woke me up for?

She said, "Sam, I'll bet you will write something about National Pickle Week."

And I said, "I'll bet I won't."

"There are possibilities," she said, relighting her cigar. "You can make fun of it, for example. You can say that it is ridiculous. I mean, stacked up against Brotherhood Week or Mental Health Week, Pickle Week seems pretty frivolous."

"I don't know about that," I said. "Year-round I see more pickles than I do Brotherhood or Mental Health. Measured by results alone, it seems to me that Pickle Week does a better job of pushing the product than Brotherhood or Mental Health weeks."

"See," said my Muse. "I knew you could write something about Pickle Week. You came up with the angle, Sam baby."

"No," I said. "I'm not going to write that Pickle Week is better than Brotherhood Week or Mental Health Week because I believe in Brotherhood and Mental Health more than I do pickles."

"Say," she said, "you might do a combination piece on Brotherhood and Pickles. It would be a catchy slogan: 'Take a Pickle to Lunch.'

"Do you know what the theme of National Pickle Week is this year?" my Muse went on.

"No, and I don't want to."

"Traffic safety," she said. "They have a bumper sticker alerting the people that for maximum safety they should drive 150 pickle lengths behind the car ahead."

"What sort of pickles," I asked, slightly interested, "gherkin, dill or watermelon?"

"Aha," she said, "you are slightly interested. That's good. The kind of pickle doesn't matter. The point is that it's in the cause of traffic safety, which is at least as important as Brotherhood and Mental Health."

"I'll take care of it during Traffic Safety Week," I said.

"People," she said, "will think you are anti-safety."

"The president of the high-level Business Council," I pointed out, "says that safety is a fad, like the hula hoop."

"There's the angle, Sam baby," said my Muse. "Get it down on paper that safety is a sometime thing, but pickles are forever."

"I won't," I said. "I won't write a thing about National Pickle Week."

And my Muse smiled.

"Want to bet, Sam baby?" she murmured drawing her draperies around her and departing, leaving only a faint aroma of cigars to be, somehow, explained.

THE FOOTBALL INTERVIEW

Kansas City Star *circa 1962*

Q Well, coach, are your Behemoths in good condition for the annual Fudge Bowl New Year's Day classic which is expected to jam the stadium to capacity besides entertaining fans across the land via the nationwide teevy?

A Thank you, Ed. First of all I might say that our team is not the Behemoths. That is the other team, I think. We are the Basset Hounds, as is well known by fans across the land who will be watching us in the Fudge Bowl classic on New Year's Day either in person or through the nationwide teevy. It is nice to be here, Ed.

Q Thank you, coach, and I do want to apologize for calling your team the Behemoths instead of the Beagle Hounds.

A Basset Hounds.

Q Basset Hounds. It's just that there are so many bowl games this year and I'm not the regular sports man around here anyway and I got stuck with this interview because the regular sports man—but never mind about him. Anyway, coach, are your charges in good condition?

A My charges are in terrible condition, due to my wife's going crazy and buying everything in the stores. Ha. Ha.

Q By charges, coach, I mean your players.

A I was being light, Ed. But seriously, Ed, are you asking about the condition of my players?

Q Yes, coach. I think we would all like to have a little insight into whether All-American Honorable Mention, the rifle-armed and game little quarterback Digby Dabbs, will be at his rifle-armed peak.

A Ed, let me say this about Digby. He is a fine boy and he will give 100 percent.

Q Well, coach, the fans will be glad to get this good news about Digby Dabbs.

A What good news? He's a fine boy who will give 100 percent, but he has a broken leg.

Q Oh, well, uh, coach, about New Year's Day. Would you care to pick a winner?

A Yes, I would care to pick a winner.

Q Well:

A Well, what?

Q Well, so pick a winner.

A O.K. Missouri in the Orange Bowl.

Q No, no, coach. I mean a winner in the Fudge Bowl.

A Frankly, Ed, between us and the Behemoths I don't think there will be a winner in the Fudge Bowl.

Q Ah, speaking of the Behemoths—have you scouted them or seen their game movies?

A We have seen their game movies and, frankly, they are the worst movies since Lionel Atwill in *The Vampire Bat.* You couldn't tell the good guys from the bad guys. Terrible movies. Half the time I didn't even know where the ball was.

Q What kind of team do the Behemoths have?

A Well, Ed, let me say this—they play hard-nosed football.

Q What kind?

A Hard-nosed.

Q What does that mean exactly?

A I don't know. I read it in the newspaper.

Q What sort of strategy have you planned?

A We will play soft-nosed football, Ed.

Q Soft-nosed?

A Yes, everybody else has played hard-nosed football this year. We figure that if we play soft-nosed football it will sort of catch the Behemoths by surprise.

Q And well it might.

A I'm the only coach left in the game who still teaches soft-nosed football, but I think the others will come back to it.

Q The pendulum—

A Will swing. Yes.

Q Speaking of coaches, how about Coach Pug Underslung of the Behemoths?

A Well, there's this to be said about the Behemoths—they play hard-nosed football, but they are very poorly coached. What's the matter, Ed? You look pale.

Q Twenty-five years in this business, and you're the first coach I ever heard say that the opposing team was badly coached.

A Don't get me wrong, Ed. As a man, as a brother-in-law, Underslung leaves a lot to be desired, but as a coach he's a mess. A hard-nosed mess.

Q Do you have any parting tips for the fans, coach?

A Yes. Watch the Rose Bowl.

THE FRIENDLY SKY, IF YOU EVER GET INTO IT

Kansas City Star *circa 1960s*

Whatever we may think of the ethics of the sixteen-year-old who flew from Kansas City to London without a ticket, apparently by picking the right flights and flashing an empty loading envelope, we must be impressed by the aplomb of the generation he represents.

It is a quality which I, at least, excessively lack. Getting on an airplane is an ordeal.

In the first place, I am obsessed with the idea that I am about to miss my flight. Sitting in the coffee shop, an hour before the scheduled departure, I see a sign on the wall: "No Flight Announcements." It has the same effect on me as though it said "Cholera." I gulp my coffee and rush outside where I can hear the loudspeaker, which I never understand.

It says, "Passenger Wahh W. Wahh for Wahhwahh, please come to the Wahh Wahh Airline ticket counter. Passenger Wahh."

I rush up to the nearest uniformed figure and surrender:

"What did I do wrong? I'm sorry. If there's anything extra I'll pay."

It turns out, of course, that it is not my airline or my destination and that the man they want is indeed a Passenger Wahh. I have never, incidently, seen any of the Passengers Wahh, but they are a family that travels extensively. At least they always seem to be going somewhere on the days when I am in an air terminal.

Even more chilling is when the horn says, "Wahhwahh Airlines announces the departure of its Slippery Elm Green Carpet Flight Wahh to Wahh. Leaving from Gate 9, Concourse G, Finger 112. All passengers please board."

Great Scott!

It doesn't sound like my flight, but it's leaving, while I'm standing here buying a newspaper. I don't dare take a chance. Gate 9, Concourse G, Finger 112 is so far away it's in a different time zone, so I gain an hour but I would have missed the plane even if it had been mine which, I discover, when I get there, it isn't.

Most air passengers only check in at the ticket counter once. I do it two or three times. How do I know that the last fellow who checked me in was really a ticket person? Anybody can fake a uniform. Maybe he was a sixteen-year-old boy trying to get to London.

I go back for reassurance. They get so they remember me and say that they don't have to check my ticket again. I beg them to do it anyway. The last man who stamped it was using a wornout ink pad. Suppose I get thrown off the plane because the stamp is illegible? They say there is really nothing more they need to do. I invite them to get my luggage back and weigh it again. They say it won't be necessary.

When the time comes to board the plane, I check in at the gate. The man says, "Thank you, sir," and I say, "Does this plane go to Chicago?" He says, "No, why?" and I say, "Well, so many of them do. Go to Chicago that is, and I'm not. I'm going to Los Angeles to see my sister's oldest boy who's in the hardware game."

He says this plane is going to Los Angeles, but I check it again at the foot of the gangplank with a man who has mufflers on his ears. I yell, "Los Angeles?" at him and he nods. At the top of the stairs I show my ticket to the stewardess and explain to her about the wornout rubber stamp. It's pretty noisy but she nods and says, "Wahh."

What does she mean by that? It worries me. I go up and down the aisle asking everybody if they are going to Chicago. They all say they aren't, which reassures me. But after I've asked them if they're going to Chicago they all take their tickets out and look at them again, which starts me worrying all over.

Well, of course, I get to Los Angeles all right. But it just proves one of the handicaps of being born before the age of air.

That kid stowed away easier than I can fly with a paid-up ticket.

Lewis Grizzard

Detractors insist that the cleverest works of Lewis Grizzard (1946–) are his book titles, i.e.: Elvis Is Dead and I Don't Feel So Good Myself. *But Grizzard (pronounced grizzARD like lard, not GRIZZard like lizard) is a folk hero to readers who also grew up poor in the South and readily recognize his stock characters, Bubba and Earl. Grizzard was born at Fort Benning, Georgia, bred in Moreland, Georgia, and attended the University of Georgia. At twenty-three, he was sports editor of the* Atlanta Journal, *moving on as a sportswriter for the* Chicago Sun-Times; *he returned to Georgia in 1977 and started his column in the* Constitution. *His books, originally published by the Peachtree Press, have become underground bestsellers. His columns are collected in* When My Love Returns from the Ladies Room, Will I Be Too Old to Care? *(New York, 1987), from which the following are taken.*

LUNCH WITH DANNY AND DUDLEY

Atlanta Constitution *circa 1980s*

Moreland, Georgia—I was having lunch here in my hometown with the folks, and Dudley Stamps and Danny Thompson, both of whom still live in these parts, dropped by.

We were boys together here, but we don't see very much of each other anymore. Twenty years ago we were inseparable. Then one day, I went my way and they went theirs.

Danny's hair is turning gray. Dudley is losing his. They both have good jobs and families. They seem happy.

So we had this idea after lunch, and that was to take a walk together. Grown men rarely take walks together, but the weather was nice and since we were in the midst of reminiscence anyway, it seemed the thing to do to take a walk around the little town from which we sprouted.

We walked slowly, and we stopped often. We told some old stories and we had us some laughs.

We were walking through what used to be my grandmother's yard where we played together before we learned our multiplication tables.

"Mama Willie's yard doesn't seem nearly as big as it did back then, does it?" said Danny.

It didn't. What, I wondered, is the shrinking agent in time?

We walked up to the Methodist church. The vacant lot in front of the church was where we played touch football.

The lot isn't vacant anymore. Somebody poured some concrete on it and put up a fence.

We walked down to Cureton and Cole's store, or to the building that used to be Cureton and Cole's store, where we met each afternoon after school and drank big orange bellywashers and ate Zagnut candy bars.

Cureton and Cole's store is now home to some sort of interior decorator. That hurt.

The post office isn't the post office anymore, either. It's a beauty salon, and they're trying to refurbish the old hosiery mill next store and make it into a museum.

We remembered the Fourth of July street dances they used to hold in front of the old hosiery mill.

"They quit having them," Danny said, "when folks got to drinkin' and fightin'."

"They're trying to bring them back, though," said Dudley. "Now, they smell your breath before they'll sell you a ticket."

We walked up what was left of the old path that leads to the schoolhouse. Danny peeked through one of the windows at the room where we spent our eighth grade year.

"Dang if that sight don't pull my stomach," he said.

We had to go to the old ballfield. Dudley was our catcher. Danny played first. I pitched.

Even the ballfield wasn't the same. They've put home plate where right field used to be, and somebody tore down the tree that provided the shade for the home team bench.

The walk was over much too quickly. Back home, we talked about the inevitability of change and how they should have left our ballfield the way it was.

Then we shook hands and said we ought to do this sort of thing more often, which we won't, of course. But at least we had this day, the day three grown men walked back through their childhood together.

I wish I had told them how much I loved them before they left. But you know how grown men are.

WHAT'S WRONG WITH SPEAKING SOUTHERN?

Atlanta Constitution *circa 1980s*

There have been several reports recently of Southerners going to special classes in an effort to learn not to speak Southern.

I read of such classes in Atlanta, where people who took the course said they were afraid if they didn't stop talking with a Southern accent, it might impede their progress toward success.

A young woman who works for IBM said, "I want to advance through the company (and) I feel I need to improve my voice. . . ."

What she really was saying is the IBM office where she works is run by a bunch of northern transplants who probably make fun of the way she talks, and she is embarrassed and wants to talk like they do.

That, in my opinion, is grounds for loss of Southern citizenship.

What are we trying to do here? Do we all want to sound like those talking heads on local television news who have tried so hard not to have an accent, their vocal cords are nervous wrecks?

What's wrong with having a Southern accent? My grandfather said "y'all" (only in the plural sense, however, as Yankees have never figured out) and my grandmother said, "I reckon." And if IBM doesn't like that kind of talk, they can just program themselves right back to where they came from.

I like accents. I like to try to emulate accents, other than my own.

I do a big-time Texas oilman: "Now, you ladyfolks just run along 'cause us menfolk got to talk about bidness."

I can even talk—or at least type—like Bostonians sound: "Where can I pock my cah?"

The wonderful thing about the way Americans treat the English language is we have sort of made it up as we have gone along, and I see absolutely nothing wrong with having different ways to pronounce different words.

New Yorkers say "mudder and fadder."

Midwesterners say, "mahmee and dee-ad."

Southerners say, "mahma and deadie."

Big deal.

If we all spoke the same, dressed the same, acted the same, thought the same, then this country would not be the unique place that it is, would not have the benefit of our spice and variety, and everybody probably would be in the Rotary Club.

What we all need to realize is the more diverse we are, the stronger we are. Being able to get second and third and even fourth and fifth opinions often will prevent the nastiest of screw-ups.

I say if you are going to classes to lose your Southern accent, you are turning your back on your heritage, and I hope you wind up working behind the counter of a convenience store with three Iranians and a former Indian holy man.

And if you happen to be from another part of the country and make fun of the way Southerners talk, may you be elected permanent program chairman of your Rotary Club.

Y'all reckon I've made my point?

Judith Martin

(Miss Manners)

George Will called Judith Martin's book on correct conduct the most formidable political work produced by an American since The Federalist. Rarely has applause been so heartfelt as for Miss Manners' Guide to Excruciatingly Correct Behavior *(New York, 1982). What Martin (1938–) did was to refurbish the moribund etiquette column for the* Washington Post *Style section, answering questions from readers with éclat and authority,* e.g.: *''What the world needs is more false cheer and less honest crabbiness.'' ''At dinner parties, married couples must be separated because they tell the same stories and tell them differently.'' ''To be deceived is the natural human condition; to read another person's mail is despicable.'' Her column started in 1979, an outgrowth of Martin's previous work as cultural and society reporter, collected in* The Name on the White House Floor *(New York, 1972). Besides sequels to her guide, she has written two novels. Miss Manners was born in Washington, attended Wellesley, and lives in Washington with her husband and two perfect children.*

ADVICE FROM MISS MANNERS

Washington Post *1979–82*

Breaking Up

Dear Miss Manners:

I am interested in knowing what is the proper method of breaking off a relationship. For the past several months, I have been exclusively dating a young lady whom I was extremely fond of. Everything seemed to be going great until about two months ago, when she suddenly seemed to lose interest in me. Every time I wanted to see her, she was busy, etc., until finally I just stopped calling her. I have not heard from her since. What do you think of breaking off a relationship in this manner, and how should it be handled?

Gentle Reader:

What you describe is your basic Kafka Romance Dissolver, and you handled it exactly correctly. Do not be offended if Miss Manners approves of the young lady's behavior, as well.

Naturally, you were hurt and bewildered when your invitations were repeatedly rejected without explanation. Miss Manners would like to point out to you, however, that there is no possible way for one person to end a romance that the other person thought was going great without causing pain and bewilderment. The chief difference between the Kafka method and those more socially approved ones that come with explanations is that the latter engender humiliation, as well as pain and bewilderment.

What, after all, can the explanation be?

"Sure I like you, but I met someone I'm really crazy about."

"I know you can't help it, but there are a lot of things about you that were beginning to get on my nerves."

"It was fun for a while, but lately I've found myself getting bored and restless."

And so on. Rarely, these days, does anyone break of an exciting, stimulating, fulfilling romance to lead a life of service or to save the family through an expedient alliance.

Therefore, all explanations can be reduced to the fact that the other person would rather do something else—sometimes anything else—than continue the romance. Attempts to obfuscate, such as "I love you, but I need room to grow," don't fool anyone.

The patronizing sweeteners customarily added to these explanations are particularly galling. It is easier to bear being denounced as a villain by someone you still love than to be told that you are a "nice person but."

Perhaps you will object that the method without explanation took some time, because its comparative subtlety confused you about what was actually happening. Granted. Nevertheless, Miss Manners maintains that the period of suffering was, in the end, shorter.

The early part, say the first two rejections, was annoyance, rather than devastation, because you did to yet believe it. Then you began to suspect and pay attention; you guessed; you tested the hypothesis by ending your calls; and then you had your proof. Indeed, that period must have hurt.

Consider what that time would have been like had you been spending it discussing the situation with the young lady. As the explanation method spuriously suggests reasons for the whims of the heart, the reaction of the rejected person is always to offer counterarguments. It would have taken just as long and, as the young lady would be forced to escalate her objections to overcome your arguments, the pain would have been more intense.

The true reward comes now. In your memory, you may set this young lady forever as a fool who didn't know how to appreciate you. You needn't carry around the certain knowledge of how little she appreciated you, nor the memory of your having made a fool of yourself trying to argue the matter with her.

Advice to the Rejectee

Falling is always easier than getting back on one's feet, as any water skier can tell you, and so it is with love. Miss Manners has observed that most people can be trusted to behave reasonably well when they have fallen in love and perfectly dreadfully when they have been dumped.

One might protest that it is unfair for the burden of proper behavior to fall on the person who is down, rather than on the one who did the pushing. But so it is. All a person has to do who wants to walk away from a love affair is to walk away. It is surprising how few rejecting lovers understand this. In this day of explanations, it is fashionable for those who no longer love to offer to talk it all over with those whom they no longer love. No worse cruelty ever disguised itself as a kindness.

In fact, the first duty (and only available pleasure) of the rejected one is to reject such offers of help. One ought to reply, as the Republican Party is said to have done to Mr. Nixon when he offered to help with the 1976 election, "Thank you, but I think you've done enough already." Few rejected lovers can bring themselves to do this, however. They always cherish the notion that if they could only make the other understand how they have been made to feel (rotten), that person would come to his or her senses (realize that the old love is true love, after all).

Unfortunately, it never works that way. Otherwise civilized people first take a grim pleasure in watching romantic sufferings that they have inflicted, and then their satisfaction turns to extreme distaste. Unhappiness that one has oneself caused always looks grotesquely exaggerated and disgusting. Therefore, the smartest thing a dumped one can do is to get out of sight, or at least to hide all traces of misery. This is not easy to do, but it is one of those rare instances in which the hardest work brings the greatest chance of success.

Success, in this case, must be defined as making the other person suffer as much as oneself. Miss Manners is sorry—this is not her usual objective in teaching manners—but it is the only way to restore equilibrium. Whether one should resume the love affair after that is another matter. Such suffering is never caused by see-how-miserable-you-made-me-feel. It is caused, as the rejectee ought to know, by the realization that a person who used to love you doesn't any longer. Thus, the proper behavior for someone whose heart is breaking is to be cheerful, not pained; ungrudgingly forgiving, not accusing; busy, not free to be comforted; mysterious, not willing to talk the situation over; absent, not obviously alone or overdoing attentions to others.

Such behavior will have two rewards. First, it will take the sufferer's mind off suffering and begin the recovery. Second, it will make the former lover worry that this supposed act of cruelty was actually a relief to the person it should have hurt. That hurts.

Who Pays

Dear Miss Manners:

I have a problem. Our divorced daughter is marrying a divorced man. They are planning on having a meal after the church ceremony and a week later is the reception. What are our responsibilities in regard to footing the bills for the meals? Do we offer to pay for both, or neither?

Gentle Reader:

Miss Manners hopes you did not neglect, on the occasion of your daughter's first wedding, the ceremony of saying, "OK, kid, that's it. From now on you're on your own." It is a practical tradition these days. No, you are not responsible for paying for festivities related to a second wedding, although it is charming to do so. For that matter, the same is true of a first wedding. These decisions are usually based on the desirability of the bridegroom, as assessed by the bride's parents.

Snooping

Dear Miss Manners:

I am heartbroken because I discovered that my husband has been receiving love letters from another woman, but he keeps changing the subject by saying I had no right to read his mail. Who is right?

Gentle Reader:

You are quite right to seek advice from an etiquette column, rather than a psychologically oriented one. Miss Manners believes that the true value in people is not what is in their murky psyches, which many keep in as shocking a state as their bureau drawers, but in how they treat one another. You are wrong, however, in your dispute with your husband. To be deceived is the natural human condition; to read another person's mail is despicable.

Civil Disobedience

Dear Miss Manners:

What is the proper conduct of demonstrators at White House demonstration, and of recipients (targets) of demonstrations?

Gentle Reader:

The first obligation of the demonstrator is to be legible. Miss Manners cannot sympathize with a cause whose signs she cannot make out even with her glasses on. The next rule of conduct is that demonstrators not vent their discontent on passersby, whom they should be impressing with their goodwill and reasonableness. As for the recipients, the proper thing is to resist the temptation to look out of a White House window. Peeking is a mistake when one may wish to declare, the next day, that one was unaware of the demonstration.

Closings

Dear Miss Manners:

I write occasional business letters. What is the proper complimentary closing? Many are to people whom I do not necessarily respect; that lets out "respectfully." "Yours" is obviously out; I am most certainly not. While I hope I am almost always sincere, "Sincerely" seems to be out of place in a business letter. So does "love," whether agape or filial. How about just "very" or how about nothing?

Gentle Reader:

How about "Yr. most humble and obedient servant"? How about not being so literal? Miss Manners signs herself "Very truly yours" to business people and "Sincerely yours" to acquaintances—believing that "sincerely" alone is as close to nothing as "very" alone—and has not yet been required to surrender herself to any of them.

Doctors and Doctorates

Dear Miss Manners:

Several women doctors I know are extremely sensitive about being addressed socially as "Doctor." One of them is married to a "Mr." and another to a Ph.D. I created a furor by sending them cards addressed to Dr. and Mr. What is the correct way to do it?

Gentle Reader:

Illicit love has given us, if nothing else, the two-line method of address, which may also be applied to married couples with different titles or names. The doctor and Mr. may be addressed as:

Dr. Dahlia Healer
Mr. Bryon Healer

and the doctor and academician, if he uses his title socially, which not all holders of doctorates do, as:

Dr. Dahlia Healer
Dr. Bryon Healer
or as:
The Doctors Healer
or as:
The Doctors Bryon and Dahlia Healer.

Interesting Arguments

Dear Miss Manners:

Do you believe it is proper for parents to fight—argue, shall we say—in front of their children? There is one theory that it is bad to burden them, and another that it's educational for them. Which do you think?

Gentle Reader:

Both. Parents should conduct their arguments in quiet, respectful tones, but in a foreign language. You'd be surprised what an inducement that is to the education of children.

On Offensive Odors

Dear Miss Manners:

We have a quarrel about married couples who order different food at restaurants. I can't stand to kiss my husband when he has garlic or onions, which I don't eat, on his breath.

Gentle Reader:

Well, Miss Manners is not going to do it for you.

Calvin Trillin

Remorselessly, Calvin Trillin (1935–) set one condition in agreeing to write a regular column for The Nation: *that he could make fun of the editor, Victor S. Navasky. The 1978 pact has been honored though Trillin's pay grew from the ''somewhere in the high two figures'' originally offered by Navasky (''What exactly to you mean by high two figures,'' I said. ''Sixty-five dollars,'' Navasky said. ''Sixty-five dollars! That sounds more like middle two figures to me.'' ''You shook on it,'' Navasky said). Trillin kept the same fresh tone as the column was syndicated and recycled in three books,* Uncivil Liberties *(New York, 1982),* With All Disrespect *(New York, 1985), and* If You Can't Say Something Nice *(New York, 1987). But then Trillin, who hails from Kansas City, Yale, and* The New Yorker, *has found a new cause—to make spaghetti carbonara the national Thanksgiving Day dinner, in honor of Columbus.*

NAVASKY AS MALE MODEL

The Nation *November 21, 1981*

Thumbing through the *New York Times Magazine* one Sunday, noticing that more and more designers like Ralph Lauren and Bill Blass and Oscar de la Renta are appearing in advertisements for their own brands of clothing, I was struck so suddenly with an inspiration that I blurted it out to my wife, Alice: "Victor S. Navasky should appear in ads for *The Nation,*" I said. "Standing with his foot on the corral fence, next to one of those blond models who look like they're too sophisticated to eat dinner."

"Victor!" Alice said. "In *that* suit?"

"He wouldn't be peddling shmatas," I said. "We're talking about a journal of opinion here. Journalists are supposed to be a bit seedy, and you have to admit that Navasky is very opinionated looking. He looks like he holds a number of opinions that other people gave up a long time ago."

"If this is another plot to embarrass Victor just because you think he's underpaying you, I want you to forget it immediately," Alice said. "It's not his fault he doesn't look like a male model."

"But that's just the point," I said. "Neither do those designers. Does Ralph Lauren strike you as someone who used to be a lifeguard?"

"Well, no," Alice acknowledged. "In fact, it's always seemed to me that Oscar de la Renta looks like a ruffle with legs."

"As it happens, Victor S. Navasky has an athletic past—he was on the basketball team at Elisabeth Irwin High School," I said, repeating an old and unconvincing Navasky claim that I figured I could pass off as true because Alice has never seen him play basketball. Years ago, Navasky and I both played in a weekly Sunday morning basketball game at the Gansevoort Street Playground in the West Village, but Alice had always declined my invitations to attend.

"I don't know much about basketball," she used to say, "and I'm afraid I'll laugh at the wrong time."

"It isn't meant to be funny at all," I would tell her—another statement I thought I might get away with because she had never seen Navasky play.

Navasky was dreadful. I don't mean he received no awards at the annual banquet we held in the spring at the Paradise Inn of the Green Olive Tree, on Forty-first Street. We had a number of awards that had nothing to do with skill—Most Improved Jewish Player, for instance, and the Richard Nixon Award for the Most Specious Out-of-Bounds Argument. His playing was such, though, that the only reason the rest of us agreed among ourselves to quit scoffing at his story of having played on the high school basketball team was to prevent him from embarrassing his family and friends by wearing his letter sweater to dinner parties.

Being unusually charitable about such matters, I finally managed to devise a theory that might explain how an American high school could have had Victor S. Navasky as a member of its basketball team—a theory based on the assumption that Elisabeth Irwin, a Greenwhich Village school with a progressive reputation, must have been full of children from left-wing families. It seems possible to me that in order to avoid sectarian disputes at PTA meetings, the basketball coach might have tried to give the teams a fair balance of all sorts of players—a few Trotskyites, a couple of Stalinist forwards, a Schachtmanite with a passable jump shot, maybe an energetic guard who couldn't hit the basket but had a grandfather who had been a Left S.R. in the old country. Navasky, I figured, must have been editing the Elisabeth Irwin newspaper, charging his writers their lunch money for the opportunity of getting into print, and he was put on the second five as the Token Exploiter.*

"How about a campus scene?" I said to Alice the next day. "Navasky's in a sweater—not his E.I. letter sweater but one of those gigantic Irish fisherman's sweaters that male models can wear to Sunday brunch in a steam-heated New York apartment once they've had their sweat glands surgically removed. He's reading *The Nation.* His foot's on a corral fence—because there happens to be a corral on this campus—and he's so engrossed he doesn't notice that all of these dynamite coeds are staring at him with a desperate longing. Some of them have long, straight hair, so they look a little left-wing, except for being such knockouts."

"I can't believe you're doing this," Alice said.

"Or maybe Navasky's on a beach," I said. "In gauzy color. He's wearing one of those alligator polo shirts, except it's got a peace sign or a Red Star or something where the alligator's supposed to be. He's reading *The Nation,* and so are all of these marvelous-looking people on the beach—everyone but the lifeguard, this wonky little guy who looks something like Ralph Lauren."

"I think you're being awful," Alice said. "You can't compare Victor to some dress designer. He's a serious person, except for that suit."

"But dress designers are not the only people appearing in their own ads now," I said. "What about Frank Perdue—with all those ads about how it takes a tough man to produce a tender chicken? Is that your idea of a glamour puss? What about that Carvel ice cream guy with the voice they make political prisoners in Central America listen to until they agree to confess to being in the pay of the Kremlin? It's just that I prefer to envision Navasky in an Irish fisherman's sweater."

"I won't have Victor discussed as if he were a ruffle," Alice said.

"Maybe you're right," I said. "We could do a sort of Perdue-style ad for *The Nation.* There's Navasky, dressed up like Edward Arnold playing the big industrialist in one of those 1940s movies, and writers are slaving away all around him—all of them dressed in dismal clothes and needing haircuts and looking undernourished."

"I don't see the point," Alice said.

"The caption, of course: 'It Takes a Real Exploiter of the Workers to Run a Left-Wing Magazine.' "

***Editor's note:* This is warmly contested by Mr. Navasky, who cites his 5.2 scoring average per game and offers to produce witnesses.

PINKO PROBLEMS

The Nation *April 17, 1982*

I've been worried lately about the possibility that *The Nation* is getting to be known around the country for being a bit pinko. I was born and brought up in Kansas City, and I'm not really keen on the folks at home getting the impression that I work for a left-wing sheet. They know I do a column for *The Nation,* of course—my mother told them—but most of them have not inquired deeply into *The Nation*'s politics, perhaps because my mother has been sort of letting on that it's a tennis magazine. She has been able to get away with that so far because *The Nation* is not circulated widely in Kansas City: in the greater Kansas City area, it goes weekly to three libraries and an unreconstructed old anarcho-syndicalist who moved to town after his release from the federal prison at Leavenworth in 1927 and set up practice as a crank. Lately, though, I've had reason to worry that *The Nation*'s political views may be revealed in the press.

My concern is not based on any notion that the people back home would react to this revelation by ostracizing my mother for having given birth to a Commie rat. Folks in the Midwest try to be nice. What I'm worried about is this: People in Kansas City will assume that no one would write a column for a pinko rag if he could write a column for a respectable periodical. They might even assume that payment for a column in a pinko rag would be the sort of money people in Kansas City associate with the summer retainer for the boy who mows the lawn. They could even go to the library—or to the home of the crank, who holds on to back issues in case they're needed for reference in sectarian disputes—and look through my columns until they find the one that revealed the luncheon negotiations in which I asked the editor of *The Nation,* one Victor S. Navasky, what he intended to pay for each of these columns and he replied, "Somewhere in the high two figures." Then the people at home, realizing that I had struggled for years in New York only to end up writing a column in a pinko rag for lawn-mowing wages, would spend a lot of time comforting my mother whenever they ran into her at the supermarket. ("There, there. Don't you worry one bit. Things have a way of working themselves out.") My mom's pretty tough, but tougher people have broken under the burden of Midwestern comforting.

Without wanting to name names, I blame all of this on Victor S. Navasky. During the aforementioned luncheon negotiations, he said nothing that would have led me to think that *The Nation* had political views I might find embarrassing. In fact, he sort of let on that it was a tennis magazine. The only reference he made to anybody who could be considered even marginally pinko was when he told me that he had reason to believe that Warren Beatty was keeping his eye on *The Nation* and might snatch up the movie rights to just about any piece for $200,000 plus 5 percent of the gross. "Always get points on the gross, never on the net," I remember Navasky saying as I dealt with the check.

When Navasky provoked a public controversy by attacking a book on the Hiss case from a position that might have been described as somewhat left of center, I tried to be understanding. I figured that Navasky was trying to pump up circulation because he lacked some of the financial resources that most people who edit journals of opinion have. Traditionally, people who run such magazines manage financially because they have a wife rich enough to have bought them the magazine in the first place. It's a good arrangement, because an editor who has his own forum for pontificating to the public every week may tend to get a bit pompous around the house, and it helps if his wife is in a position to say, "Get off your high horse, Harry, or I'll take your little magazine away from you and give it to the cook." I haven't made any detailed investigation into the finances of Navasky's wife, but it stands to reason that if she had the wherewithal to acquire entire magazines she would by now have bought him a new suit.

There is no excuse, though, for Navasky's latest caper, particularly after I had specifically said to him only last year, "I don't care what your politics are, Navasky, but I hope you'll have the good taste to keep them to yourself." A couple of months ago, around the time Susan Sontag charged in her Town Hall speech that left-wing journals had never really faced up to the basic evil of the Soviet regime, *The Nation* began advertising a tour it was co-sponsoring—a "Cruise up the Volga." The other co-sponsors were outfits with names like National Council of American-Soviet Friendship. The advertised attractions included a visit to Lenin's birthplace. I suppose some long-buried smidgen of restraint kept Navasky from simply headlining the ad "Pinko Tours Inc. Offers a Once-in-a-Lifetime Opportunity to Fellow Travelers." Didn't he know he would be attacked by *The New Republic*? Didn't he know that he would provoke an argument about the possibility that what *Nation* tourists need to examine is not the GUM department store but the loony bins where dissidents are stashed? Didn't he know this sort of thing could get back to Kansas City?

I implored Navasky to cut his losses by explaining the tour in humanitarian terms. Where, after all, are old Commies supposed to go for their vacations—Palm Springs?

"The tour is oversubscribed," he said.

I could see that he was getting defensive. He had the look I noticed when I asked him if it was really true that two elderly Wobbly copyreaders at *The Nation* had been told that their salaries were still being diverted straight into a defense fund for Big Bill Haywood. I knew that in his mood he might even launch some preventive strikes—maybe write an editorial suggesting that *Commentary* give its readers a New York tour that included the Brooklyn street where Norman Podhoretz was first taunted by black kids and the office where Daniel Bell got his first government grant.

"I'm going," I said.

"To Russia?" he asked.

"To call my mother," I said.

THE MAILBAG

The Nation *May 29, 1982*

Sir:

I don't know who to believe anymore. First the *Washington Post* and the *New York Times Magazine* have to admit that they ran phony stories; now Ann Landers and Abigail Van Buren have both been caught recycling used letters from secondhand kvetches. It occurs to me that you may be making up quotations and letters yourself and they may explain why your column has always sounded a little—if you'll forgive me for saying so—cockamamie. Can I trust you? Please try to be truthful in your response.

R.F.G., Pelham, N.Y.

Trust me. I know some misunderstanding might have been caused by my admitting in public print recently that the editor of The Nation, *one Victor S. Navasky, once asked me whether a certain quotation I had attributed to John Foster Dulles ("You can't fool all of the people all of the time, but you might as well give it your best shot") was authentic, and I answered, "At these rates, you can't always expect real quotes." However, everyone involved understood that to be a negotiating tactic rather than an admission of guilt. As to concocting letters, I can assure you that yours is the first.*

* * *

Sir:

Ten years ago, I wrote a letter to Ann Landers about my husband, who thought he was Katherine Hepburn. I signed it with a pseudonym ("Concerned"), but everyone knew it was me, of course, because I am the only Swedenborgian in Luke's Crossing, Montana. Miss Landers gave me what I thought was sensible advice ("Tell him his slip is showing"). Recently, though, she reprinted the letter as if it had just been sent, and, as it happens, I have a different husband now. He's very upset because he thinks he's Wallace Beery. If Abigail Van Buren prints it, can I sue?

R.V.M., Luke's Crossing, Mont.

Pull up your socks.

* * *

Sir:

Is it true that employees of *The Nation* are forced to sell flowers and candy in airports and turn the proceeds over to Victor S. Navasky?

G.B., Dayton, Ohio

Not exactly.

* * *

Sir:

I have been a loyal subscriber to *The Nation* for forty-five years, and I simply can't imagine what possessed them to begin running a column by a bubblehead such as yourself. Where are your politics, young man? Why don't you ever write about the Scottsboro Boys? What's your position on the withering away of the state?

S.A.S., Boston, Mass.

I was once intensely political. I had a position on everything. If something new came up, I'd get a position on it by the end of that week at the latest. Then a lot of new countries came into being, and a political person had to have a position on every one of them. One day, someone said to me, "I don't think much of your position on Guinea."

"My position on Guinea is impeccable," I said. By chance, I had that very morning taken out my position on Guinea and tidied it up a bit. "What is your objection to my position on Guinea?" I thought about calling him a crypto-fascist pig straight off, but I decided to keep that card up my sleeve.

"To start with, you don't have it in the right continent," he said. "It's in Africa."

"Well, no wonder some people thought that the Indonesian territorial claims were historically dubious," I said. "Still, as a Third World country, Indonesia —"

"That's New Guinea," he said.

"Is it possible we're talking about Ghana?" I said. "Or maybe British Guiana?"

"There isn't any British Guiana," he said. "It's Guyana now. What's your position on Guyana?"

"I'd rather not say," I told him.

I tried to catch up, of course, but I found that I no longer had time to do the dishes, and at the office I was falling behind in my work. Finally, I decided I could no longer be political. I kept a position on British unilateral disarmament, as a sort of souvenir, but nobody ever asks me about it—which is a shame, because it's impeccable.

* * *

Sir:

Did Victor S. Navasky name names?

S.F., Kansas City, Mo.

It's all very complicated.

* * *

Sir:

Don't you think CBS should have allowed the White House equal time to reply to that Bill Moyers program about three families that slipped through the social safety net? How can the press expect people to take it seriously when the networks just tell one side of the story and Ann Landers and Abigail Van Buren both keep running the same letter about some wacko in Montana who still thinks he's Katharine Hepburn?

M.N.V., Chicago, Ill.

The President's advisers differed as to how to reply. Some thought the reply should be an upbeat program showing three millionaires who were happy as clams and about to leave for extended vacations on the Continent, where they intended to depreciate Luxembourg. The hard-liners wanted to show that the parents involved were acting irresponsibly, since it was obvious that children as skinny and undernourished as theirs would slip through any net. The debate finally had to be settled by the President himself, and the program CBS rejected was to show a welfare chiseler driving to the supermarket in a Cadillac, buying vodka with foodstamps, and giving birth to a child at the checkout counter just to qualify for Aid to Families with Dependent CHildren and the Wednesday coupon special.

* * *

Sir:

Who said, "You can't fool all of the people all of the time, but you might a well give it your best shot"?

L.V., Troy, N.Y.

John Foster Dulles.

Molly Ivins

A columnist for the Dallas Times Herald, *Molly Ivins (1944–   ) says she writes "about Texas politics and other bizarre happenings." Born in Houston, Ivins attended Smith College and earned a master's degree at Columbia's School of Journalism before starting out as the Complaint Department of the* Houston Chronicle. *She cubbed for the* Minneapolis Tribune, *returning to Texas as co-editor of the peppery* Texas Observer. *From 1976 to 1982, she reported local politics for the* New York Times, *then covered nine Rocky Mountain states from Denver. Since then, her down-home liberalism has won a local audience in a conservative town, while her commentary on Texas has gained her a national audience on public radio and television.*

THE VIETNAM MEMORIAL

Dallas Times Herald *November 30, 1982*

She had known, ever since she first read about the Vietnam Veterans Memorial, that she would go there someday. Sometime she would be in Washington and would go and see his name and leave again.

So silly, all that fuss about the memorial. Whatever else Vietnam was, it was not the kind of war that calls for some "Raising the Flag at Iwo Jima" kind of statue. She was not prepared, though, for the impact of the memorial. To walk down into it in the pale winter sunshine was like the war itself, like going into a dark valley and damned if there was ever any light at the end of the tunnel. Just death. When you get closer to the two walls, the number of names start to stun you. It is terrible, there in the peace and the pale sunshine.

The names are listed by date of death. There has never been a time, day or night, drunk or sober, for 13 years she could not have told you the date. He was killed on Aug. 13, 1969. It is near the middle of the left wall. She went toward it as though she had known beforehand where it would be. His name is near the bottom. She had to kneel to find it. Stupid cliches. His name leaped out at her. It was like being hit.

She stared at it and then reached out and gently ran her fingers over the letters in the cold black marble. The memory of him came back so strong, almost as if he were there on the other side of the stone, she could see his hand reaching out to touch her fingers. It had not hurt for years and suddenly, just for a moment, it hurt again so horribly that it twisted her face and made her gasp and left her with tears running down her face. Then it stopped hurting but she could not stop the tears. Could not stop them running and running down her face.

There had been a time, although she had been an otherwise sensible young woman, when she had believed she would never recover from the pain. She did, of course. But she is still determined never to sentimentalize him. He would have hated that. She had thought it was like an amputation, the severing of his life from hers, that you could live on afterward but it would be like having only one leg and one arm. But it was only a wound. It healed. If there is a scar it is only faintly visible now at odd intervals.

He was a biologist, a t.a. at the university getting his Ph.D. They lived together for two years. He left the university to finish his thesis and before he could line up a public school job—teachers were safe in those years—the draft board got him. They had friends who had left the country, they had friends who had gone to prison, they had friends who had gone to Nam. There were no good choices in those years. She thinks now he unconsciously wanted to go even though he often said, said in one of his last letters, that it was a stupid, f—-in' war. He felt some form of guilt about a friend of theirs who was killed during the Tet offensive. Hubert Humphrey called Tet a great victory. His compromise was to refuse officer's training school and go as an enlisted man. She had thought then it was a dumb gesture and they had a half-hearted quarrel about it.

He had been in Nam less than two months when he was killed, without heroics, during a firefight at night, by a single bullet in the brain. No one saw it happen. There are some amazing statistics about money and tonnage from that war. Did you know that there were more tons of bombs dropped on Hanoi during the Christmas bombing of 1971 than in all of World War II? Did you know that the war in Vietnam cost the United States $123.3 billion? She has always wanted to know how much that one bullet cost. Sixty-three cents? $1.20? Someone must know.

The other bad part was the brain. Even at this late date, it seems to her that was quite a remarkable mind. Long before she read C. P. Snow, the ferociously honest young man who wanted to be a great biologist taught her a great deal about the difference between the way scientists think and the way humanists think. Only once has she been glad he was not with her. It was at one of those bizarre hearings about teaching "creation science." He would have gotten furious and been horribly rude. He had no patience with people who did not understand and respect the process of science.

She used to attribute his fierce honesty to the fact that he was Yankee. She is still prone to tell "white" lies to make people feel better, to smooth things over, to prevent hard feelings. Surely there have been dumber things for lovers to quarrel over than the social utility of hypocrisy. But not many.

She stood up again, still staring at his name, stood for a long time. She said, "There it is," and turned to go. A man to her left was staring at her. She glared at him resentfully. The man had done nothing but make the mistake of seeing her weeping. She said, as though daring him to disagree, "It was a stupid, f—-in' war," and stalked past him.

She turned again at the top of the slope to make sure where his name is, so whenever she sees a picture of the memorial she can put her finger where his name is. He never said goodbye, literally Whenever he left he would say, "Take care, love." He could say it many different ways. He said it when he left for Vietnam. She stood at the top of the slope and found her hand half-raised in some silly gesture of farewell. She brought it down again. She considered thinking to him, "Hey, take care, love," but it seemed remarkably inappropriate. She walked away and was quite entertaining for the rest of the day, because it was expected of her.

She thinks he would have liked the memorial O.K. He would have hated the editorials. He did not sacrifice his life for his country or for a just or noble cause. There just were no good choices in those years and he got killed.

LUBBOCK: "HER TEETH ARE STAINED, BUT HER HEART IS PURE"

Dallas Times Herald *February 3, 1987*

Austin—When in the course of human events one is called upon to explain Lubbock art, the oxymoron, to San Francisco, the city of sophisticates, one might well take a dive.

Duty called. Some outfit in San Francisco is putting on a program in April entitled "The Texans' Project," which consists of a bunch of Lubbock artists and musicians doing art and music. In my semi-official capacity as a person supposedly capable of explaining Texas to normal people, this theater asked me to explain to San Franciscans why so many great musicians come from Lubbock. How fortunate that I know:

Because there is nothing else to do in Lubbock.

Except perhaps for Wednesday Fellowship Night.

This is not a cheap shot at Lubbock. This is fact.

Another reason there is so much music in Lubbock is because people there know what sin is. Lubbock is a godly place, so it follows as night the day that there should be a lot of country music and down-home rock, with their consequent and probably inevitable accompaniment of drinking and dancing and other forms of enjoyable sin. The sheer beauty of having something as clear-cut as Lubbock to rebel against is almost enough to make me move there.

What is a teenager in San Francisco to rebel against, for pity's sake? Their parents are all so busy trying to be non-judgmental, it's no wonder they take to dyeing their hair green. But Lubbock will by-God let you know what sin *is*. So you can go and do it, and enjoy it.

Lubbock is full of They-Sayers and They Say that if you sin long enough, you will become An Example to Us All. However, if you should become sufficiently rich and famous in the process, they will also put you in the Lubbock Music Hall of Fame—along with Buddy Holly, Bob Wills, Waylon Jennings, Mac Davis, Joe Ely and the like.

Also, we should give a little credit to the musical history of Lubbock, which for many years harbored the late lamented Cotton Club, the finest honky-tonk in all of West Texas. Kent Hance, the former congressman from Lubbock who fell into Republicanism and came to electoral grief, reports, "When I was in college we went to the Cotton Club to dance, to pick up girls and to drink beer out of Coca-Cola cups in case a minister came in, and it would embarrass him and you both. Outside they had soap and water so you could wash that stamp off your hand when you left at the end of the night, so it wouldn't show Sunday morning at church.

"The dancin' started at 9, and everybody'd dance until 11:30 and then everybody's go out to the parkin' lot and fistfight or somethin' else and then go back into dance again until 2. Which is how a whole lot of great cheatin' songs and tragedy love songs came to be written."

Good Lord, Lubbock, Texas. Well about 88.3 percent of the world there is sky, and if you are used to that, it feels like freedom and everywhere else feels like jail. I have no idea what that has to do with Lubbock art, but maybe someone can think of a connection. I should probably explain more to the people in San Francisco about Lubbock weather, but I reckon hearing Joe Ely is close to a Lubbock tornado, and he'll be out there.

It is extremely difficult to develop either pretensions or affectations in Lubbock. Without getting laughed out of town. Which probably does account for a lot about the music. Lubbock is sometimes called "The Hub

City of the Plains"—actually the Lubbock Chamber of Commerce is the only thing that ever calls it that—and I think it was Jimmie Gilmore who once observed, "Plain is the opposite of fancy."

Hance says the failures of Lubbock music deserve some attention, too. "Not everybody from Lubbock has been able to sing well. I knew a guy from Matador named Robin Dorsey who wanted to be a country-western writer. He wrote a song about his girl friend, who was from Muleshoe, called 'Her Teeth Are Stained But Her Heart Is Pure.' And she quit him. And he had her name, Patty, tattooed on his arm which ever after cut down on his social life because we always had to fix him up with girls named Patty. So he wrote a tragedy love song about it, 'I Don't Know Whether to Kill Myself or Go Bowling.' "

One of my favorite Lubbock songs that never made it big is "I Wish I Was in Dixie Tonight, But She's Out of Town." The others are all too dirty to print.

So, come April, San Francisco will meet Lubbock. Lord have mercy. I just don't think those people out there are *ready*.

Dave Barry

In one of his columns for the Miami Herald, *Dave Barry (1948–   ), daringly burlesqued the totem of totems, the Pulitzer Prize. The following year he received the award, for Distinguished Commentary in his syndicated column. Barry, born in Armonk, New York, grew up in an upper middle-class neighborhood, played Little League baseball, and spent summers at Camp Sharparron. (''There is some kind of rule that says summer camps have to have comical sounding Indian names.'') He attended Haverford and worked for the* Daily Local News *in West Chester, Pennsylvania, before moving south. He began his column as a free-lancer, joined the* Herald *in 1983, and assembled his sketches in* Dave Barry's Greatest Hits *(New York, 1988).*

GOING FOR A PULITZER

Miami Herald   *March 29, 1987*

The burgeoning Iran-contra scandal is truly an issue about which we, as a nation, need to concern ourselves, because . . .

(Secret Note to Readers: Not really! The hell with the Iran-contra affair! Let it burgeon! I'm just trying to win a journalism prize here. Don't tell anybody! I'll explain later. Shhhh.)

. . . when we look at the Iran-contra scandal, and for that matter the mounting national health-care crisis, we can see that these are, in total, two issues, each requiring a number of paragraphs in which we will comment, in hopes that . . .

(. . . we can win a journalism prize. Ideally a Pulitzer. That's the object in journalism. At certain times each year, we journalists do almost nothing except apply for the Pulitzers and several dozen other major prizes. During these times you could walk right into most newsrooms and commit a multiple ax murder naked, and it wouldn't get reported in the paper because the reporters and editors would all be too busy filling out prize applications. "Hey!" they'd yell at you. "Watch it! You're getting blood on my application!")

. . . we can possibly, through carefully analyzing these important issues—the Iran-contra scandal, the mounting national health-care crisis and (while we are at it) the federal budget deficit—through analyzing these issues and mulling them over and fretting about them and chewing on them until we have reduced them to soft, spit-covered globs of information that you, the readers, can . . .

(. . . pretty much ignore. It's okay! Don't be ashamed! We here in journalism are fully aware that most of you skip right over stories that look like they might involve major issues, which you can identify because they always have incomprehensible headlines like "House Parley Panel Links NATO Tax Hike to Hondurans in Syrian Arms Deal." Sometimes we'll do a whole SERIES with more total words than "The Brothers Karamazov" and headlines like: "The World Mulch Crisis: A Time to Act." You readers don't bother to wade through these stories, and you feel vaguely guilty about this. Which is stupid. You're not SUPPOSED to read them. We JOURNALISTS don't read them. We use modern computers to generate them solely for the purpose of entering them for journalism prizes. We're thinking about putting the following helpful advisory over them: "Caution! Journalism Prize Entry! Do Not Read!")

. . . again, through a better understanding of these very important issues—the Iran-contra scandal; the health-care crisis (which as you may be aware is both national AND mounting); the federal budget deficit; and yes, let's come right out and say it, the Strategic Defense Initiative—you readers can gain a better understanding of them, and thus we might come to an enhanced awareness of what they may or may not mean in terms of . . .

(. . . whether or not I can win a Pulitzer Prize. That's the one I'm gunning for. You get $1,000 cash, plus all the job offers the mailperson can carry. Unfortunately, the only category I'd be eligible for is called "Distinguished Social Commentary," which is a real problem, because of the kinds of issues I generally write about. "This isn't distinguished social commentary!" the Pulitzer judges would say. "This is about goat boogers!" So today I'm trying to class up my act a little by writing about prize-winning issues. Okay? Sorry.)

. . . how we, as a nation, can, through a deeper realization of the significance of these four vital issues—health care in Iran, the strategic federal deficit, the contras, and one other one which slips my mind at the moment, although I think it's the one that's burgeoning—how we can, as a nation, through distinguished social commentary such as this, gain the kind of perspective and foresight required to understand . . .

(. . . a guy like noted conservative columnist George Will. You see him on all those TV shows where he is always commenting on world events in that way of his with his lips pursed together, and you quite naturally say to yourself, as millions have before you: "Why doesn't somebody just take this guy and stick his bow tie up his nose? Huh?" And the answer is: Because a long time ago, for reasons nobody remembers anymore, GEORGE WILL WON A PULITZER PRIZE. And now he gets to be famous and rich and respected FOR EVER AND EVER. That's all I want! Is that so much to ask?!)

. . . what we, and I am talking about we as a nation, need to have in order to deeply understand all the issues listed somewhere earlier in this column. And although I am only one person, one lone distinguished social commentator crying in the wilderness, without so much as a bow tie, I am nevertheless committed to doing whatever I can to deepen and widen and broaden and lengthen the national understanding of these issues in any way that I can, and that includes sharing the $1,000 with the judges.

TUNING IN THE RUSH-HOUR BLUES

Miami Herald *October 16, 1988*

Monday morning. Bad traffic. Let's just turn on the radio here, see if we can get some good tunes, crank it up. Maybe they'll play some early Stones. Yeah. Maybe they . . .

(Power On)

". . . just reached the end of 14 classic hits in a row, and we'll be right back after we . . ."

(Scan)

". . . send Bill Doberman to Congress. Because Bill Doberman agrees with us. Bill Doberman. It's a name we can trust. Bill Doberman. It's a name we can remember. Let's write it down. Bill . . ."

(Scan)

". . . just heard 19 uninterrupted classic hits, and now for this . . ."

(Scan)

". . . terrible traffic backup caused by the . . ."

(Scan)

". . . EVIL that cameth down and DWELLETH amongst them, and it DID CAUSETH their eyeballs to ooze a new substance, and it WAS a greenish color, but they DID not fear, for they kneweth that the . . ."

(Scan)

". . . followingisbasedonan800-yearleaseanddoesnotinclude taxtagsinsuranceoranactualcarwegetyourhouseandyourchildrenandyour kidneys."

"NINE THOUSAND DOLLARS!!! BUD LOOTER CHEVROLET OPEL ISUZU FORD RENAULT JEEP CHRYSLER TOYOTA STUDEBAKER TUCKER HONDA WANTS TO GIVE YOU, FOR NO GOOD REASON . . ."

(Scan)

". . . Bill Doberman. He'll work for you. He'll FIGHT for you. If people are rude to you, Bill Doberman will KILL them. Because Bill Doberman . . ."

(Scan)

". . . enjoyed those 54 classic hits in a row, and now let's pause while . . ."

(Scan)

". . . insects DID swarm upon them and DID eateth their children, but they WERE NOT afraid, for they trustedeth in the . . ."

(Scan)

". . . listening audience. Hello?"

"Hello?"

"Go ahead."

"Steve?"

"This is Steve. Go ahead."

"Am I on?"

"Yes. Go ahead."

"Is this Steve?"

(Scan)

"This is Bill Doberman, and I say convicted rapists have NO BUSINESS serving on the Supreme Court. That's why, as your congressman, I'll make sure that . . ."

(Scan)

". . . a large quantity of nuclear waste has been spilled on the interstate, and police are trying to . . ."

(Scan)

". . . GIVE YOU SEVENTEEN THOUSAND DOLLARS IN TRADE FOR ANYTHING!!! IT DOESN'T EVEN HAVE TO BE A CAR!!! BRING US A ROAD KILL!!! WE DON'T CARE!!! BRING US A CANTALOUPE-SIZED GOB OF EAR WAX!!! BRING US . . ."

(Scan)

". . . huge creatures that WERE like winged snakes EXCEPT they had great big suckers, which DID cometh and pulleth their limbs FROM their sockets liketh this, 'Pop,' but they WERE not afraid, nay they WERE joyous, for they had . . ."

(Scan)

". . . just heard 317 uninterrupted classic hits, and now . . ."

(Scan)

"Bill Doberman will shrink your swollen membranes. Bill Doberman has . . ."

(Scan)

". . . glowing bodies strewn all over the road, and motorists are going to need . . ."

(Scan)

". . . FORTY THOUSAND DOLLARS!!! WE'LL JUST GIVE IT TO YOU!!! FOR NO REASON!!! WE HAVE A BRAIN DISORDER!!! LATE AT NIGHT, SOMETIMES WE SEE THESE GIANT GRUBS WITH FACES LIKE KITTY CARLISLE, AND WE HEAR THESE VOICES SAYING . . ."

(Scan)

"Steve?"

"Yes."

"Steve?"

"Yes."

"Steve?"

(Scan)

"Yea, and their eyeballs DID explode like party favors, but they WERE NOT sorrowful, for they kneweth . . ."

(Scan)

"Bill Doberman. Him good. Him heap strong. Him your father. Him . . ."

(Scan)

". . . finished playing 3,814 consecutive classic hits with no commercial interruptions dating back to 1978, and now . . ."

(Scan)

". . . the radiation cloud is spreading rapidly, and we have unconfirmed reports that . . ."

(Scan)

". . . liquefied brain parts did dribbleth OUT from their nostrils, but they WERE not alarmed, for they were . . ."

(Scan)

". . . getting sleepy. Very sleepy. When you hear the words 'Bill Doberman,' you will . . ."

(Power Off)

Okay, never mind. I'll just drive. Listen to people honk. Maybe hum a little bit. Maybe even, if nobody's looking, do a little singing.

(Quietly)

I can't get nooooooo
Sa-tis-FAC-shun . . .

Anna Quindlen

Anna Quindlen (1950–   ) is a graduate of Barnard College who joined the New York Times *in 1977 as a general assignment and City Hall reporter. She was deputy metropolitan editor when A. M. Rosenthal, then executive editor, proposed her "Life in the '30's" column. From 1986 until the birth of her third child she wrote weekly about the intensely felt with a deftness that won her a national following. In 1990, Ms. Quindlen returned to the* Times *as an Op-Ed essayist. Her columns are collected in* Living Out Loud *(New York, 1988).*

PREGNANT IN NEW YORK

New York Times   *March 27, 1986*

I have two enduring memories of the hours just before I gave birth to my first child. One is of finding a legal parking space on Seventy-eighth Street between Lexington and Park, which made my husband and me believe that we were going inside the hospital to have a child who would always lead a charmed life. The other is of walking down Lexington Avenue, stopping every couple of steps to find myself a visual focal point—a stop sign, a red light, a pair of $200 shoes in a store window—and doing what the Lamaze books call first-stage breathing. It was 3:00 A.M. and coming toward me through a magenta haze of what the Lamaze books call discomfort were a couple in evening clothes whose eyes were popping out of their perfect faces. "Wow," said the man when I was at least two steps past them. "She looks like she's ready to burst."

I love New York, but it's a tough place to be pregnant. It's a great place for half sour pickles, chopped liver, millionaires, actors, dancers, akita dogs, nice leather goods, fur coats, and baseball, but it is a difficult place to have any kind of disability and, as anyone who has filled out the forms for a maternity leave lately will tell you, pregnancy is considered a disability. There's no privacy in New York; everyone is right up against everyone else and they all feel compelled to say what they think. When you look like a hot-air balloon with insufficient ballast, that's not good.

New York has no pity: it's every man for himself, and since you are yourself-and-a-half, you fall behind. There's a rumor afoot that if you are pregnant you can get a seat on the A train at rush hour, but it's totally false. There are, in fact, parts of the world in which pregnancy can get you a seat on public transportation, but none of them are within the boundaries of the city—with the possible exception of some unreconstructed parts of Staten Island.

What you get instead are rude comments, unwarranted intrusions and deli countermen. It is a little-known fact that New York deli countermen can predict the sex of an unborn child. (This is providing that you order, of course. For a counterman to provide this service requires a minimum order of seventy-five cents.) This is how it works: You walk into a deli and say, "Large fruit salad, turkey on rye with Russian, a large Perrier and a tea with lemon." The deli counterman says, "Who you buying for, the Rangers?" and all the other deli countermen laugh.

This is where many pregnant women make their mistake. If it is wintertime and you are wearing a loose coat, the preferred answer to this question is, "I'm buying for all the women in my office." If it is summer and you are visibly pregnant, you are sunk. The deli counterman will lean over the counter and say, studying your contours, "It's a boy." He will then tell a tedious story about sex determination, his Aunt Olga, and a clove of garlic, while behind you people waiting on line shift and sigh and begin to make Zero Population Growth and fat people comments. (I once dealt with an East Side counterman who argued with me about the tea because he said it was bad for the baby, but he was an actor waiting for his big break, not a professional.) Deli countermen do not believe in amniocentesis. Friends who have had amniocentesis tell me that once or twice they tried to argue: "I already know it's a girl." "You are wrong." They gave up: "Don't forget the napkins."

There are also cabdrivers. One promptly pulled over in the middle of Central Park when I told him I had that queasy feeling. When I turned to get back into the cab, it was gone. The driver had taken the $1.80 on the meter as a loss. Luckily, I never had this problem again, because as I grew larger, nine out of ten cabdrivers refused to pick me up. They had read the tabloids. They knew about all those babies christened Checker (actually, I suppose now most of them are Plymouths) because they're born in the back seat in the Midtown Tunnel. The only way I could get a cabdriver to pick me up after the sixth month was to hide my stomach by having a friend walk in front of me. The exception was a really tiresome young cabdriver whose wife's due date was a week after mine and who wanted to practice panting with me for that evening's childbirth class. Most of the time I wound up taking public transportation.

And so it came down to the subways: men looking at their feet, reading their newspapers, working hard to keep from noticing me. One day on the IRT I was sitting down—it was a spot left unoccupied because the rainwater had spilled in the window from an elevated station—when I noticed a woman standing who was or should have been on her way to the hospital.

"When are you due?" I asked her. "Thursday," she gasped. "I'm September," I said. "Take my seat." She slumped down and said, with feeling, "You are the first person to give me a seat on the subway since I've been pregnant." Being New Yorkers, with no sense of personal privacy, we began to exchange subway, taxi, and deli counterman stories. When a man sitting nearby got up to leave, he snarled, "You wanted women's lib, now you got it."

Well, I'm here to say that I did get women's lib, and it is my only fond memory of being pregnant in New York. (Actually, I did find pregnancy useful on opening day at Yankee Stadium, when great swarms of people parted at the sight of me as though I were Charlton Heston in *The Ten Commandments*. But it had a pariah quality that was not totally soothing.)

One evening rush hour during my eighth month I was waiting for a train at Columbus Circle. The loudspeaker was crackling unintelligibly and ominously and there were as many people on the platform as currently live in Santa Barbara, Calif. Suddenly I had the dreadful feeling that I was being surrounded. "To get mugged at a time like this," I thought ruefully. "And this being New York, they'll probably try to take the baby, too." But as I looked around I saw that the people surrounding me were four women, some armed with shoulder bags. "You need protection," one said, and being New Yorkers, they ignored the fact that they did not know one another and joined forces to form a kind of phalanx around me, not unlike those that offensive linemen build around a quarterback.

When the train arrived and the doors opened, they moved forward, with purpose, and I was swept inside, not the least bit bruised. "Looks like a boy," said one with a grin, and as the train began to move, we all grabbed the silver overhead handles and turned away from one another.

MOTHERS

New York Times *July 9, 1986*

The two women are sitting at a corner table in the restaurant, their shopping bags wedged between their chairs and the wall: Lord & Taylor, Bloomingdale's, something from Ann Taylor for the younger one. She is wearing a bright silk shirt, some good gold jewelry; her hair is on the long side, her makeup faint. The older woman is wearing a suit, a string of pearls, a diamond solitaire, and a narrow band. They lean across the table. I imagine the conversation: Will the new blazer go with the old skirt? Is the dress really right for an afternoon wedding? How is Daddy? How is his ulcer? Won't he slow down just a little bit?

It seems that I see mothers and daughters everywhere, gliding through what I think of as the adult rituals of parent and child. My mother died when I was nineteen. For a long time, it was all you needed to know about me, a kind of vestpocket description of my emotional complexion: "Meet you in the lobby in ten minutes—I have long brown hair, am on the short side, have on a red coat, and my mother died when I was nineteen."

That's not true anymore. When I see a mother and a daughter having lunch in a restaurant, shopping at Saks, talking together on the crosstown bus, I no longer want to murder them. I just stare a little more than is polite, hoping that I can combine my observations with a half-remembered conversation, some anecdotes, a few old dresses, a photograph or two, and re-create, like an archaeologist of the soul, a relationship that will never exist. Of course, the question is whether it would have ever existed at all. One day at lunch I told two of my closest friends that what I minded most about not having a mother was the absence of that grown-up woman-to-woman relationship that was impossible as a child or adolescent, and that my friends were having with their mothers now. They both looked at me as though my teeth had turned purple. I didn't need to ask why; I've heard so many times about the futility of such relationships, about women with business suits and briefcases reduced to whining children by their mothers' offhand comment about a man, or a dress, or a homemade dinner.

I accept the fact that mothers and daughters probably always see each other across a chasm of rivalries. But I forget all those things when one of my friends is down with the flu and her mother arrives with an overnight bag to manage her household and feed her soup.

So now, at the center of my heart there is a fantasy, and a mystery. The fantasy is small, and silly: a shopping trip, perhaps a pair of shoes, a walk, a talk, lunch in a good restaurant, which my mother assumes is the kind of place I eat all the time. I pick up the check. We take a cab to the train. She reminds me of somebody's birthday. I invite her and my father to dinner. The mystery is whether the fantasy has within it a nugget of fact. Would I really have wanted her to take care of the wedding arrangements, or come and stay for a week after the children were born? Would we have talked on the telephone about this and that? Would she have saved my clippings in a scrapbook? Or would she have meddled in my affairs, volunteering opinions I didn't want to hear about things that were none of her business, criticizing my clothes and my children? Worse still, would we have been strangers with nothing to say to each other? Is all the good I remember about us simply wishful thinking? Is all the bad self-protection? Perhaps it is at best difficult, at worst impossible for children and parents to be adults together. But I would love to be able to know that.

Sometimes I feel like one of those people searching, searching for the mother who gave them up for adoption. I have some small questions for her and I want the answers: How did she get her children to sleep through the night? What was her first labor like? Was there olive oil in her tomato sauce? Was she happy? If she had it to do over again, would she? When we pulled her wedding dress out of the box the other day to see if my sister might wear it, we were shocked to find how tiny it was. "My God," I said, "did you starve yourself to get into this thing?" But there was no one there. And if she had been there, perhaps I would not have asked in the first place. I suspect that we would have been friends, but I don't really know. I was simply a little too young at nineteen to understand the woman inside the mother.

I occasionally pass by one of those restaurant tables and I hear the bickering about nothing. You did so, I did not, don't tell me what you did or didn't do, oh, leave me alone. And I think that my fantasies are better than any reality could be. Then again, maybe not.

Appendix

Other Columns and Columnists

"And where is . . . ?" These capsules are an inadequate answer to a reasonable question. They are necessarily suggestive and selective, not comprehensive. Injustices abound, alas. I have omitted all celebrity columnists, with the glowing exception of Eleanor Roosevelt. Only deceased gossip columnists are included, and only the best known. Book columnists and cultural critics are too numerous to mention, but I allow three exceptions: Vincent Starrett, Ada Louise Huxtable, and John Crosby. Some magazine columns and columnists seemed too good to pass over: TRB, A. J. Liebling, Charles Erskine Scott Wood. A single service column is listed, Henry Mitchell's Sunday gardening column in the *Washington Post.* Of omissions, I am particularly conscious of an entire phalanx of first-rate syndicated political columnists, among them William Pfaff, Richard Reeves, Jules Witcover, John Hess, Milton Viorst, and Mark Shields. These and others deserve better. Foreign language newspapers are unrepresented, save for Bintel Brief, from the Yiddish *Daily Forward.* Europeans, Latin Americans, and Canadians are left out, which means no references to Ireland's Myles na Gopaleen (Brian Nolan), France's Raymond Aron, Colombia's Gabriel García Márquez, Canada's Robertson Davies, Britain's Peregrine Worsthorne, and numerous others. Only one editorial writer makes the roll, Hal Borland, late author of the Sunday nature editorials in the *New York Times,* who overcame the anonymity of a hapless calling. The idea, in short, is simply to suggest missing riches.

About New York

Running column on local affairs in *New York Times,* inaugurated (1939–40) by the renowned local reporter, Meyer Berger, who relaunched it in 1953 and continued until his death in 1959. See Meyer Berger's *New York*

(New York, 1960). Revived in 1979 on a rotating basis, moving to prime space on second front page; among the outstanding writers have been William Geist, Anna Quindlen, Francis Cline.

Anderson, Jack (1922–   )

Born in Long Beach, Cal., brought up in Salt Lake City, a Mormon missionary (1941–43). Worked on Shanghai edition of *Stars and Stripes,* joined Drew Pearson as staff reporter for "Washington Merry Go-Round" in 1947, taking over column in 1969 after Pearson's death. A thorn to successive administrations, liberal by inclination but ecumenical in pursuit of scandal. See *Confessions of a Muckraker* (with James Boyd, New York, 1979).

Baer, Arthur "Bugs" (1886–1969)

Born in Philadelphia, worked as artist for Philadelphia *Ledger,* moved to *Washington Times* as sports columnist and artist known for his cartoons of a baseball-shaped insect, hence his nickname. After Baer moved to *New York World,* Hearst so delighted by a baseball quip ("His head was full of larceny but his feet were too honest") that he hired Baer in 1914 to write syndicated humor column, "One Word Led to Another." This led to a Sunday feature, "The Family Album," whose cynical yokelisms made it a favorite of highbrows like Gilbert Seldes. See Baer, *The Family Album* (New York, 1925).

Bintel Brief

A feature begun in 1906 in Yiddish-language, Socialist-inclined *Daily Forward* in which readers, most of them immigrants, could discuss personal problems and seek advice. As editor Abraham Cahan remarked, "The name of the feature became so popular that it was often used as part of American Yiddish. When we speak of an interesting event in family life, you can hear a comment like, 'A remarkable story—just for Bintel Brief.' " Selections translated in Isaac Metzker, *Bintel Brief* (New York, 1972).

Borland, Hal (1900–1978)

Born in Nebraska, raised in Colorado, settled in New England in 1945. Wrote 1,750 Sunday nature editorials for *New York Times,* from 1941 to 1977. Though unsigned, Borland's work became nationally known through republication, as in *Hal Borland's Twelve Moons of the Year* (New York, 1979).

Brisbane, Arthur (1864–1936)

Born in Buffalo, N.Y., son of eminent Utopian Socialist, educated in France and Germany, reporter and London correspondent of *New York Sun,* managing editor of *New York World* before joining Hearst organization in 1897 as editor of *New York Journal.* Celebrated for his sententious "Today" column, long a fixture in Hearst papers.

Broder, David (1929–)

Born in Chicago Heights, Ill., attended University of Chicago (B.A. and M.A.), reporter, *Bloomington* (Ill.) *Pantagraph, Washington Star, New York Times* Washington bureau; joined *Washington Post* in 1966 as political reporter and columnist, associate editor since 1976; author of numerous books. The doyen of political analysts.

Browne, Charles Farrar (1834–67)

Known as Artemus Ward; born Waterford, Me., learned printer's trade at *Lancaster* (N.H.) *Weekly Democrat,* broke in as writer in 1852 on (Boston) *Carpet Bag,* made his reputation on Cleveland *Plain Dealer* where his pen-name first appeared February 3, 1858. Found a trans-Atlantic audience for his broadly written vernacular lampoons, became lecturer, meeting young Sam Clemens in Virginia City, Nev. Sailed in 1866 for England, where he was feted, fell ill, and died. A liberating figure in his time. See D.C. Seitz, *Artemus Ward* (New York, 1919).

Buchanan, Patrick (1928–)

Born in Washington, D.C., attended Georgetown University, earned master's in journalism at Columbia University, editorial writer, *St. Louis Globe-Democrat* (1962–66); White House aide to President Nixon (1966–73); syndicated columnist (1975–85); director of communications for President Reagan (1985–87). Resumed hardline conservative column in 1988.

Burdette, Robert Jones (1844–1914)

Born in Pennsylvania, raised in Cumminsville, Ohio, and Peoria, Ill.; worked on *Peoria Review* until 1874, when he joined staff of *Burlington* (Iowa) *Hawkeye.* His "Hawkeyetems" won him a wider reputation as "the Burlington Hawkeye Man." His collected sketches, variations on home-spun themes, were bestsellers; see *The Rise and Fall of the Mustache and other Hawkems* (New York, 1877).

Childs, Marquis (1903–1990)

Born in Clinton, Iowa, attended University of Wisconsin; spent three years with United Press; syndicated by United Features (1944–54); special correspondent for *Post-Dispatch* (1954–62); chief Washington correspondent and syndicated columnist (1962–68). Fluent left-of-center columnist on national and foreign affairs; received first Pulitzer Prize for commentary, in 1969; prolific author. See Childs, *Witness to Power* (New York, 1975).

Clapper, Raymond (1892–1944)

Born in La Cygne, Kansas, worked as journeyman printer in Kansas City; attended University of Kansas; cubbed at *Kansas City Star;* joined United Press bureau in Chicago and moved in 1917 to Washington; started column with advent of New Deal in 1934, first for Scripps-Howard chain, then United Features Syndicate, appeared in 180 papers as a "think" columnist. Killed in wartime air crash in Pacific. Described himself as a "seventy-five percent New Dealer."

Cohen, Richard (1941–)

New York-born, attended New York University, earned a master's in journalism at Columbia University, joined the *Washington Post* as reporter in 1968, syndicated columnist since 1976. Unabashedly liberal general commentary.

Considine, Robert (1906–75)

Born in Washington, D.C., attended George Washington University, broke in as weekly tennis columnist ($5 a column) for *Washington Post* in 1929; moved to *Washington Herald,* where his sportswriting caught eye of W. R. Hearst. Hired by *New York Journal-American* in 1937 and paired with Damon Runyon as sports columnist; a year later his "On the Line" column was syndicated and Considine emerged as Hearst's chief color writer for all occasions. Carried column on for four decades, wrote or co-authored 25 books.

Crosby, John (1912–)

Born in Milwaukee, attended Yale, broke in as reporter for *Milwaukee Sentinel,* joined *New York Herald Tribune* in 1935. After service in U.S. Army (1941–46) rejoined paper as radio and television critic. Wrote a literate, irreverent column on broadcasting. Moved to Paris for *Tribune* in 1963, then to London as weekly columnist for the *Observer.* See *Out of the Blue* (New York, 1952) and *With Love and Loathing* (New York, 1963).

Drummond, Roscoe (1902–)

Born in Theresa, N.Y., attended Syracuse University, joined *Christian Science Monitor* in 1924, serving as executive editor (1934–40) and head of Washington bureau (1940–53); author of syndicated "State of the Nation" column for *New York Herald Tribune* and *Los Angeles Times.* Longtime barometer of moderate Republicanism.

Fleeson, Doris (1901–70)

Born in Sterling, Kansas, attended University of Kansas; reporter and Washington correspondent, *New York Daily News* (1927–42); war correspondent, *Woman's Home Companion* (1943–44); New Dealish Washington columnist for United Feature Syndicate from 1945 to 1967. Teamed with husband John O'Donnell for nine years as author of "Capitol Stuff" column in *Daily News;* marriage ended when O'Donnell (and paper) broke with Roosevelt.

Frazier, George (1911–74)

Boston-born, attended Boston Latin School and Harvard. Hired as a writer for *Life* magazine in 1942, free lanced after World War II, joining *Boston Herald* as daily columnist in 1961; moved to *Globe* in 1971. Wrote lyrics for Count Basie's "Harvard Blues." Once reported Red Sox-Yankee game in Latin (*Globe* headline: "Tibialibus Rubris XV, Eboracum Novum V"). Haughty, idiosyncratic: "I'm a lonely man. The column precludes friendship." See Charles Fountain's biography, *Another Man's Poison* (Chester, Conn., 1984).

Gould, John (1908–)

Born in Freeport, Me., attended Bowdoin College, became farmer and professional guide living in Lisbon Falls, Me. Has written weekly column on rural themes since 1952 for the *Christian Science Monitor.* See Gould, *The Shag Bag* (Boston, 1972).

Graham, Sheilah (1908–88)

Born in London, England, performed in musical comedies on London stage; began writing Hollywood column (1935–40); war correspondent in London, (1940–45); resumed Hollywood column in 1945, eventually rivaling Hopper and Parsons as doyenne. Wrote extensively about her four-year affair with F. Scott Fitzgerald, who modeled Kathleen, heroine of *The Last Tycoon,* on Graham.

Greene, Robert Bernard, J. (Bob) (1947–)

Born in Columbus, Ohio, attended Northwestern University, reporter on *Chicago Sun-Times* (1969–71); columnist (1971–78). Field Syndicate columnist (1976–81); columnist, *Chicago Tribune,* since 1978. Professional baby-boomer; see *The Best of Bob Greene* (New York, 1976).

Greenfield, Meg (1930–)

Born in Seattle, Wash., attended Smith College and Newnham College, Cambridge University. Staff writer, *Reporter* magazine (1957–68), Washington editor (1965–68). Joined *Washington Post* editorial page in 1968, named its editor of page editor in 1979; awarded Pulitzer for editorial writing, 1978. Columnist for *Newsweek* since 1974.

Harris, Sydney (1917–87)

Born in London, England; attended University of Chicago; joined *Chicago Daily News* as reporter and feature writer in 1941; started "Strictly Personal" column in 1944, moving to *Chicago Sun-Times* in 1963. General interest column collected in a dozen books.

Harsch, Joseph C. (1905–)

Born in Toledo, Ohio, attended Williams College and Corpus Christi College, Cambridge University. Joined *Christian Science Monitor* in 1929; in Washington bureau (1931–39); Rome and Berlin (1939–41); chief editorial writer (1971–74); foreign affairs columnist since 1952.

Hellinger, Mark (1903–47)

Born in New York, hired in 1923 as nightlife reporter for newly founded *New York Daily News;* started "About Broadway" column mixing gossip and tough/sentimental tales. Joined *New York Daily Mirror* in 1930 and launched "All in a Day" column combining short stories and show business reportage. Moved to Hollywood in 1937 as writer-producer for Warner Brothers, forming own production company in 1945; films include *Naked City, The Killers, Two Mrs. Carrolls.* An inventor of the Broadway column and streetsmart fable. See Hellinger, *Moon Over Broadway* (New York, 1931), *Ten Million* (New York, 1934), and *The Mark Hellinger Story* (New York, 1949) by his former legman, Jim Bishop.

Hers

Innovative *New York Times* column edited by Nancy Newhouse to provide a forum for "serious, funny, factual, nostalgic, reportorial, inward-looking" pieces by women of interest to both sexes. Begun as weekly feature in 1977, changed to twice monthly and moved from news section to Sunday magazine in 1988. Outstanding among seventy-odd contributors were Joyce Maynard, Mary Cantwell, and Phyliss Theroux. See Nancy Newhouse, ed., *Hers* (New York, 1986).

Higgins, Marguerite (1920–66)

Born in Hong Kong, attended the University of California, and earned a master's in journalism at Columbia University; joined the *New York Herald Tribune* in 1942 and in two years was a war correspondent in France, later covering the Berlin blockade and the Korean War; from 1955, Washington columnist for the *Tribune,* after 1963 for *Newsday.*

Hitchins, Christopher (1949–)

Born in Malta, attended Balliol College, Oxford; staff writer, various Fleet Street popular dailies; contributor and foreign editor, *New Statesman* of London (1960s); Washington correspondent, *Spectator* of London (1979–85); author of combative "Minority Report" column for *The Nation* since 1981. See Hitchens, *Prepared for the Worst* (New York, 1988).

Hopper, Hedda (1890–1966)

Born Elda Furry in Hollidaysburg, Pa., of Quaker parentage; a chorus girl, stage and film actress; from 1936 to 1966, a daily Hollywood gossip columnist, syndicated by the *Chicago Tribune* and known for lavish headgear and her feud with archrival Louella Parsons.

Huxtable, Ada Louise

Born in New York, attended Hunter College and Institute of Fine Arts, New York University; assistant curator, architecture and design, Museum of Modern Art (1946–50). Began as free-lance writer in 1950, named first full-time architectural critic on *New York Times* in 1963. Received Pulitzer for criticism, 1970. Joined *Times*'s editorial board, 1973, while continuing Sunday column; retired in 1982 after receiving MacArthur Fellowship. A sorely missed voice. See Huxtable, *Will They Ever Finish Bruckner Boulevard?* (New York, 1976); and Lawrence Woodhouse, *Ada Louiser Huxtable: An Annotated Bibliography* (New York, 1981).

Kaul, Donald (1934–   )

Born in Detroit, attended Wayne State University, joined *Des Moines Register* as reporter in 1960, assigned to Washington where his "Over the Coffee" column was more popular with readers than with his editor. After quarrel in 1983, Kaul announced his resignation in a column, which he continued to syndicate elsewhere. Rehired in 1989 when new editor took over, to applause of devotees of his offbeat ruminations.

Kieran, John (1892–1981)

Born in New York, attended Fordham University, taught himself French and Italian, developed skills at identifying birds and plants; in 1915, joined sports department of *New York Times;* on January 1, 1927, started paper's first daily signed column; conducted "Sports of the Times" until 1943. Displayed erudition and encyclopedic recall on radio quiz program, "Information, Please." See his memoirs, *Not Under Oath* (Boston, 1964).

Kinsley, Michael (1951–   )

Born in Detroit, attended Harvard, Harvard Law School, and Magdalen College, Oxford. Managing editor, *Washington Monthly* (1975); editor, *Harper's* magazine (1981–83); managing editor, *New Republic* (1976–79), editor (1985–89). Essayist for *Time* and fluent author TRB column (*q.v.*) since 1983. See Kinsley, *The Curse of the Giant Muffin* (New York, 1987).

Kraft, Joseph (1925–86)

Born in South Orange, N.J., served as U.S. Army cryptographer in World War II, attended Columbia University; editorialist, *Washington Post* (1951–52); with Sunday department, *New York Times* (1953–57); Washington correspondent, *Harper's Magazine* (1962–65); syndicated Washington columnist (1965–86). Long a fixture on *Washington Post*'s Op Ed page; reflective middle roader, aptly credited with coining term "Middle America."

Krauthammer, Charles (1950–)

Born New York, attended McGill University and Balliol College, Oxford; earned M.D. at Harvard. Resident in psychiatry, Massachusetts General Hospital (1975–78); speech writer, Vice President Walter Mondale (1980–81); senior editor, *New Republic,* since 1981; essayist for *Time* since 1983; syndicated columnist since 1984. Conservative commentary, with interventionist fervor, won Pulitzer Prize for commentary, 1988. See Krauthammer, *The Cutting Edge* (New York, 1985).

Krock, Arthur (1886–1974)

Reporter (1927–32), Washington bureau chief (1932–53), wrote "In the Nation" column (1953–67) for the *New York Times.* Winner of multiple Pulitzers, friend of Presidents, Krock was known to successive generations for cumbrous syntax and Olympian tone. See Krock, *Memoirs* (New York, 1968) and *In the Nation* (New York, 1966).

Kupcinet, Irv (1912–)

Born in Chicago, attended Northwestern University, columnist on *Chicago Daily Times,* (1935–43); author of "Kup's Column" for *Chicago Sun-Times,* since 1943. Perky local columnist and civic-minded host of a long-running Chicago TV program, "Kup's Show."

Lardner, John (1912–59).

Born in Chicago, a son of Ring Lardner, attended Harvard, reporter, *New York Herald Tribune* (1931–33); sports columnist for North American Newspaper Alliance until 1948; World War II correspondent principally for *The New Yorker;* much-admired sports columnist for *Newsweek* (1950s). See Roger Kahn, *The World of John Lardner* (New York, 1961).

Lawrence, David (1883–1973)

Born in Philadelphia, attended Princeton, Washington correspondent for *New York Evening Post,* (1916–19); began syndicated column in 1919; founded *U.S. News and World Report* (1946), for which he also wrote weekly column. Continued almost to the end with a newspaper column notable for intrepid conservatism and anti-Soviet ardor. See Lawrence, *Diary of a Washington Correspondent,* (New York, 1942).

Lewis, Anthony (1927–)

Born in New York, attended Harvard, deskman for *New York Times* Sunday department (1948–52); Pulitzer prize-winning reporter, *Washington Daily News,* (1952–55); joined *New York Times* Washington bureau (1955), chief London correspondent (1965–72), thereafter conducted "At Home Abroad" and "Abroad at Home" column, a left-of-center institution, outstanding on constitutional themes.

Lewis, Flora

Born in Los Angeles, attended U.C.L.A. and earned master's in journalism at Columbia University; foreign correspondent for *Washington Post* (1958–66); columnist for *Newsday* (1967–72); acclaimed Paris bureau chief, *New York Times* (1972–80), roving foreign affairs columnist since 1980.

Liebling, A. J. (1904–63)

Born in New York, attended Dartmouth, cubbed on *Providence Journal*, hired sandwich-man in 1930 to walk in front of *New York World* offices with sign reading "Hire Joe Liebling," winning a berth as paper was sinking; his signed features for *World-Telegram* led to offer to join *New Yorker* as staff writer in 1930. In 1945 he relaunched magazine's "Wayward Pressman" column, which rapidly established itself as wittiest running critique of U.S. journalism. Selections in *The Press* (New York, 2nd rev. ed., 1975).

Locke, David Ross (1833–88)

Known to readers as Petroleum V. Nasby, Locke was born in Vestal, N.Y., and became famous as the anti-slavery satirist in Ohio, where he edited the *Findlay Jeffersonian* and the *Toledo Blade*. His Nasby letters ridiculed pro-slavery Copperheads. Paid a rare compliment when Lincoln read Nasby letters to visitors; their popularity however, perished with the period.

Lyons, Leonard (1906–76)

A New Yorker who started his Broadway column "The Lyons Den" in 1934 for the *New York Post*. A column with less malice or bite than Winchell's, its long run ended in 1974.

McCabe, Charles (1915–83)

Born in New York, attended Manhattan College, started as police reporter for *New York American* in 1936; managing editor, *Puerto Rico World Journal;* war correspondent, United Press. Moved to San Francisco in mid-1950s, joined the *Chronicle* where his episodic "Fearless Spectator" column turned into a daily confessional column entitled "Himself," running from 1960s until his death.

McCormick, Anne O'Hare (1880–1954)

Born Yorkshire, England, of American parents, grew up in Columbus, Ohio, where she attended Academy and College of Saint Mary of the Springs. Her husband, an engineer, frequently visited Europe and she wrote to the managing editor of *New York Times* offering to send dispatches. In 1922, she was hired as a regular correspondent; began solid centrist foreign affairs column "Abroad" in 1935, continuing until her death; was first woman on *Times*'s editorial board.

McGill, Ralph (1898–1969)

Born in Soddy, Tenn., attended Vanderbilt University, covered police, politics and sports for *Nashville Banner* until 1929, when he joined *Atlanta Constitution* as sports editor; named executive editor in 1938, publisher in 1960. For 40 years wrote daily column that became local and regional institution; an outstanding voice of decency on racial issues. Awarded Pulitzer Price in 1958. See Michael Strickland, ed., *The Best of Ralph McGill* (Atlanta, 1980); Harold H. Martin, *Ralph McGill, Reporter* (Boston, 1973).

McIntyre, Oscar Odd (1884–1938)

Born in Plattsburg, Mo., broke in as local reporter for *Gallipolis* (Ohio) *Daily Sun and Journal,* writer and editor for *Dayton Herald* and *Cincinnati Post* until 1911, when he joined *New York Evening Mail.* Discharged for incompetence, launched "O. O. McIntyre's New York Letter," sentimental tales about the Great White Way and the Sidewalks of New York, phrases he helped popularize in syndicated column (1912–38). See his *The Big Town: New York Day-by-Day* (New York, 1935), and Charles Driscoll, *Life of O. O. McIntyre* (New York, 1938).

Mitchell, Henry (1923–   )

Grew up in Memphis, Tenn., attended University of Vermont; worked for *Memphis Commercial-Appeal* and *Delta Review* before joining *Washington Post* as reporter in 1970. A backyard cultivator, he began notably literate Sunday "Earthman" column in 1974. See Mitchell, *The Essential Earthman* (Bloomington, Ind., 1981).

Morley, Christopher (1890–1957)

A Philadelphian educated at Haverford College and a Rhodes Scholar at Oxford; he was well-known novelist and critic when he started "The Bowling Green" in the *New York Evening Post* in 1920, moving in 1930s to *Saturday Review of Literature.* His column launched and publicized the Baker Street Irregulars, a Sherlock Holmes society whose charter, as drafted by Elmer Davis, was published in *SRL,* February 17, 1934.

Murray, Jim (1919–   )

Born in Hartford, Conn., where he attended Trinity College. Reported for *New Haven Register* (1943–44), moving to *Los Angeles Examiner* (1944–48), thence to *Time* as Los Angeles correspondent (1948–59). Helped found *Sports Illustrated,* serving as West Coast editor before joining *Los Angeles*

Times as sports columnist in 1961. See *Best of Jim Murray* (New York, 1978)

Parker, Dan (1893–1967)

Born in Waterbury, Conn., worked as sportswriter for *Waterbury American* (1913–24); joined *New York Daily Mirror* in 1924 when Gene Fowler was sports editor; after Fowler departed in 1926, became sports editor and seven-day-a-week columnist, jobs he held until paper's demise in 1963. A stylist in the Runyon mode, he once remarked, "I'd rather write about the characters you run into in sports than the games they play."

Parsons, Louella (died 1972)

Born Freeport, Ill., attended Dixon College, started motion picture column for the *Chicago Herald,* wrote for *New York Morning Telegraph* before joining Hearst Publications in 1922; Hollywood columnist from 1925 to 1965. Known for her feud with Hedda Hopper and for using column to punish those out of favor with William Randolph Hearst.

Pearson, Drew (1897–1969)

Born in Evanston, Ill., attended Swarthmore College; foreign editor, *United States Daily* (1926–29); Washington correspondent, *Baltimore Sun* (1929–33). In collaboration with Robert S. Allen, he anonymously published *The Washington Merry-Go-Round* in 1931, a *succès de scandale* that led to a syndicated column of the same title. Its launching in 1933 coincided with advent of New Deal, and column was the first and most successful of its kind. Pearson won readers with his often malicious leaks and muckraking populism, oddly commingled with Quaker piety.

Povich, Shirley (1905–   )

Born in Bar Harbor, Me., attended Georgetown University; joined *Washington Post* in 1923 as reporter; sports editor (1926–33), superlative sports columnist (1933–1974), thereafter editor emeritus.

Nye, Edgar Wilson (1850–96)

Born in Shirley, Me., "Bill" Nye made his name as editor of the *Laramie* (Wyoming) *Boomerang* in the 1880s. He joined the *New York World* as a columnist in 1887, became a popular platform performer and fashioned a broad, hayseed style in best-selling books whose hilarity has waned.

Raspberry, William (1935–   )

Born in Okolona, Miss., attended Indiana Central College, reporter for *Indianapolis Recorder* (1956–60); joined *Washington Post* in 1962. The author of an urban affairs column since 1966, written from a black perspective.

Roosevelt, Eleanor (1881–1962)

Began writing a weekly column and also a page in *Woman's Home Companion* soon after taking residence in White House in 1933; contracted with United Features syndicate in 1936 to write six columns a week. When her husband disparaged columnists, a reporter reminded the President of "My Day." He responded, "She is in an entirely different category; she simply writes a daily diary." Continued column after FDR's death, shifting in 1957 from *New York World-Telegram* to *New York Post.* Dictated daily from 2 to 4 p.m., sometimes on picnics, in trains, cars, ships, planes, once on a destroyer. Only once, she wrote proudly in 1949, did she fail to meet copy deadline. Changed over years from diary-style entries to whole columns on controversial topics. See E. Roosevelt, *This I Remember* (New York, 1949); Joseph Lash, *Eleanor: The Years Alone* (New York, 1972); and *Eleanor's Roosevelt's My Day,* 1936–1945 (New York, 1989).

Rosenthal, A. M. (1922–   )

Born in Sault St. Marie, Canada, grew up in New York, attended City College of N.Y., joined *New York Times* in 1944; won Pulitzer Prize in 1960 for coverage of Poland; executive editor (1977–86); began national and foreign affairs column, "On My Mind," in 1986.

Rowan, Carl (1925–   )

Born in Ravenscoft, Tenn., attended Tennessee State University, Washburn University, Oberlin College, earned master's in journalism, University of Minnesota. Reporter, *Minneapolis Tribune* (1950–61); from 1961 to 1965, served as a deputy assistant secretary of state, U.S. ambassador to Finland, and director, U.S.I.A.; syndicated columnist since 1965. A pioneer in breaking color line in white journalism, and a centrist political commentator.

Royster, Vermont (1914–   )

Born in Raleigh, N.C., attended University of North Carolina, joined Washington bureau of *Wall Street Journal* in 1936, becoming chief correspondent in 1946, editor, (1958–1971), columnist since 1971. Won Pulitzer Price for commentary, 1984. See Royster, *A Pride of Prejudices* (New York, 1967).

Schanberg, Sydney (1934–   )

Born in Clinton, Mass., attended Harvard, joined *New York Times* as reporter in 1959, known for his prize-winning coverage of fall of Phnom Penh in 1975; wrote column on New York politics (1981–85). After internal differences with publisher, joined *Newsday* as local columnist in 1986.

Shannon, William V. (1927–88)

Born Worcester, Mass., attended Clark University, earned M.A. at Harvard, Washington correspondent and columnist, *New York Post* (1951–64); editorial board, *New York Times* (1964–77); U.S. ambassador to Ireland (1977–81); contributor, *Boston Globe* (1982–88). Remembered for developing a loyal following when he filled in for vacationing *New York Times* pundits.

Shaw, Henry Wheeler (1818–85).

Known as Josh Billings, born in Lanesboro, Mass., attended Hamilton College, adventured through Great West before debut as journalist in *New Ashford* (Mass.) *Eagle* and in Poughkeepsie newspapers. His "Allminix" used grotesque misspellings for labored cracker-barrel witticisms. Acclaimed, erroneously, as a Yankee Rochefoucauld. See Cyril Clemens, *Josh Billings, Yankee Humorist* (Webster Grove, Mo., 1932).

Sidey, Hugh (1927–   )

Born in Greenfield, Iowa, attended Iowa State University, reporter for *Adair County* (Iowa) *Free Press* and *Omaha World-Herald* before joining *Life* in 1955, moving over in 1958 to *Time* magazine. Column, "The Presidency," has appeared in both magazines since 1966. Chief, *Time*'s Washington bureau (1969–78). Known to readers and viewers as fervent celebrator of the Presidency.

Sokolsky, George (1893–1962)

Born in Utica, N.Y., grew up in New York City, attended Columbia University; went to Russia in 1917 to edit English-language *Daily News* in Petrograd; departed in 1918 a convinced anti-Bolshevik, settling in Shanghai, where he served as correspondent for *New York Post, Philadelphia Ledger,* and *London Express.* Returning to U.S. in 1932, became broadcaster for National Association of Manufacturers and columnist, successively for *New York Herald Tribune, New York Sun,* and King Features (Hearst). A tireless if bromidic upholder of free market capitalism.

Starrett, Vincent (1856–1974)

Born in Toronto, Canada; broke in on *Chicago Inter-Ocean* (1905–06); on staff of *Chicago Daily News* (1906–16); wrote Sunday "Books Alive" column for *Chicago Tribune* from 1942. A fine essayist in old belles-lettres mode, kept alive memory of Chicago's literary renaissance; esteemed by devotees for his *Private Life of Sherlock Holmes* (Chicago, 1960). See Starrett, *The Column Book* (Chicago, 1957).

Stokes, Thomas L. (1898–1958)

Born in Atlanta, attended University of Georgia; after apprenticeship on three Georgia papers, hired by United Press in Washington in 1921, becoming known for solid coverage of national politics. Hired in 1933 as Washington correspondent by Scripps-Howard, winning a Pulitzer five years later and becoming a United Features columnist in 1944. Distinguished for his equable tone, New Dealish views. See Stokes, *Chip Off My Shoulder* (New York, 1940)

Strout, Richard (1898–1990)

New Englander who attended Harvard; sailed to England in 1919, where he broke in on *Sheffield Independent.* Returning to Boston, worked briefly for *Post;* joined Washington bureau, *Christian Science Monitor,* in 1925. In 1943, took over "TRB" column in *New Republic,* doubling as *Monitor* reporter and liberal columnist for four decades; steadying rudder as magazine lurched port to starboard. See Strout, *TRB* (New York, 1979).

Sullivan, Ed (1902–74)

Born in Manhattan, spent childhood in Port Chester, N.Y., where he worked for the *Item;* became sports writer for *New York Evening Mail* in 1920, shifted to *World,* then *Morning Telegraph* columnist, in 1932; he joined *Daily News* where his "Little Old New York" column ran until 1974. He hosted *Daily News*'s Harvest Moon Ball and in 1948, CBS tried him out on "Toast of the Town," a variety program that became "The Ed Sullivan Show" and ran until 1971.

Sulzberger, C. L. (1912–   )

Born in New York, attended Harvard, conducted the Paris-based "Foreign Affairs" column for *New York Times* from 1954 until 1978. A prolific writer and proficient cultivator of influential figures, he published three volumes of an extensive diary.

TRB

Long-running *New Republic* column, launched by liberal weekly on January 7, 1925, over cryptic initials said to be inspired by those of Brooklyn Rapid Transit subway line, read backward. Initially conducted by *Baltimore Sun*'s Frank Kent, then by *Newsweek*'s Kenneth Crawford, whose successor in 1943 was Richard Strout (*q.v.*), who fixed its chatty, outspoken character in 40-year run; notable "TRBs" since 1983 have been Michael Kinsley (*q.v.*) and Hendrik Hertzberg.

Topics of the Times

One of the oldest, continuously published editorial page features in U.S.; started in *New York Times* in 1860s by William Livingstone Alden as "Minor Topics"; changed to "Topics of the Times" in 1896 by Frederick Craig Mortimer, who conducted this miscellany column until his retirement in 1926; succeeded by Simeon Strunsky (1879–1948), erudite all-rounder, who yielded in turn in 1947 to rotating staff contributors; after 1976, became a portmanteau feature with unrelated items bundled together. See selections in Herbert Mitgang, ed., *American at Random* (New York, 1970); and Harold Phelps Stokes, ed., *Simeon Strunsky's America* (New York, 1956).

Tyrrell, R. Emmett, Jr. (1944–)

Attended University of Indiana at Indiana at Bloomington, where he turned a campus conservative journal, the *Alternative,* into national magazine, renaming it *American Spectator* in 1974. A widely syndicated columnist heavy on wordy sarcasm; hailed, precipitately, as a new Mencken.

Wechsler, James (1915–83)

Born in New York, attended Columbia University, assistant editor and then Washington bureau chief, *PM* (1940–44); joined *New York Post* as Washington correspondent in 1946, editor from 1949 to 1961, editor of editorial page and an admired columnist (1961–80).

Wicker, Tom (1926–)

Born in Hamlet, N.C., attended the University of North Carolina, wrote for *Winston-Salem Journal* and *Nashville Tennessean* before joining *New York Times* Washington bureau in 1960; general political columnist since 1968.

Wills, Garry (1934–)

Born in Atlanta, attended Xavier University and earned doctorate at Yale; Henry R. Luce Professor of American Culture and Public Policy, Northwestern University, since 1980. Author of dozen books on politics past and present, he has been syndicated newspaper columnist since 1970. See Wills, *Lead Time: A Journalist's Education* (New York, 1983).

Wood, Charles Erskine Scott (1892–1944)

A Pennsylvanian who took up law in Oregon and writing in California's Bay Area. A leonine figure and eminent regional poet, he wrote 40 dialogues during World War I for *The Masses,* a Socialist magazine suppressed by U.S. Government on sedition charges. In Wood's heaven, God, Jesus, and St. Peter chat about humanity's failings with Rabelais, Voltaire, Paine, and other freethinkers. See Wood, *Heavenly Discourse* (New York, 1927).

Bibliographic Note

Those interested in the newspaper column will look in vain for a history or extended essay on the form. I have found two books about columnists. Charles Fisher's *The Columnists* (New York, 1944) offers opinionated sketches of political and gossip columnists (Lippmann, Winchell, Thompson, Pegler et al.); Neil A. Grauer's *Wits and Sages* (Baltimore, 1984) is altogether superior, combining biography and commentary on Jack Anderson, Russell Baker, Erma Bombeck, Jimmy Breslin, David Broder, Art Buchwald, William F. Buckley, Jr., Ellen Goodman, James J. Kilpatrick, Carl T. Rowan, Mike Royko, and George Will. Women journalists are the subject of Ishbel Ross, *Ladies of the Press* (New York, 1936) and Barbara Belford, *Brilliant Bylines* (New York, 1986), the latter including selections as well as biographies of Anne O'Hare McCormick, Dorothy Thompson, Doris Fleeson, Mary McGrory, Marguerite Higgins, Ada Louise Huxtable, and Ellen Goodman. Going backwards, there is an extensive literature of the Algonquin literati; I liked best James R. Gaines's dryly written *Wit's End: Days and Nights of the Algonquin Round Table* (New York, 1977). Carl Van Doren's *Many Minds* (New York, 1924) has first-rate essays on George Ade, H. L. Mencken, Ring Lardner, Franklin P. Adams, Don Marquis, Heywood Broun, and Christopher Morley. *The Seven Lively Arts* (New York, 1924) by Gilbert Seldes has chapters on Ade, Finley Peter Dunne, Lardner, and on columns in general (F.P.A., B.L.T., Arthur "Bugs" Baer, Baird Leonard, Marquis, Morley, and Broun). Of many books on American humor, Walter Blair and Hamlin Hill, *America's Humor* (New York, 1978) is notable for dealing with modern-day columnists as well as forebears like George Horatio Derby, Mark Twain, and Seba Smith. Anthologies of humor that I found useful included Leonard C. Lewin, ed., *A Treasury of American Political Humor* (New York, 1964); E. B. White and Katharine White, eds., *A Subtreasury of American Humor* (New York, 1941); J. B.

Mussey, ed., *The Cream of the Jesters* (New York, 1935). Sound selections of journalism are hard to find, but Louis L. Snyder and Richard B. Morris, eds., *A Treasury of Great Reporting* (New York, 1949) can be recommended. On the origins of American journalism, besides standard academic histories, Leonard W. Levy, ed., *Freedom of the Press from Zenger to Jefferson* (Indianapolis, 1966) has special value. Similarly, Ralph Ketchum, ed., *The Anti-Federalist Papers and the Constitutional Convention Debates* (New York, 1986), throws needed light on a press polemics overshadowed by *The Federalist.* General histories of American literature, past and present, tend to give short shrift to journalism; yet many of the writers represented in this collection are discussed in William Peterfield Trent et al., *The Cambridge History of American Literature* (Cambridge, 1917); Robert Spiller et al., *Literary History of the United States* (New York, 1948), and in Van Wyck Brooks, *The Times of Melville and Whitman* (New York, 1947) and *The Confident Years* (New York, 1952).

Index

The index includes proper names of people and publications in introduction, headnotes, and appendices. The columns themselves are not indexed.